FLUORESCENT POLYMERS

FLUORESCENT POLYMERS

N. N. BARASHKOV
Karpov Institute of Physical Chemistry
Moscow, and
Visiting Professor, Department of Physics
Texas Tech University

and

O. A. GUNDER
Institute for Single Crystals
Kharkov
Ukraine

Translation Editor:
Professor T. J. Kemp
Department of Chemistry
University of Warwick

ELLIS HORWOOD
NEW YORK   LONDON   TORONTO   SYDNEY   TOKYO   SINGAPORE

This English edition first published 1994 by
Ellis Horwood Limited
Market Cross House, Cooper Street
Chichester
West Sussex, PO19 1EB
A division of
Simon & Schuster International Group

Translated by E. Sakharov from *Fluorescent Polymers* published in 1987 by Khimiya, Moscow

Printed and bound in Great Britain by
Hartnolls, Bodmin

Library of Congress Cataloging-in-Publication Data
Available from the Publisher

British Library Cataloguing in Publication Data
A catalogue record for this book is available from the British Library

ISBN   0-13-323510-6 (hbk)

1 2 3 4 5   97 96 95 94

Table of contents

Preface

Of the great variety of synthetic materials, specialists are giving growing attention to tailor made polymers.

In a number of fields of application, polymers are required to have well defined optical characteristics, e.g. transparency, bright colour, etc. The ability to convert radiant energy of one spectral composition to that of another is another essential feature of polymers used in plastic scintillators, active elements for laser equipment, and fluorescent pigments and paints.

It is known that many organic compounds are able to luminesce when exposed to various forms of radiation. Ultraviolet rays and light of the short wavelength visible spectrum (photoluminescence), high energy radiation (radioluminescence), chemical reactions (chemiluminescence), etc., may serve as the sources of excitation energy.

Compounds capable of luminescing when exposed to light or any other forms of exciting radiation are called luminophors. The present book is restricted to discussion of luminescent compounds of the greatest practical utility such as photoluminophors. A polymer will luminesce when its molecule has a definite structure, e.g. a branched system of conjugated bonds, a sufficiently rigid structure, etc. [1–7].

Most natural and synthetic polymers do not have such properties and, consequently, they do not luminesce. To impart emissive powers to polymers, the analogues of low molecular organic luminophors such as luminophoric moieties are introduced into the macromolecules (in the main or side chain).

At present there is an abundant original literature on the synthesis and study of the spectral and luminescent properties of luminophor-containing (fluorescent) polymers. The reviews published are far fewer [8–11], most of them generalizing data pertaining to fluorescent polymerized polymers. Meanwhile knowledge of the spectral and luminescent properties of polymerized, polycondensed and natural polymers is of great theoretical importance because it opens the possibilities of studying processes of intra- and intermolecular transfer and the migration of electronic excitation energy in the polymeric medium.

In recent years several trends have taken shape in the extensive application of the marked fluorescence capacity of a number of polymers. These include the application of fluorescent polymers in the paint and varnish industry, the production of decorative plastics and textile materials, materials for monitoring ionizing radiation, the manufacture of plastic scintillators for conversion of polymeric compositions [12], and the production of materials for laser equipment [13].

Additionally, the rapid advance of modern polymer science and the wide use of polymeric materials in the economy have become instrumental in attracting specialists' attention to the physical and chemical methods of studying polymers, especially spectral methods. Great interest in these methods is expressed not only by those concerned with polymer physics but also by synthetic chemists and specialists in the treatment and application of polymeric materials.

Meanwhile, the available monographs [14, 15] and reviews [16–19] consider the potentialities of luminescence spectroscopy, mainly as applied to the study of structures and properties of polymers. At the same time, the problems of preparing polymers with prescribed fluorescent properties, and studying the processes of synthesis and chemical conversion of macromolecules by luminescence methods have not been given sufficient attention. In view of this fact, in this book we have made an attempt to generalize the material accumulated thus far on these topics.

The subject of light fastness and photoconversions of fluorescent polymers is only outlined in this book as most of the data on the above have been analysed by Allen [20] and Shlyapintokh [21].

In this book we place particular emphasis on the chapter covering the prospects of practical application of fluorescent polymers and materials based on them.

We would like to express our sincere gratitude to Professor D. N. Shigorin and Dr R. N. Nurmukhametov for valuable suggestions and comments made during the discussion of the separate topics covered in this book, and to Professor T. J. Kemp who kindly consented to be the translation editor of this English edition.

1

Photophysical processes in polymers

The photophysical processes occurring in polymers, such as the nonradiative and radiative deactivation of electronically excited states in macromolecules, compete with radical formation processes caused by bond rupture. The occurrence of a particular process is determined by the correlation between the possibility of the localization of excitation energy in a specified chemical bond and the dissipation of excitation energy in a macromolecule, the emission of the excitation energy, or the transfer of energy to molecules of impurities. We consider the special features of deactivation of excited states in polymeric systems.

1.1 FORMATION OF EXCITED STATES IN POLYMERS AND ENERGY TRANSFER

On interaction with electromagnetic radiation of sufficient energy, elementary polymeric units, like low molecular organic compounds, pass into an excited state which leads to a considerable change in their electronic structure and increase of their reactivity. Furthermore, the excited states are either deactivated or take part in photochemical reactions.

The photophysical processes taking place during the absorption and emission of light by organic molecules can be conveniently presented as an energy level diagram (Fig. 1.1). At room temperature, practically all molecules are in the ground unexcited state (level S_0). After absorption of light, a molecule passes into an excited singlet state with two unpaired electrons whose spins are antiparallel (levels S_1^x and S_2^x). We draw attention to the difference between *fluorescence*, associated with the electronic transition $S_1^x \rightarrow S_0$ and the electronic transition from the triplet state, in which the electron spins are parallel to the ground state $T \rightarrow S_0$, known as *phosphorescence*.

In greater detail the photophysical processes occurring in organic molecules are analysed in [1–5]. As applied to polymers, such processes may be presented in the following kinetic scheme:

$$1. \quad M_0 + h\nu_0 \overset{k_0}{\rightarrow} M^* \qquad \text{Excitation of monomer unit}$$

2. $M^* \xrightarrow{k_d} M_0$ Nonradiative deactivation of monomer unit

3. $M^* \xrightarrow{k_1} M_0 + h\nu$ Luminescence of monomer unit

4. $M_0 + M^* \xrightarrow{k_e} \left(M_0 M^* \right)$ Formation of excimer[†]

5. $M^* + Q_0 \xrightarrow{k_q} Q^* + M_0$ Transfer of energy from monomer unit to impurity molecules

6. $Q^* \to Q_0$ Nonradiative deactivation of impurity

7. $M^* + A_0 \to A^* + M_0$ Transfer of energy from monomer unit to luminescent compound

8. $\left(M_0 M^* \right) + A_0 \to A^* + 2M_0$ Transfer of energy from excimer to luminescent
compound

9. $Q^* + A_0 \to A^* + Q_0$ Transfer of energy from impurity to luminescent
compound

10. $A^* \to A_0 + h\nu_1$ Fluorescence of luminescent compound

11. $A^* \to A_0$ Nonradiative deactivation

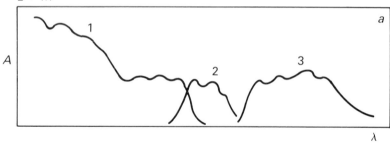

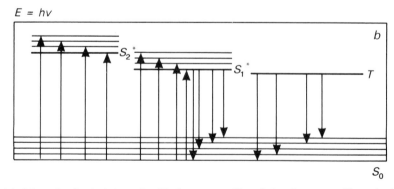

Fig. 1.1. Schematic of typical absorption (1), fluorescence (2), and phosphorescence (3) spectra (a) and
energy levels (b) of complex organic molecules.

<hr />

[†] Excimer—Unstable dimer formed as a result of association of a singlet excited and a ground state molecule.

After the transition into its excited state (stage 1), the monomer unit of a polymer undergoes radiative or nonradiative deactivation proceeding by intramolecular (stages 2 and 3) and intermolecular (stages 5, 7) mechanisms. Obviously the contribution of each of the above processes depends on the structure of the elementary unit of the polymer chain. When studying intermolecular polymeric processes it is necessary to take into account the presence in polymers of impurities introduced by the production technology, and in the form of low molecular additives such as light- and thermo-stabilizers.

As with liquid low molecular organic compounds, electronic excitation energy in polymers is transferred from the donor (polymer) to the acceptor (molecules of impurity). One of the factors determining the efficiency of energy transfer in polymers is the migration of energy, i.e. the transfer of energy between identical chromophores. In organic liquids the role of migration is revealed by the discrepancy between the theoretical and experimental values of fluorescence quenching rates [22]. The migration factor is calculated on the basis of the dipole–dipole interaction between the molecules, and the donor and acceptor overlap integral is determined by the spectral overlap of the natural absorption and fluorescence [23].

Unlike liquids, in which, owing to the random arrangement of molecules, the migration is always isotropic, the presence of a chain in the polymer may result in a preferred direction of the exciton along the macromolecule. Thus, in a polymer solution the migration is expected to be one-dimensional and intramolecular. In the bulk of a polymer, the anisotropy may disappear, as on approaching each other the chromophores of different chains are likely to switch over from one-dimensional intramolecular to three-dimensional intermolecular migration.

The possibility of changing from intramolecular to intermolecular migration after an increase in the polymer concentration of the solution was first suggested in [24]. The multi-dimensional migration occurring in block poly(vinyltoluene) was studied in [25]. The results obtained support three-dimensional migration. The experimental study of migration (intra- and intermolecular) in polyvinylcarbazole [26] enabled its authors to conclude that the possibility of migration of excitation energy is at least by an order of magnitude greater between the monomer units within a polymer chain than between the units of different macromolecules.

Excimers play an important role in the migration of electronic excitation energy in polymers. The fluorescence spectrum of an excimer is a structureless band with the maximum red shifted relative to that of monomeric emission by 5000 to 6000 cm^{-1}. The excimer has no intrinsic absorption because it does not exist in the ground (unexcited) state. Excimer emission is quite typical of solid polymers. In dilute solutions, monomeric (I_M) and excimeric (I_E) fluorescence can be observed. The ratio $I_E{:}I_M$ varies according to the nature of the solvent, strength of the solution, and RMM of the polymer.

The formation of excimers in polymers is determined by the conformation of the polymer chain and occurs with a parallel (sandwich) arrangement of the chromophoric groups [27, 28]. Accordingly, the possibility of excimer formation by chromophoric groups (excimer centres) is not very high.

For polystyrene, the formation of excimer centres is determined by the *trans-trans* configuration of the elementary units of the polymer chain [29]. Excimer fluorescence is detected in practically all aromatic vinyl polymers such as polystyrene, polyvinyltoluene, polyvinylxylene, polyvinyldiphenyl, polyvinylnaphthalene, as well as in heterocyclic vinyl polymers, especially polyvinylcarbazole.

Excimers may be formed as a result of molecular interactions of two types [30]: exciton resonance which is essentially the dipole–dipole or multipole–multipole interaction between two similar molecules in the excited and unexcited states, and charge resonance, occurring on Coulombic interaction between the positive and negative states of the molecular field. In good agreement with experiment data are calculations based on a configurational combination of these mechanisms.

As shown in [31], on interaction of molecules in excited and unexcited states, favourable conditions develop for the formation of a charge transfer complex. So far it has been singlet excimers that have been discovered in vinyl chromophore-containing polymers.

Along with the main process of complex formation, excimer emission,

$$X_{S_1^*} + X_{S_0} \rightarrow (X_{S_1} X_{S_0})^* \rightarrow 2X_{S_0} + h\nu_1$$

the deactivation via triplet–triplet annihilation occurs between two excited triplet states to yield an excited singlet state (which fluoresces) and an unexcited molecule [32].

$$X_{T_1} + X_{T_1} \longrightarrow X_2^{**} \rightleftharpoons (X_{S_1} + X_{S_0})^*$$

$$\Big\downarrow{+h\nu_1} \qquad\qquad \Big\updownarrow$$

$$2X_{S_0} \qquad\qquad 2X_{S_0} + h\nu$$

The difference between these two mechanisms is determined by the lifetime of the excimer state. If $\tau < 10^{-4}$ s, a structureless band appears, solely due to complex formation. Excimer fluorescence caused by triplet–triplet annihilation is slow ($\tau > 10^{-3}$ s), and the intensity of such fluorescence is very low (approximately 10% of the total fluorescence) [32].

Study of the concentration dependence of the I_E/I_M ratio on going from highly dilute solutions of polystyrene to concentrated solutions and solid polymers, as well as the concentration dependence of the excimer fluorescence quenching factor after the introduction of a quencher (bromostyrene) in the polymer chain, enabled the authors of [33] to conclude that the increase in concentration of excimers in dilute solutions is caused by a change in conformation of the macromolecules resulting in the formation of intramolecular excimers. In concentrated polymer solutions, the I_E/I_M ratio increased smoothly, which confirms the identical nature of the (intramolecular) excimers.

A sudden change in the I_E/I_M ratio during the transition from concentrated solutions to films is explained by the transition from one-dimensional to three-dimensional migration of electronic excitation energy. It is confirmed by comparison of theoretical calculations with experimental data on the dependence of the excimer fluorescence quenching factor of polystyrene with/without a quencher in the chain on the polymer concentration in the solution [33, 34].

On the whole, the question about the relationship between the conformation changes in aromatic vinyl polymers of the polystyrene type and the migration of electronic excitation energy is ambiguous. It has been suggested [35, 36] that the migration of energy in polymers accounts for depolarization of the excimer emission. The dependence of the I_E/I_M ratio on the concentration of polystyrene solutions is attributed [37] to the compression of polymer helices due to the mutual repulsion of polystyrene chain segments.

The authors of [38] point out that excimer formation in polystyrene is effectively quenched by atmospheric oxygen, owing to the formation of a charge transfer complex between the aromatic nucleus and the oxygen molecule. A similar effect was noticed during the photo-oxidation of a mixture of polystyrene and poly(p-hydroxystyrene) [39]. It was shown that the fluorescence spectrum of the latter polymer has a vibronic structure and cannot be regarded as an excimer. In the authors' opinion, the presence in poly(p-hydroxystyrene) of bulky hydroxyl groups does not allow for the parallel orientation of the aromatic nuclei in the polymer chain necessary to form excimers.

The conformation of the copolymer macromolecules of styrene and methyl methacrylate has a distinct bearing on the efficiency of transfer of electronic excitation energy with the participation of such fluorescence probes as pyrene and fluorene [40].

A number of studies discuss the relationship between the migration of energy and the formation of excimers in the copolymers of styrene and acrylate [41], vinylcarbazole [42], N-alkylcarbazole [43], and diacetylene [44, 45]. Polymers of the first type serve as an example to show that energy migration is limited by intramolecular excimer formations. At the same time, the existence of a relationship between the efficiency of energy migration and the formation of excimers is open to question [35].

The transfer of electronic excitation energy from the polymer (donor) to impurity molecules (acceptor) is affected both by the radiative and nonradiative mechanisms. At low acceptor concentrations ($C < 10^{-5}$ mol dm^{-3}), the transfer is only radiative, while at higher concentrations nonradiative energy transfer dominates.

The radiative transfer mechanism suggested by Birks in [46] consists of the absorption by the acceptor of the photon emitted by the donor, with the subsequent emission of photons corresponding to acceptor fluorescence. The necessary condition for operation of this mechanism is overlap of the acceptor absorption and donor fluorescence spectra, as well as high optical transmission of the medium in the spectral region corresponding to the absorption and fluorescence of the donor and the acceptor. In view of the fact that the radiative transfer of energy is effected by means of photons, the scope of this mechanism is infinite.

In the case of nonradiative transfer, travelling photons do not act as means of transferring energy, and the mechanisms are wholly different. The theory of excitation energy transfer by an inductive resonance mechanism, associated with the names of Förster and Galanin, is based on quantum mechanical concepts. The dipole–dipole interaction is effected through the presence of appropriate allowed transitions in the molecules of donor and acceptor. The probability of energy transfer is in inverse proportion to the distance between the molecules of donor and acceptor (R_0^{-6}), and the transfer is possible when R_0 is from 2 to 10 nm. In a number of cases, when the dipole–dipole transfer of electronic excitation energy is impossible (the transition is forbidden either in the acceptor or donor), the multipole interaction between the donor and acceptor becomes significant. If the probability of multipole interactions is low, the exchange resonance interaction prevails, as described by Dexter. At small distances between donor and acceptor, the contributions made by the dipole–quadrupole and quadrupole–quadrupole mechanisms are comparable with that of the dipole–dipole mechanism. Favourable conditions for the exchange-resonance transfer of excitation energy result from the molecules being in structural contact (when the donor and acceptor shells overlap) and on the collision of molecules.

The exchange-resonance mechanism is illustrated by the transfer of energy between the triplet states of the donor and acceptor discovered by Terenin and Ermolaev [47]. The theory of electronic excitation energy transfer is presented in detail in [1, 46, 48] and for polymer scintillation systems in [49, 50].

As shown above, the occurrence of energy migration in polymers should be taken into account when studying the mechanism of electronic excitation energy transfer in polymer scintillation compositions. Excitation energy migration accelerates energy transfer by inductive resonance and leads to the transfer of excitation energy by collision [48, 49, 51]. The simplest and quite exact formulae for calculating the rates of energy transfer, with due account made for the effect of migration and diffusion on excitation energy transfer by inductive resonance, are presented in [52]. The formulae derived in [52] make a simultaneous allowance for excitation energy transfer by collision. In polymers, the migration coefficient is largely dependent on the formation of excimers, which are essentially traps for excitation energy [25]. In a number of studies [22, 53, 54], energy transfer has been analysed with account taken of the formation of donor excimers. Thus, proceeding from the assumption that the excimer formation rate is much higher than that of excitation energy transfer, the authors of [53] kept in mind only the excitation energy transferred from the donor excimers. Assuming that the transfer of energy occurs before the formation of excimers, the authors of [22] calculated energy transfer only from a monomer unit. A more consistent approach to excitation energy transfer in polymers was developed by David and co-workers in [54] and Powell [25]. When studying excitation energy transfer in polyvinyltoluene with *p*-terphenyl and diphenylstilbene, Powell showed that excitation energy is transferred from the monomer unit in the first case and from excimers in the second.

The authors of [55, 56] suggested a new way of calculating the efficiency of electronic excitation energy transfer with account made of the contribution to this process of monomer (elementary) units, excimer formation by the donor, and different types of both fluorescent and quenching impurity molecules. At the same time, it was believed that the transfer of energy from the monomer units proceeded by the inductive resonance and collision mechanisms accelerated by migration, while in the case of excimers formed as a result of migration, excitation energy transfer was effected solely by inductive resonance. Allowance was also made for the non-steady formation of excimers and impurity molecules involved in the process, particularly of residual monomer.

The calculations made on real systems show good agreement with experimental data (Fig. 1.2). Fig. 1.2 illustrates the efficiency of electronic excitation energy transfer in polystyrene and poly(2,4-dimethylstyrene) in the presence of the luminophor, 2-diphenyl-5-phenyloxadiazole-1,3,4. The higher value of the total energy transfer in poly(2,4-dimethylstyrene) at similar acceptor concentrations is explained by a lower concentration of excimer centres in this polymer ($C_y = 0.03$ mol dm^{-3}) as compared to polystyrene ($C_y = 0.2$ mol dm^{-3}).

The topic of energy migration and trapping in fluorescent polymers has been reviewed recently [56A, 56B].

1.2 POLYMER FLUORESCENCE QUENCHING BY ADDITIVES

Fluorescence quenching by various additives is the most widely used method for studying the deactivation of excited states in polymers [57]. This method, as will be shown in

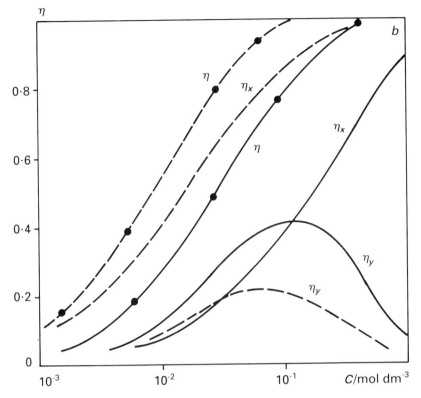

Fig. 1.2. Efficiency of electronic excitation energy transfer from the monomer unit η_x, excimer η_y, and total η in polystyrene and poly(2,4-dimethylstyrene) vs concentration of 2-diphenyl-5-phenyloxadia-zole-1,3,4 (points—experimental data).

Chapter 5, is widely used for investigating the structure and relaxation properties of hydrogels, diffusion phenomena in polymer solutions and gels, the kinetics of polyelectrolyte exchange in polymer complexes, etc.

On the basis of the kinetic scheme of photophysical processes (section 1.1), it is possible to derive kinetic equations describing the rates of different stages in terms of the concentration of donor molecules in the ground state $[D_0]$, excited state $[D^*]$, and the concentration of quencher molecules $[Q]$. In this case, the rates of stages 1 to 3 and 5 will be equal to:

$$v_1 = k_0[D_0]; \quad v_2 = k_d[D^*]; \quad v_3 = k_1[D^*]; \quad v_5 = k_q[D^*][Q].$$

In the excited or stationary state as well, as in the absence of irreversible photochemical reaction, the concentration of donor molecules is constant and is determined by [57]:

$$d[D^*]/dt = v_1 - (k_1 + k_d)[D^*] = 0.$$

In the presence of quencher molecules Q:

$$d[D^*]/dt = v_1 - (k_1 + k_d + k_q[Q])[D^*] = 0$$

from which: $v_1 = \left(k_1 + k_d + k_q[Q]\right)[D^*]$.

In the absence of quencher, the quantum emission efficiency of the donor molecule is equal to:

$$\varphi_0 = \frac{k_1[D^*]}{v_a} = \frac{k_1}{k_1 + k_d} \qquad (1.1)$$

In the presence of quencher molecules:

$$\varphi_Q = \frac{k_1[D^*]}{v_a} = \frac{k_1}{k_1 + k_d + k_q[Q]} \qquad (1.2)$$

On dividing (1.1) by (1.2), we obtain:

$$\frac{\varphi_0}{\varphi_Q} = \frac{k_1 + k_d + k_q[Q]}{k_1 + k_d} \quad \text{or} \quad \frac{\varphi_0}{\varphi_Q} = 1 + \frac{k_q}{k_1 + k_d}[Q]$$

Bearing in mind that the lifetime of excited state τ is determined by the inverse value of the sum of the rate constants of monomolecular deactivation processes $\tau = 1/\sum_i k_1 = 1/(k_1 + k_d)$, we obtain:

$$\frac{\varphi_0}{\varphi_Q} = 1 + \tau k_q[Q] \qquad (1.3)$$

where τ is the lifetime of the excited state of the donor molecule measured in the absence of quencher. Eqn. (1.3) is referred to as the Stern–Volmer equation. In graphical form it may be linearized in the form of the plot of φ_0/φ_Q vs [Q] (Fig. 1.3).

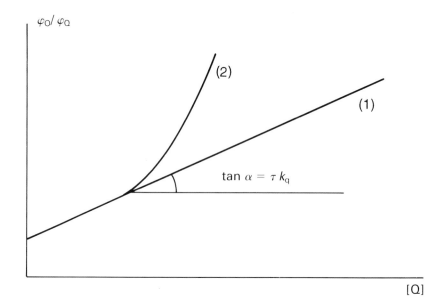

Fig. 1.3. Typical Stern–Volmer relationship (1) and possible deviation (2).

Having determined τ in the absence of quencher, the bimolecular quenching rate constant may be calculated for the transfer of electronic energy.

The Stern–Volmer equation is often used in the following forms:

$$f = \frac{k_q \tau [Q]}{1 + k_q \tau [Q]}$$ (1.4)

where f is the efficiency of energy transfer from the excited singlet or triplet state

and

$$\frac{\varphi_0}{\varphi_Q} = \frac{1 + \tau k_q [Q]}{1 + \alpha u}$$ (1.5)

where α is the efficiency of energy transfer from a triplet donor molecule to the quencher during the lifetime of the collision complex ($\alpha = 1$ in the case of diffusion-controlled triplet energy transfer) and; u is the fraction of donor molecules which have at least one quencher molecule in their immediate neighbourhood.

Eqn (1.5) is used when the φ_0/φ_Q ratio goes up rapidly with increase in quencher concentration [Q]. In this event, account should be taken of the statistical distribution of the reagents.

When $\alpha u \ll 1$ (the concentration of quencher is low or energy transfer occurs with low efficiency), eqn (1.5) takes the form of the normal Stern–Volmer eqn (1.3). As $\alpha u \to$ 1, the linear relationship between φ_0/φ_Q and [Q] is not obeyed, and the plot of φ_0/φ_Q against [Q] has the appearance illustrated in Fig. 1.3 (curve 2).

2

Monomers used in production of fluorescent polymers

Most fluorescent polymers obtained by polymerization are essentially the copolymers of low molecular luminophors, containing vinyl groups, and of conventional unsaturated compounds. The first type of copolymer is usually represented by derivatives of naphthalene, anthracene, pyrene, other condensed aromatic hydrocarbons and some heteroaromatic compounds such as carbazole, pyrazoline, etc. The second type of copolymer in general use includes styrene and derivatives of acrylic and methacrylic acids.

2.1 SPECTRAL LUMINESCENT PROPERTIES OF MONOMERS

The monomers used in the synthesis of fluorescent polymers are complex molecules whose electronic spectra lack individuality in comparison with the highly distinctive spectra of simple molecules. Ref. [1] suggested a scheme for the formation of electronic bands in the spectra of such molecules. According to this scheme, an electronic band is the result of the overlap of two distributions: (i) configurational, i.e. the distribution of probabilities of transition from each vibration level of the initial state to all vibration levels of the final state, and (ii) thermal, i.e. the distribution of molecules in the vibration levels of the initial electron state.

The theory of formation of absorption and emission bands in complex molecules is discussed in [47]. It should only be noted here that one of the deductions of this theory is as follows: the position of the electronic band maximum of a polyatomic molecule depends, in a complex way, on the distribution of Franck–Condon probabilities of separate vibronic transitions, the density of levels in the final electronic state, and the distribution of molecules in the vibration levels of the initial electronic state. Considering the above, the frequencies of absorption and emission band maxima appear to be ambiguous from the physical point of view.

As shown in ref. [2] (p. 121), most experimental studies deal primarily with the positions of absorption and fluorescence spectral peaks, which complicates interpretation of the data obtained owing to the absence of clear-cut concepts about the physical nature of these peaks. It would be more correct to consider the frequency of the electron transfer, corresponding to the difference in energy of the zero point electronic levels for both states.

The experimental determination of the frequency of the purely electron transfer is exemplified in Fig. 2.1 with styrene and 9-vinylanthracene which are often used for the synthesis of fluorescent polymers (v_{00} = 34 500 and 25 100 cm^{-1}, respectively). However, this method for determining v_{00} is not always possible. The point is that for the styrene and 9-vinylanthracene molecules, like most fluorescent compounds, the red shift of the emission peak, relative to that of absorption, equals 20–70 nm. Cases are known [5] when after absorption of a light quantum, the excited molecule undergoes a number of changes corresponding to a new distribution of the electron density, and finds itself at another lower excited level. After emission of a light quantum, it again undergoes a number of changes, this time in the opposite direction. Representative examples of such molecules are internally H-bonded compounds such as 2-(2'-hydrox-yphenyl)benzimidazole, for which the Stokes shift is as much as 100–200 nm and the absorption and emission spectra do not intersect. The excited state intramolecular pro-ton transfer of this molecule [2A] provides the basis of its reasonably efficient laser action [2B, 2C].

It is well known, ([2] p. 115), that the appearance of broad structureless bands of complex molecules is associated with intramolecular processes or, more exactly, with the strong interaction of vibrations rather than with intermolecular processes. This inter-action makes impossible the localization of energy at a certain vibrational degree of freedom: on excitation of a vibration by light, the respective energy is redistributed very quickly around the entire system of vibration levels. Such a situation is explained by the high density of vibration levels in each electronic state of a polyatomic molecule. For this reason, the time required for the establishment of equilibrium between the vibra-tional levels of the given excited state is one of the major characteristics of the complex molecule.

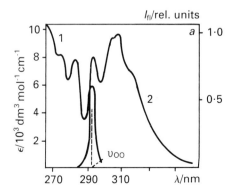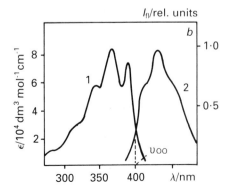

Fig. 2.1. Absorption (1) and emission (2) spectra of cyclohexane solutions of styrene (a) and 9-vinyl-anthracene (b).

Table 2.1. Classification of molecules

Main classes	Sub-classes	Groups			
		I	II	III	IV
1	2	3	4	5	6
I	σ_0	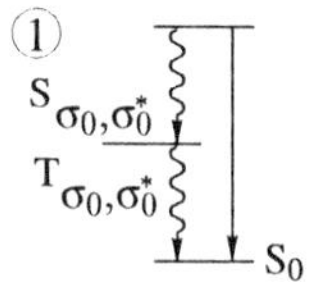 $(\sigma_0)_{\alpha\beta\mu\chi}$ ① S_{σ_0,σ_0^*} T_{σ_0,σ_0^*} S_0			
II	v_0 σ_0	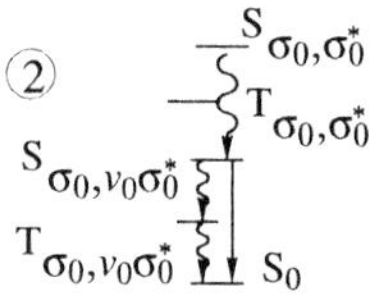 $(\sigma_0 v_0)_{\alpha\beta\mu\chi}$ ② S_{σ_0,σ_0^*} T_{σ_0,σ_0^*} $S_{\sigma_0,v_0\sigma_0^*}$ $T_{\sigma_0,v_0\sigma_0^*}$ S_0			
III	n_0 σ_0	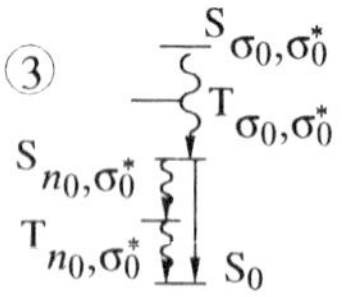 $(\sigma_0 n_0)_{\alpha\beta\mu\chi}$ ③ S_{σ_0,σ_0^*} T_{σ_0,σ_0^*} S_{n_0,σ_0^*} T_{n_0,σ_0^*} S_0			
IV	v_0 n_0 σ_0	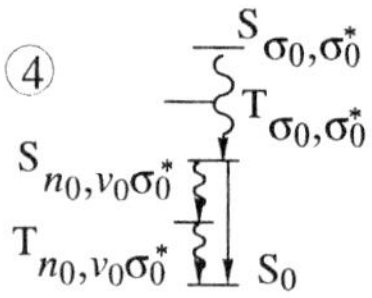 $(\sigma_0 n_0 v_0)_{\alpha\beta\mu\chi}$ ④ S_{σ_0,σ_0^*} T_{σ_0,σ_0^*} $S_{n_0,v_0\sigma_0^*}$ $T_{n_0,v_0\sigma_0^*}$ S_0			
V	π σ		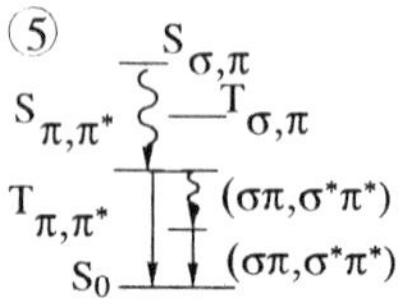 $(\sigma\pi)_{\alpha\beta\mu\chi}$ ⑤ $S_{\sigma,\pi}$ $T_{\sigma,\pi}$ S_{π,π^*} T_{π,π^*} $(\sigma\pi,\sigma^*\pi^*)$ S_0 $(\sigma\pi,\sigma^*\pi^*)$		
VI	v π σ		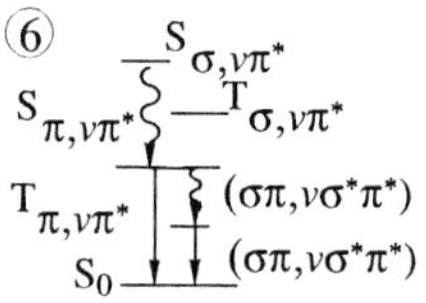 $(\sigma\pi v)_{\alpha\beta\mu\chi}$ ⑥ $S_{\sigma,v\pi^*}$ $T_{\sigma,v\pi^*}$ $S_{\pi,v\pi^*}$ $T_{\pi,v\pi^*}$ $(\sigma\pi,v\sigma^*\pi^*)$ S_0 $(\sigma\pi,v\sigma^*\pi^*)$		

by spectral luminescent properties

Groups			Subgroups			Homo-logical series	Elements	
V	VI	VII	A	B	C		Periods	Groups
7	8	9		10		11	12	13
					1	I, 1C $(\sigma_0)\alpha\beta\mu\chi$	1–2	I(H) IV(C)
					2	I, 2C $(\sigma_0\nu_0)\alpha\beta\mu\chi$	1–6	I–IV
					3	I, 3C $(\sigma_0 n_0)\alpha\beta\mu\chi$	1–2	I(H) IV(C) V(N) VI(O)
					4	I, 4C $(\sigma_0 n_0 \nu_0)\alpha\beta\mu\chi$	1–6	I–VII
				1		II, 1B $(\sigma\tau)\alpha\beta\mu\chi$	1–2	I(H) IV(C)
				2		II, 2B $(\sigma\tau\nu)\alpha\beta\mu\chi$	1–6	I–IV

Table 2.1.

Main classes	Sub-classes	Groups			
		I	II	III	IV
1	2	3	4	5	6
VII $l\ \pi\ \sigma$	$(\sigma\pi l)_{\alpha\beta\text{шх}}$		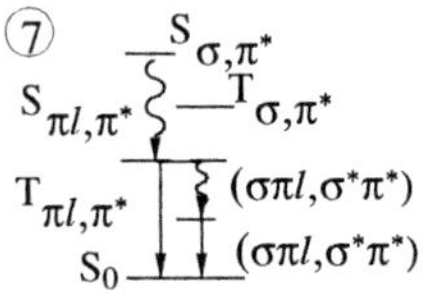 ⑦ $S_{\pi l,\pi^*}$ — S_{σ,π^*}, T_{σ,π^*}; $T_{\pi l,\pi^*}$ $(\sigma\pi l,\sigma^*\pi^*)$; S_0 $(\sigma\pi l,\sigma^*\pi^*)$		
VIII $v\ l\ \pi\ \sigma$	$(\sigma\pi l v)_{\alpha\beta\text{шх}}$		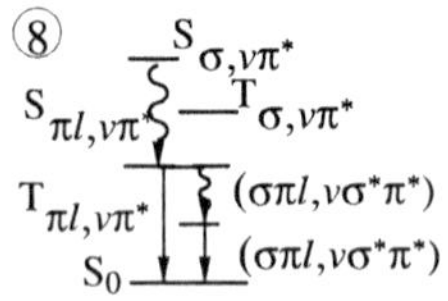 ⑧ $S_{\pi l,v\pi^*}$ — $S_{\sigma,v\pi^*}$, $T_{\sigma,v\pi^*}$; $T_{\pi l,v\pi^*}$ $(\sigma\pi l,v\sigma^*\pi^*)$; S_0 $(\sigma\pi l,v\sigma^*\pi^*)$		
IX $n\ \pi\ \sigma$	$(\sigma\pi n)_{\alpha\beta\text{шх}}$			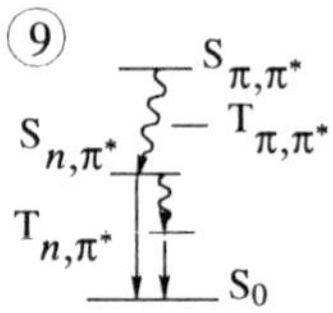 ⑨ S_{n,π^*}, S_{π,π^*}, T_{π,π^*}, T_{n,π^*}, S_0	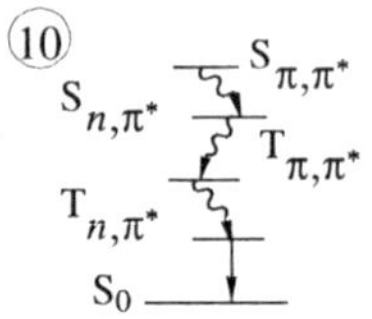 ⑩ S_{n,π^*}, S_{π,π^*}, T_{π,π^*}, T_{n,π^*}, S_0
X $v\ n\ \pi\ \sigma$	$(\sigma\pi n v)_{\alpha\beta\text{шх}}$			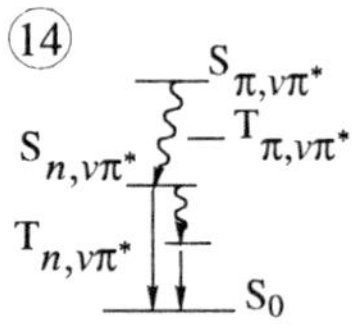 ⑭ $S_{n,v\pi^*}$, $S_{\pi,v\pi^*}$, $T_{\pi,v\pi^*}$, $T_{n,v\pi^*}$, S_0	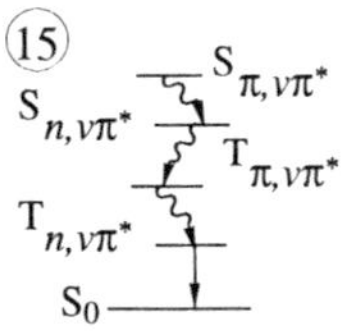 ⑮ $S_{n,v\pi^*}$, $S_{\pi,v\pi^*}$, $T_{\pi,v\pi^*}$, $T_{n,v\pi^*}$, S_0
XI $l\ n\ \pi\ \sigma$	$(\sigma\pi n l)_{\alpha\beta\text{шх}}$			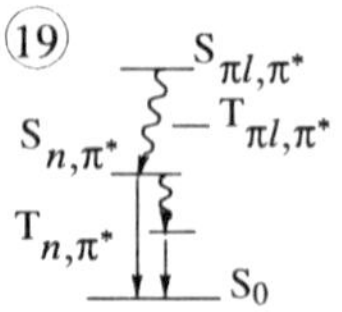 ⑲ S_{n,π^*}, $S_{\pi l,\pi^*}$, $T_{\pi l,\pi^*}$, T_{n,π^*}, S_0	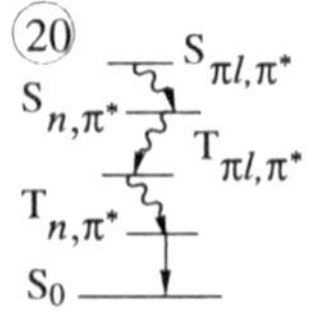 ⑳ S_{n,π^*}, $S_{\pi l,\pi^*}$, $T_{\pi l,\pi^*}$, T_{n,π^*}, S_0
XII $v\ l\ n\ \pi\ \sigma$	$(\sigma\pi n l v)_{\alpha\beta\text{шх}}$			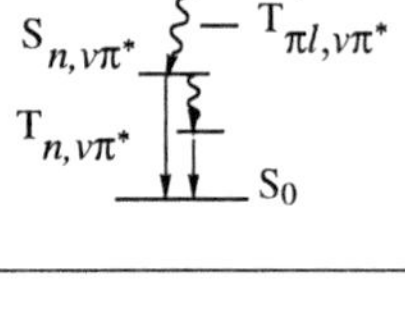 ㉔ $S_{n,v\pi^*}$, $S_{\pi l,v\pi^*}$, $T_{\pi l,v\pi^*}$, $T_{n,v\pi^*}$, S_0	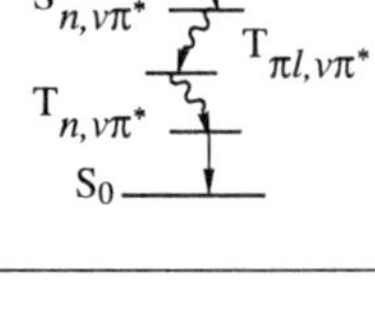 ㉕ $S_{n,v\pi^*}$, $S_{\pi l,v\pi^*}$, $T_{\pi l,v\pi^*}$, $T_{n,v\pi^*}$, S_0

Continued

Groups			Subgroups			Homo-logical series	Elements	
V	VI	VII	A	B	C		Periods	Groups
7	8	9	10			11	12	13
			3			II, 3B $(\sigma\pi l)_{\alpha\beta\mu\chi}$	1–2	I(H) IV(C) V(N) VI(O)
			4			II, 4B $(\sigma\pi l v)_{\alpha\beta\mu\chi}$	1–6	I–VII
⑪ 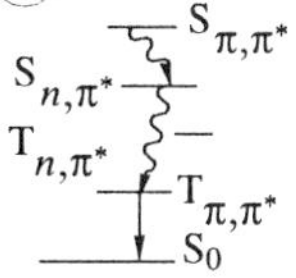 S_{π,π^*} S_{n,π^*} T_{n,π^*} — T_{π,π^*} S_0	⑫ 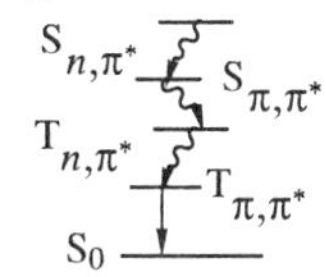 S_{n,π^*} S_{π,π^*} T_{n,π^*} — T_{π,π^*} S_0	⑬ S_{n,π^*} — S_{π,π^*} T_{π,π^*} S_0	1			III–VII, 1A $(\sigma\pi n)_{\alpha\beta\mu\chi}$	1–2	I(H) IV(C) V(N) VI(O)
⑯ 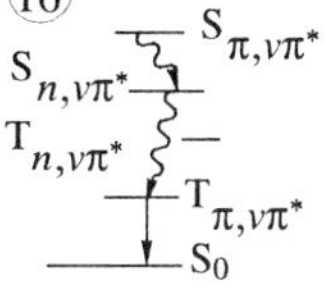 $S_{\pi,v\pi^*}$ $S_{n,v\pi^*}$ $T_{n,v\pi^*}$ — $T_{\pi,v\pi^*}$ S_0	⑰ $S_{n,v\pi^*}$ $S_{\pi,v\pi^*}$ $T_{n,v\pi^*}$ — $T_{\pi,v\pi^*}$ S_0	⑱ $S_{n,v\pi^*}$ — $S_{\pi,v\pi^*}$ $T_{\pi,v\pi^*}$ S_0	2			III–VII, 2A $(\sigma\pi v n)_{\alpha\beta\mu\chi}$	1–6	I–VI
㉑ 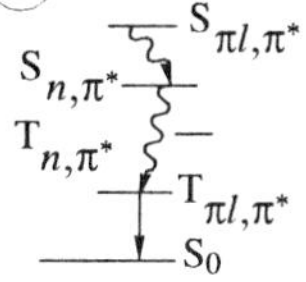 $S_{\pi l,\pi^*}$ S_{n,π^*} T_{n,π^*} — $T_{\pi l,\pi^*}$ S_0	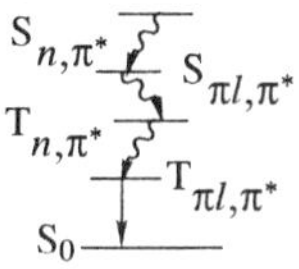 ㉒ S_{n,π^*} $S_{\pi l,\pi^*}$ T_{n,π^*} — $T_{\pi l,\pi^*}$ S_0	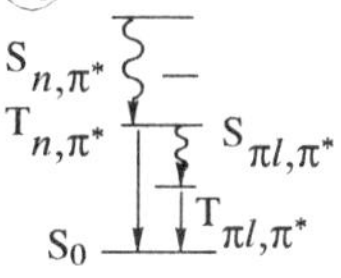 ㉓ S_{n,π^*} — $S_{\pi l,\pi^*}$ $T_{\pi l,\pi^*}$ S_0	3			III–VII, 3A $(\sigma\pi l)_{\alpha\beta\mu\chi}$	1–2	I(H) IV(C) V(N) VI(O)
㉖ 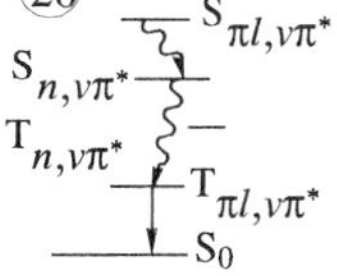 $S_{\pi l,v\pi^*}$ $S_{n,v\pi^*}$ $T_{n,v\pi^*}$ — $T_{\pi l,v\pi^*}$ S_0	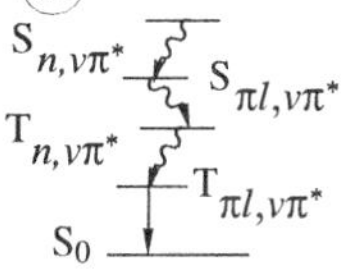 ㉗ $S_{n,v\pi^*}$ $S_{\pi l,v\pi^*}$ $T_{n,v\pi^*}$ — $T_{\pi l,v\pi^*}$ S_0	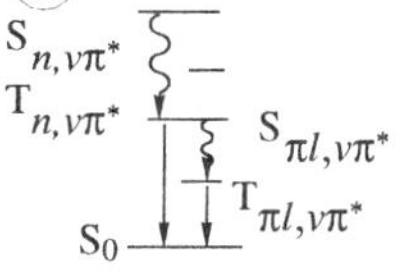 ㉘ $S_{n,v\pi^*}$ — $S_{\pi l,v\pi^*}$ $T_{\pi l,v\pi^*}$ S_0	4			III–VII, 4A $(\sigma\pi l v)_{\alpha\beta\mu\chi}$	1–6	I–VII

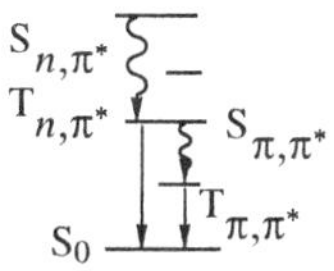

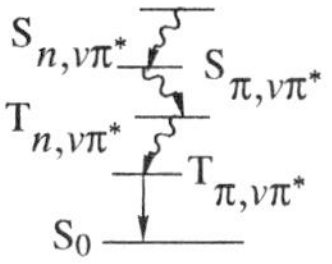

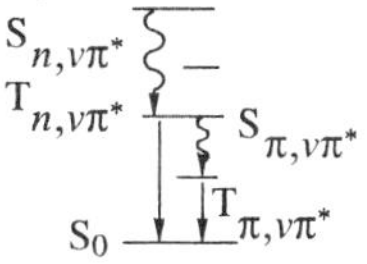

The classification or systematization of molecules plays an important role in arriving at a correct understanding of the electronic structure, spectral, luminescent, and photo-chemical properties of organic molecules [59]. The classification is based on the general rule: the spectral luminescent properties of molecules are determined by the relative positions of electronically excited state energy levels of different orbital nature and multiplicity, which changes in response to various structural factors and intermolecular interactions [3, 60]. Based on the principle of the relative positions of the lowest electronically excited state energy levels, a homologous series of molecules with similar properties would stand out very distinctly [61].

The electronic structure of organic molecules containing no heteroatoms is characterized in the ground excited state by the aggregate of molecular orbitals of the σ, π, π^*, σ^*-types. The result of introducing heteroatoms is that, after using the appropriate number of spin valencies for the establishment of chemical bonds, molecules come into possession of active orbitals involved in the formation of the electron shell, especially when they are in the excited state.

Free and occupied orbitals of heteroatoms involved in π-delocalization are generally quasi-autonomous and marked by letters v and l, respectively. In molecules undergoing sp^2- or sp-hybridization of heteroatom orbitals, the electron pair-occupied orbitals of valence heteroatoms cannot participate in π-delocalization of the system and are marked as n. So, five types of orbitals, namely σ, π, n, l, v, can be used as the basis for classifying molecules. The energy levels of the orbitals l, v, and n are determined by the ordinal number of the element, the basic quantum number, and their hybridization.

The classification of molecules comprising 12 classes of compounds distinguished by the orbital nature (σ, π, n, l, v) of the ground singlet state is given in Table 2.1. Additionally, the principle of the relative positions of electronically excited state energy levels of different orbital nature and multiplicity can be applied to explain the properties of molecules whose ground state is not singlet (for instance, with participation of d-elements, radicals, etc.). Table 2.1 presents four classes (I to IV) of molecules containing only σ_0-bonds (and orbitals n_0, v_0) and eight classes (V to XII) of molecules containing σ- and π-bonds whose heteroatoms (with orbitals l, v, n) are included directly in the conjugate or α-substituent chain.

As shown in [62], classes V to XII, characterized by a similar relative position of electronically excited state energy levels of different orbital nature and multiplicity, form five spectral luminescent groups (SLG). In turn, each SLG consists of four subgroups or analogues differentiated by the orbital nature of the state of corresponding multiplicity. The classification of molecules by spectral luminescent properties makes it possible to explain and predict any change in these properties due to various structural factors and intermolecular interaction. Such factors may cause a transition from one homologous series to another within one SLG as a result of which the molecular properties change in a monotonic way and remain similar. The above factors may be responsible for the inversion of energy levels and the transition between the homologous series of different SLGs accompanied by abrupt quantitative and qualitative changes in the molecular properties.

The overwhelming majority of monomers used for the synthesis of fluorescent polymers belong to SLG V. They are characterized by a high quantum fluorescence yield, especially for molecules of class $\sigma\pi nl$ (up to 0.7–0.9) and a very low phosphorescence yield. At the same time, molecules of class $\sigma\pi nl$ show good photochemical resistance, which is also very important.

Discussed below are the methods for synthesis of some typical monomers required for the synthesis of fluorescent polymers of two types: unsaturated compounds capable of polymerization and luminophors with two functional groups capable of polycondensation.

2.2 SYNTHESIS OF MONOMERS

2.2.1 Polymerization of monomers with vinyl groups

Naphthalene derivatives
α-Vinylnaphthalene (I). Prepared from α-naphthylmagnesiumbromide and acetaldehyde (US Patent 1985844).

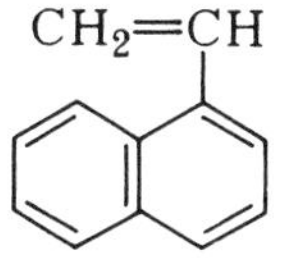

$$CH_2{=}CH$$

I

α-Naphthyldiazomethane (II). Prepared from α-naphthaldehyde and hydrazine hydrate with subsequent oxidation of α-naphthalhydrazone by mercury(II) oxide [63].

$$CHN_2$$

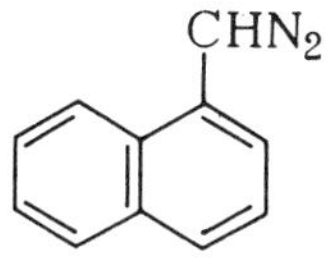

II

α-Naphthyl methacrylate (III). Prepared by the Schotten–Baumann reaction from sodium naphthylate and methacryloyl chloride [64].

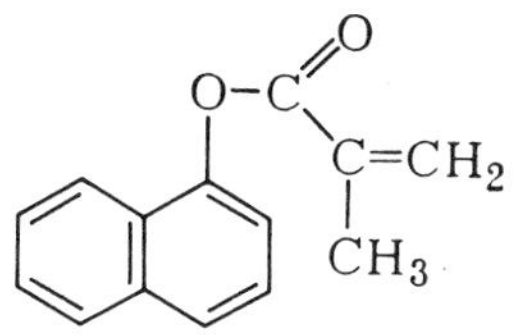

III

Anthracene derivatives
9-Anthryldiazomethane (IV). Prepared from 9-anthraldehyde and hydrazine hydrate with the subsequent oxidation of 9-anthral hydrazone by mercury(II) oxide [63].

$$CHN_2$$

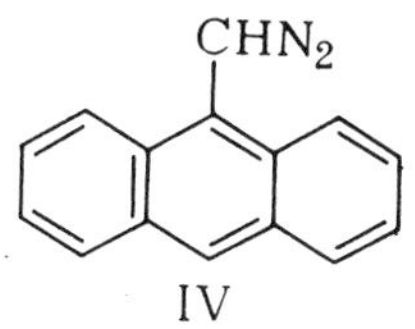

IV

9-Vinylanthracene (V). Prepared by reduction of 9-acetylanthracene by lithium tetrahydro-aluminate to obtain 1,9-anthrylethanol and subsequent heating of the latter with potassium hydrogen sulphate [65].

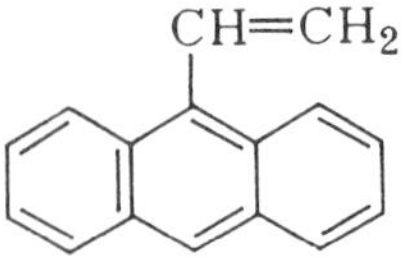

$$CH{=}CH_2$$

V

9-Methacryloylhydroxymethylanthracene (VI). Prepared from 9-anthryldiazomethane and methacrylic acid [66] or 9-chloromethylanthracene and methacrylic acid [67].

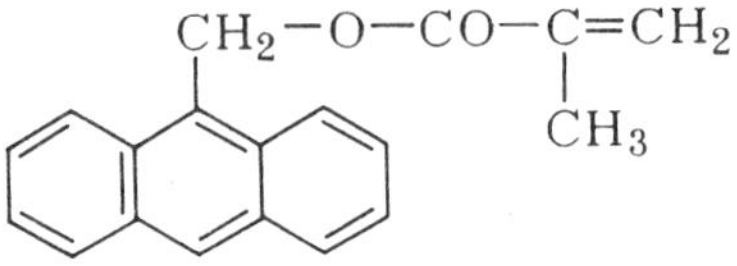

$$CH_2{-}O{-}CO{-}C{=}CH_2$$
$$CH_3$$

VI

9-Methacryloylhydroxymethyl-10-methylanthracene (VII). Prepared in the same way as 9-methacryloylhydroxymethylanthracene from 9-(10-methyl)anthryldiazomethane [66] and 9-chloromethyl-10-methylanthracene [67].

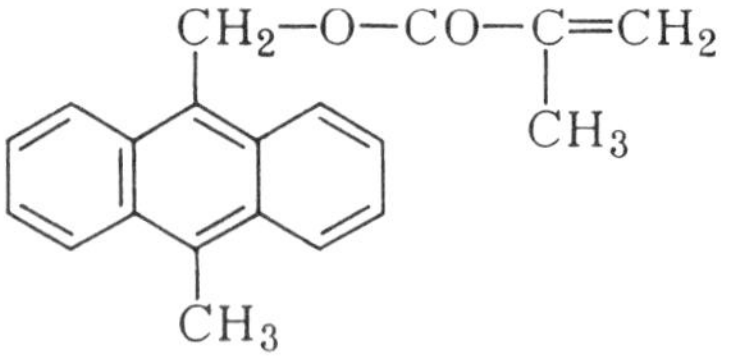

$$CH_2{-}O{-}CO{-}C{=}CH_2$$
$$CH_3$$

$$CH_3$$

VII

9,10-Bis(methacryloylhydroxymethyl)anthracene (VIII) and 9,10-bis(acryloylhydroxy-methyl)anthracene (VIII). Prepared from 9,10-bis(chloromethyl)anthracene and acrylic acid (R—H) or methacrylic acid (R = CH$_3$) [67].

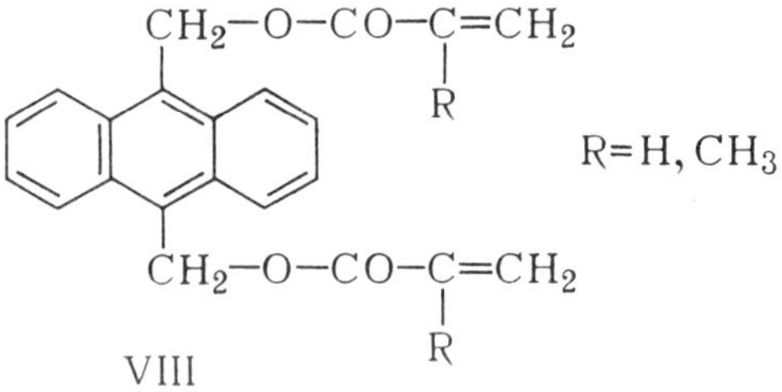

$$CH_2{-}O{-}CO{-}C{=}CH_2$$
$$R$$

$$R{=}H, CH_3$$

$$CH_2{-}O{-}CO{-}C{=}CH_2$$
$$R$$

VIII

4-Vinyl-9,10-diphenylanthracene (IX). Prepared from 9,10-anthraquinone and phenyl-magnesium bromide with subsequent treatment of 9-phenyloxanthranol with 4-magne-siumbromostyrene and reduction of the intermediate product with hydrogen iodide [68].

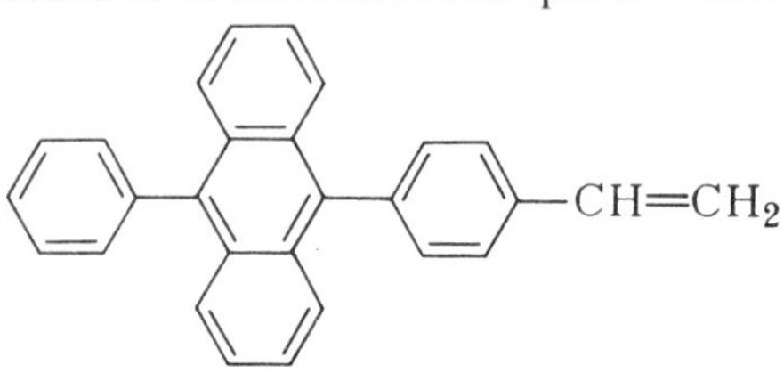

$$-CH{=}CH_2$$

IX

1-Vinylpyrene (X). Prepared from pyrene and *N,N*-dimethylformamide by the Vilsmeier reaction to obtain pyrene-1-aldehyde with subsequent treatment of the latter with methylenetriphenylphosphorane by the Wittig reaction [69].

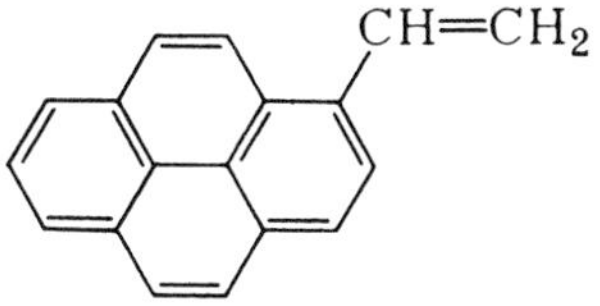

$CH=CH_2$

X

1-Pyrenylmethyl methacrylate (XI). Prepared from 1-hydroxymethylpyrene and methacryloyl chloride [70].

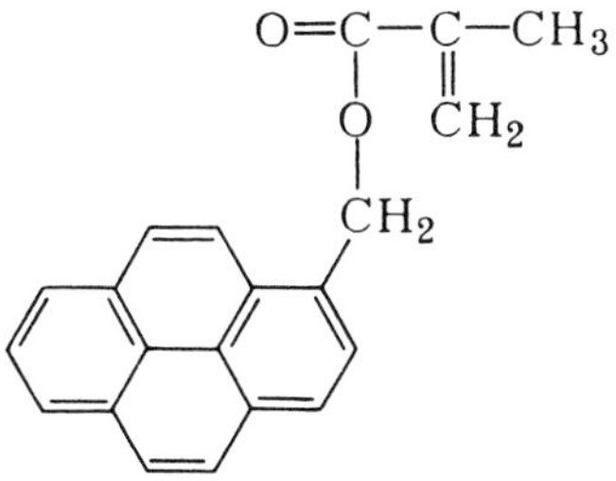

$O=C-C-CH_3$
O CH_2
CH_2

XI

3-Vinylphenanthrene (XII). Prepared from phenanthrene-9-aldehyde and methyltriphosphonium bromide [71].

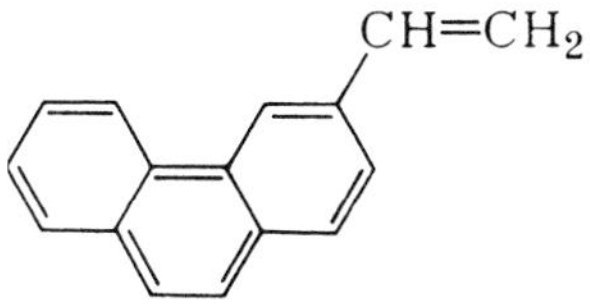

$CH=CH_2$

XII

N-Vinylcarbazole (XIII). Prepared from carbazole and β-chloroethyl ester of *p*-toluenesulfonic acid with subsequent treatment of 9-(*N*-chloroethyl)carbazole with alcoholic alkali [72].

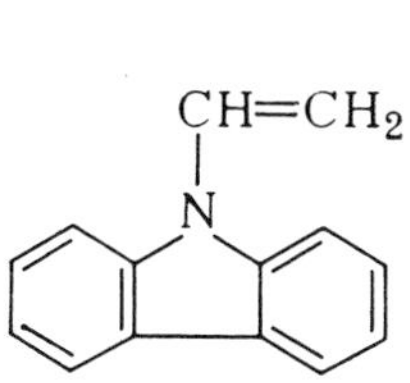

$CH=CH_2$
N

XIII

9-Vinylacridine (XIV). Prepared from 9-methylacridine by the Mannich reaction with subsequent methylation and decomposition by the Hofmann reaction of 9-(2-dimethyl-aminoethyl)acridine formed at the first stage [73].

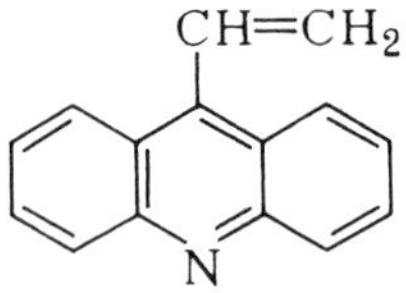

CH=CH₂

XIV

In their spectral luminescent properties, the vinylated derivatives of condensed aro-matic hydrocarbons and their heteroaromatic analogues are much like the corresponding compounds containing no vinyl groups, especially when the double bond is not in close proximity with the π-system of the aromatic rings. By contrast, when the vinyl sub-stituent is in the immediate vicinity of the aromatic ring, the extension of the conjugated system causes a red shift of the absorption and emission bands and normally increases the fluorescence quantum yield. Thus, the introduction of a vinyl group in the 9-position of the anthracene ring leads to a bathochromic shift of the fluorescence band by 20 nm and an increase in the fluorescence quantum yield from 0.22 to 0.60 [74].

The effect of substituents on the fluorescent properties of aromatic compounds is cov-ered in greater detail in ref. [5] p. 16.

2.2.2 Bifunctional monomers used in polycondensation

Naphthalene derivatives
2,6-Bis(N-methyl-aminomethyl) naphthalene (XV). Prepared from 2,6-bis(bromo-methyl)naphthalene and methylamine [75].

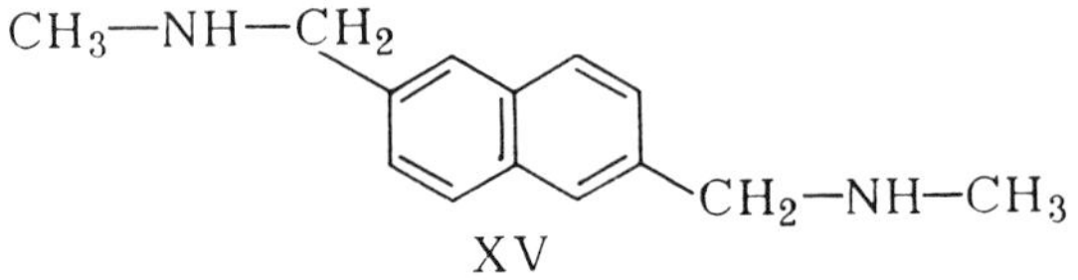

CH₃—NH—CH₂

CH₂—NH—CH₃

XV

2-[(1-Naphthyl)methyl]-1,3-propanediol (XVI). Prepared by condensation of 1-chloro-methylnaphthalene and diethyl malonate with subsequent reduction of diethyl [(1-naph-thyl)methyl] malonate with lithium tetrahydroaluminate [76].

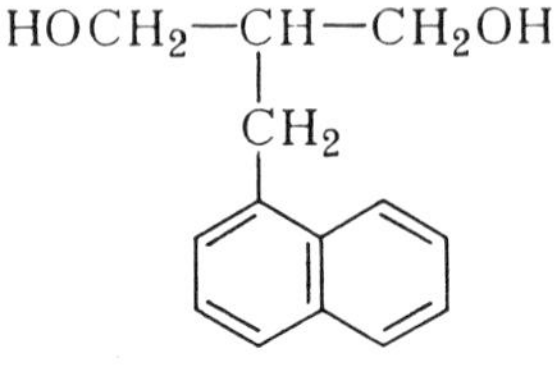

HOCH₂—CH—CH₂OH
|
CH₂

XVI

Anthracene derivatives

9,10-Di(aminomethyl)anthracene (XVII). Prepared by condensation of anthracene with *N*-bromomethylphthalimide and subsequent hydrolysis of the compound formed with aqueous alkali (US Patent 1873402).

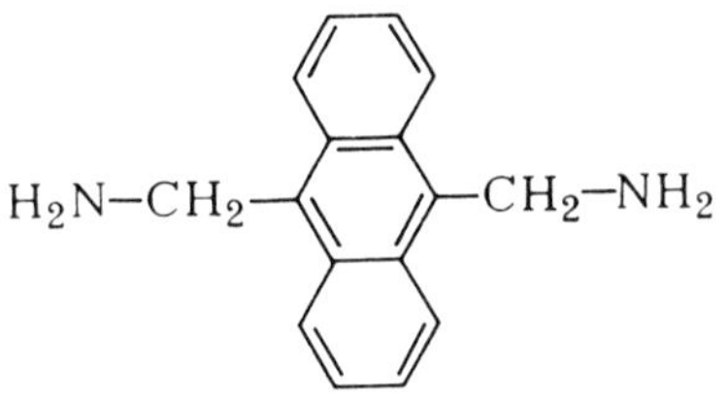

$$H_2N-CH_2- \quad -CH_2-NH_2$$

XVII

Bis(p-aminophenyl)-9,10-anthracene (XVIII). Prepared by condensation of 9,10-anthra-quinone and aniline hydrochloride in boiling aniline [77].

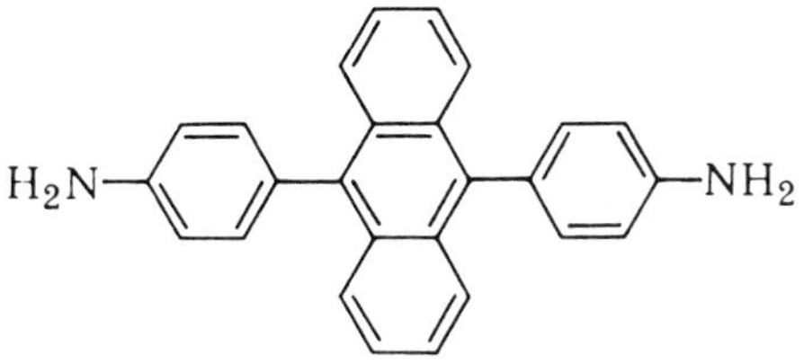

$$H_2N- \quad -NH_2$$

XVIII

9,10-Dianilinoanthracene (XIX). Prepared during the interaction of 9,10-anthraquinone and aniline in the presence of aluminium chloride and zinc powder ([5] p. 32).

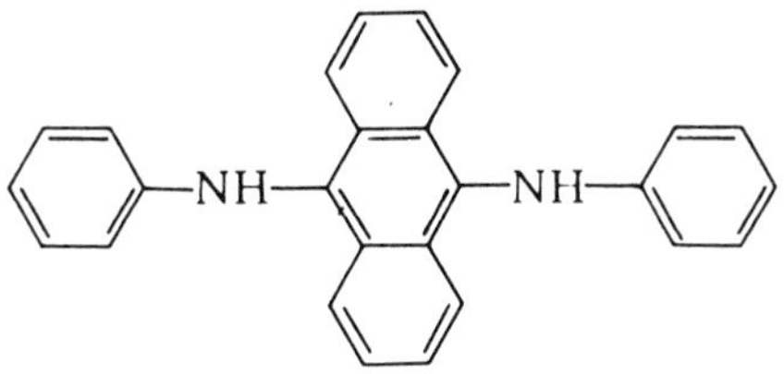

$$-NH- \quad -NH-$$

XIX

2-(9-Anthryl)methyl-1,3-propanediol (XX). Prepared in the reaction of 9-chloromethylan-thracene and diethyl malonate with subsequent reduction of diethyl(9-anthryl)methyl malonate with lithium tetrahydroaluminate [78].

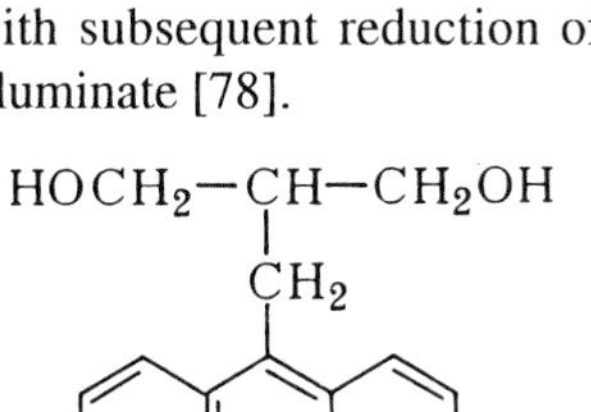

$$HOCH_2-CH-CH_2OH$$
$$CH_2$$

XX

Bis(2-hydroxyethyl)-9-anthrylmethyl malonate (XXI). Prepared by the reaction of diethyl-(9-anthryl)methyl malonate with ethylene glycol [79].

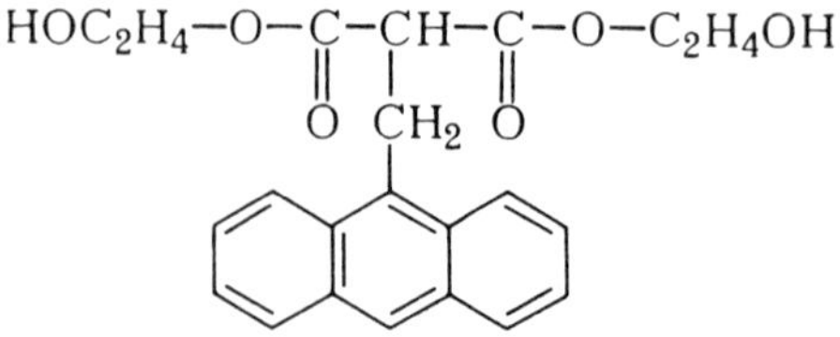

$$HOC_2H_4-O-\underset{\underset{\displaystyle CH_2}{O}}{\overset{\overset{\displaystyle }{||}}{C}}-CH-\overset{O}{\overset{||}{C}}-O-C_2H_4OH$$

XXI

9,10-Bis-4-(hydroxybutylene oxycarbonyl)dianthracene (XXII). Prepared from the chloro-anhydride of 9-carboxyanthracene and 1,4-butylene glycol with the subsequent photo-dimerization of 4-hydroxybutyl-9-anthroate in tetrahydrofuran solution on irradiation (Hg lamp, $\lambda < 320$ nm) [80].

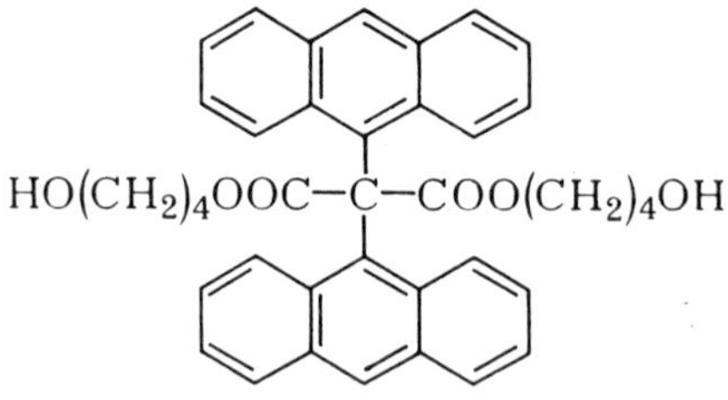

$$HO(CH_2)_4OOC-\overset{}{\underset{}{C}}-COO(CH_2)_4OH$$

XXII

9,10-Anthracene dipropionic acid (XXIII). Prepared by the reaction of 9,10-dichloro-methylanthracene and sodiomalonic ester with subsequent saponification of 9,10-diethyl-1,1,1-tetracarbethoxyanthracene with the ethanolic alkali solution and decarbo-xylation [81].

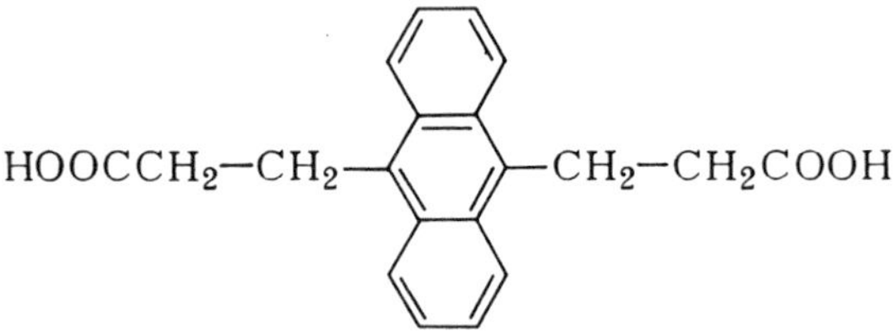

$$HOOCCH_2-CH_2--CH_2-CH_2COOH$$

XXIII

9,10-Bis(p-carboxyphenyl)anthracene (XXIV). Prepared by the treatment of anthracene with denitrated *p*-aminobenzoic acid (French Patent 1085860).

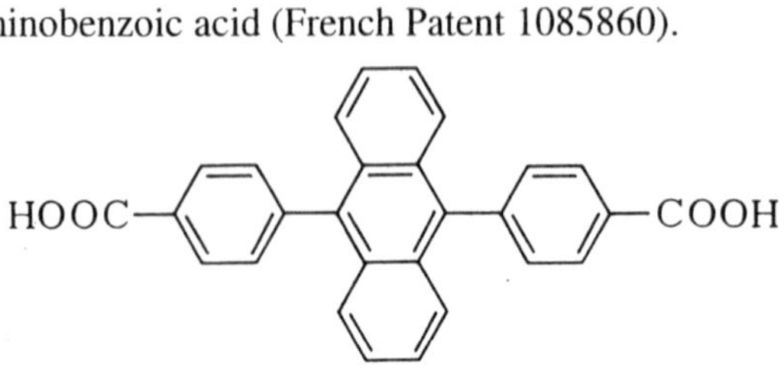

$$HOOC---COOH$$

XXIV

9,10-Bis(p-carbomethoxystyryl)anthracene (XXV). Prepared from 9,10-diformyl anthra-cene [82].

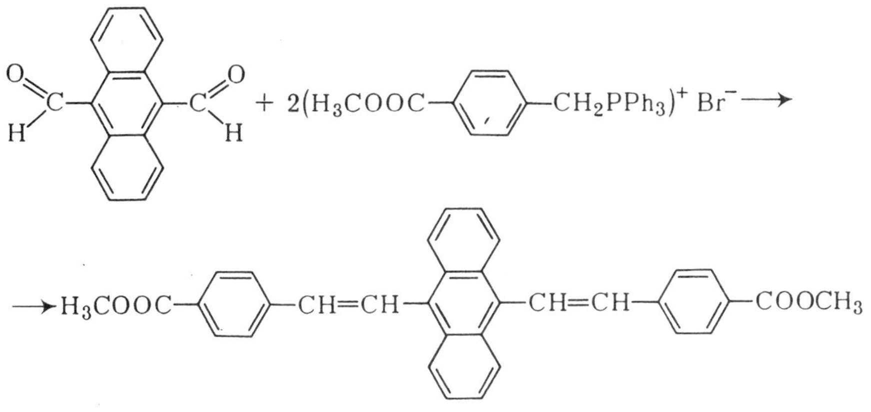

$$\text{(anthracene-9,10-dicarbaldehyde)} + 2(\text{H}_3\text{COOC}-\!\!\bigcirc\!\!-\text{CH}_2\text{PPh}_3)^+ \text{ Br}^- \longrightarrow$$

$$\longrightarrow \text{H}_3\text{COOC}-\!\!\bigcirc\!\!-\text{CH}=\text{CH}-\text{(anthracene)}-\text{CH}=\text{CH}-\!\!\bigcirc\!\!-\text{COOCH}_3$$

XXV

Derivatives of pyrene and other aromatic hydrocarbons

1-Pyrenyl-1,3-propane diol (XXVI). Prepared by condensation of 1-chloromethylpyrene and diethyl malonate with subsequent reduction of diethyl [(3-pyrenyl)methyl] malonate with lithium tetrahydroaluminate [76].

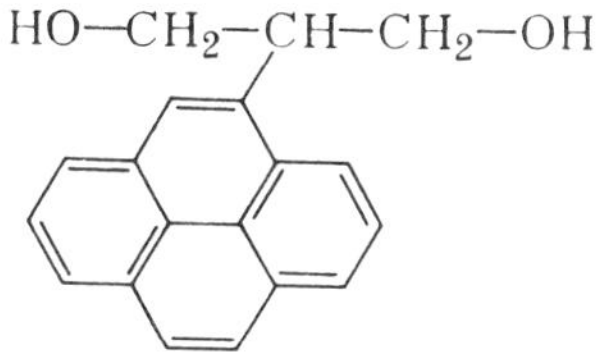

$$\text{HO}-\text{CH}_2-\text{CH}-\text{CH}_2-\text{OH}$$

XXVI

Disodium 4,4'-diamino-2,2'-stilbenedisulfonate (XXVII). Prepared by oxidation of 4-nitrotoluene-2-sulfonic acid with atmospheric oxygen and subsequent reduction of the dinitro derivative ([5] p. 49).

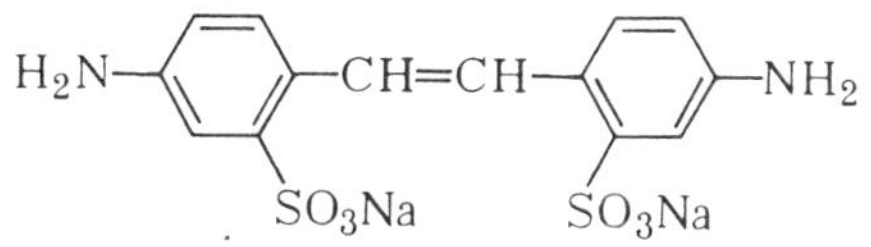

$$\text{H}_2\text{N}-\!\!\bigcirc\!\!-\text{CH}=\text{CH}-\!\!\bigcirc\!\!-\text{NH}_2$$
$$\overset{|}{\text{SO}_3\text{Na}} \qquad \overset{|}{\text{SO}_3\text{Na}}$$

XXVII

1,4-Bis(p-carboxystyryl)benzene (XXVIII). Prepared in the same way as 9,10-bis(p-carboxystyryl)anthracene from terephthalaldehyde [82].

$$\text{HOOC}-\!\!\bigcirc\!\!-\text{CH}=\text{CH}-\!\!\bigcirc\!\!-\text{CH}=\text{CH}-\!\!\bigcirc\!\!-\text{COOH}$$

XXVIII

Derivatives of heteroaromatic compounds

3,6-Diamino-N-methylphthalimide (XXIX). Prepared by nitration of 3-acetylamino-*N*-methylphthalimide and subsequent reduction of 3-acetylamino-6-nitro-*N*-methylphthalimide with tin(II) chloride and the hydrolysis of 3-acetylamino-6-amino-*N*-methylphthalimide with muriatic acid [83].

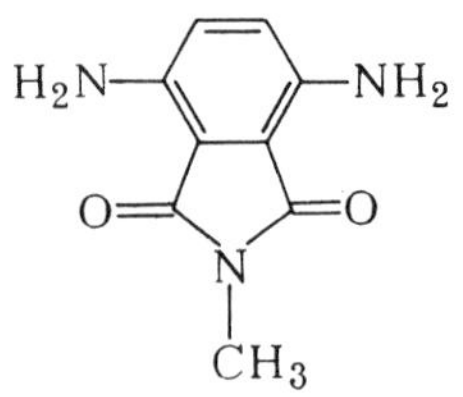

XXIX

2,8-Diamino-3,7-dimethylacridine (XXX). Prepared by condensation of 2,4-diaminotoluene with oxalic acid [84].

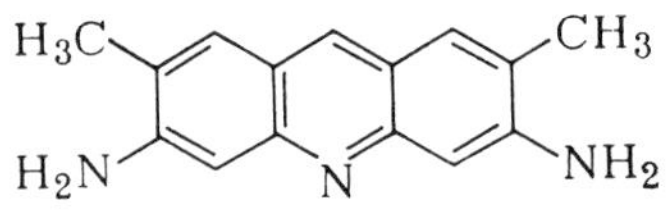

XXX

Unsubstituted rhodamine (base) (XXXI). Prepared by heating 3-aminophenol with phthalic anhydride in the presence of sulfuric acid [85].

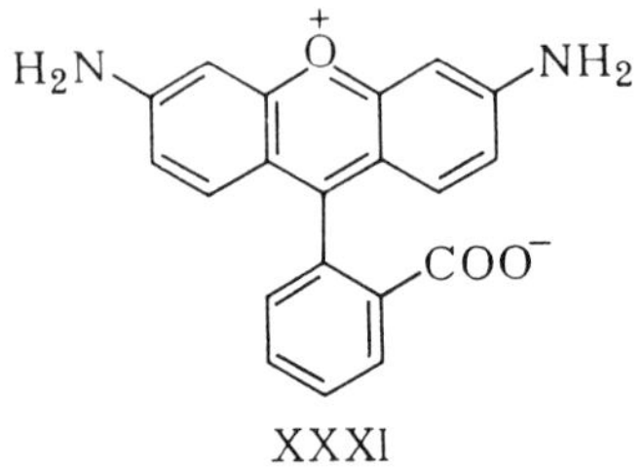

XXXI

3,6-Diamino-9-ethylcarbazole (XXXII). Prepared by nitration of 9-ethylcarbazole with nitric acid and subsequent reduction of 3,6-dinitro-9-ethylcarbazole with tin(II) chloride [86].

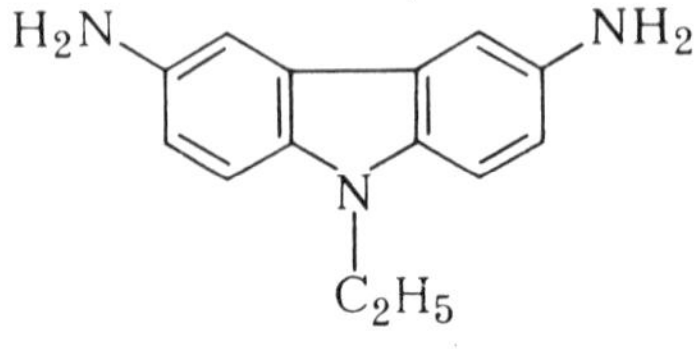

XXXII

6-Carbazyl-2,2-dihydroxymethylhexane (XXXIII). Prepared by treatment of 9-(4-bromo-butyl)carbazole with diethyl methylmalonate sodium salt and subsequent reduction of the reaction product with lithium tetrahydroaluminate [87].

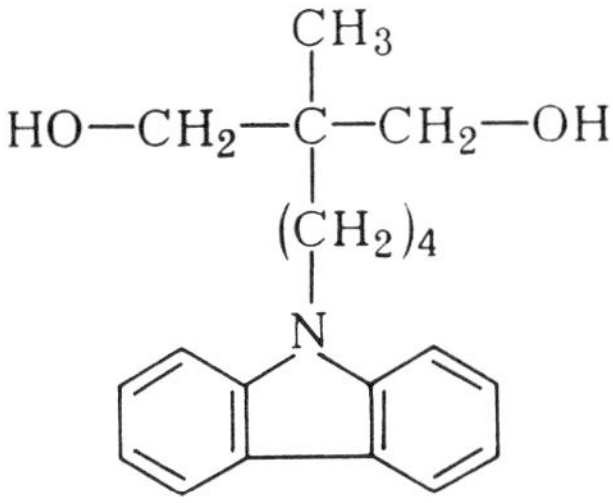

$$CH_3$$
$$HO-CH_2-\underset{\underset{\displaystyle N}{\underset{|}{(CH_2)_4}}}{\overset{|}{\underset{|}{C}}}-CH_2-OH$$

XXXIII

Many of the above bifunctional derivatives capable of polycondensation are organic dyes. In [4] p. 191 this class of compounds is characterized as having the following features.

(i)   An extremely high intensity of the absorption band in the visible region [$\varepsilon \sim 10^5$ dm^3 mol^{-1} cm^{-1}].

(ii) The highest intensity is usually displayed by the 0,0-band of the $S_1 \leftarrow S_0$ electronic transition. The absorption corresponding to this transition is the superposition of two or three vibronic bands. On the average, the vibrational states of S_1 are well separated (some 1200 to 1600 cm^{-1}).

(iii) The intensity of the vibronic bands falls rapidly on moving away from the 0,0-transition. Analysis of these data was made in terms of the Franck–Condon principle (according to which the deactivation of electronic excitation via a change to vibrational energy is possible only when the nuclear configuration of the molecule in its excited state is similar to that of the ground state). This analysis enabled the conclusion to be made that the potential energy curves of states S_1 and S_0 are similar and arranged vertically, one above the other.

(iv) The high extinction coefficient (transition $S_1 \leftarrow S_0$) is explained by the fact that the natural fluorescence lifetime is roughly 10^{-9} s. This short lifetime increases the fluorescence efficiency compared with other processes leading to the nonradiative deactivation of the S_1 state.

(v)   The fluorescence spectrum is usually mirror symmetric about the absorption spectrum, and the Stokes shift of the fluorescence band (relative to that of absorption) is *ca.* 500 cm^{-1}.

(vi) Owing to their dipolar nature, the molecules of dyes with strong polarizability are prone to association and aggregation at high concentrations.

(vii) Dyes are characterized by a low $S_1 - T_1$-splitting (4500–1400 cm^{-1}).

(viii) Dyes have a low fluorescence quantum yield (less than 0.1) and a long phosphorescence lifetime (up to several seconds in rigid glasses at 77 K).

(ix) Many dyes exhibit delayed 'thermally activated' fluorescence, i.e. the $S_1 \rightarrow S_0$ fluorescence which follows kinetically the thermal excitation $S_1 \leftsquigarrow T_1$.

Many luminescent dyes with two functional groups, such as unsubstituted rhodamine and acridine yellow, are inclined towards association. These compounds dissolve only in polar solvents and exist in them mainly in the form of associates, and, only at low concentrations, as monomers [88]. The susceptibility of dyes to form associates varies according to their structural features and is also dependent on the nature of the solvent. The association of dye molecules in solution usually occurs owing to van der Waals' forces and hydrogen bond formation.

During the formation of associates, the deformation of the absorption spectra depends on the intermolecular force in the associates, which is in many respects determined by their mutual orientation and distance. As the molecules approach each other, interaction of the resonance type is sure to happen, leading to splitting of the electronic levels including S_1.

The dimer may have two transitions, with the longer wavelength absorption bands being shifted relative to that of the monomer to both short and long wavelength regions. For instance, the authors of [89] found that with increase in the concentration of fluoresceine in aqueous solution, the intensity of the long wavelength absorption maximum rapidly decreases. Simultaneously a dimer band is formed which is shifted towards shorter wavelengths.

When studying the spectral and luminescent properties of unsubstituted rhodamine, uranine and other xanthene dyes, it should be noted that the quinoid structure attributed to these by most authors does not correspond to the spectral characteristics of these compounds. According to [89, 90], the electronic system of the xanthene ring is very much like the π-system of the anthracene ring. The derivatives of anthracene, acridine, and xanthene dyes share the following spectral properties:

(i)   The lowest $S_1 \rightarrow S_0$ transition responsible for the long wavelength absorption and fluorescence of the β-derivatives of anthracene, acridine, and xanthene dyes, is polarized along the long axis of the molecule. The energy of this transition is much the same for similar derivatives of the above compounds.

(ii) The spectra of the acridine and xanthene dyes have an absorption band similar in intensity and position to the ^1L_a-band of anthracene. The transition moment responsible for the appearance of this band is polarized along the short molecular axis in all three types of compound.

(iii) The T_1-level energy responsible for the phosphorescence of these compounds is much the same.

The features common to the polarized spectra of acridine and xanthene dyes are also covered in [91].

Thus, it may be concluded that in their most essential spectral characteristics, the compounds of these three classes are very much alike. There are strong grounds to believe that the ionic form of the xanthene ring is π-isoelectronic in relation to the anthracene and acridine rings (XXXIV). The spectral properties of these compounds are distinguished only by the difference in symmetry of the rings and the nature of the heteroatom.

Of the xanthene dyes, along with unsubstituted rhodamine, rhodamine 6 G (XXXV) may serve as a monomer for polycondensation purposes owing to its two secondary amine groups. The substitution of hydrogen by the ethyl group results in red shifts of the bands in the absorption and emission spectra and a minor increase in the fluorescence yield.

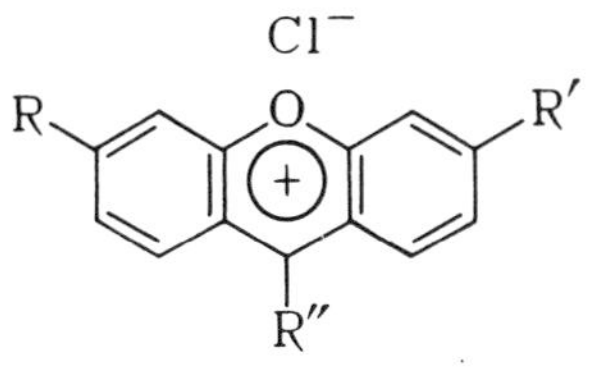

$$\text{XXXIV}$$

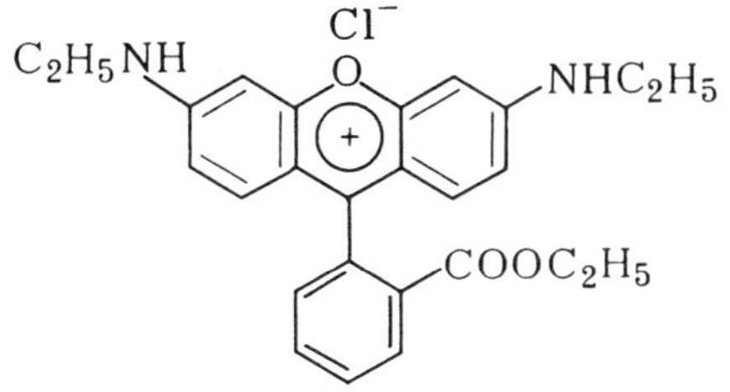

$$\text{XXXV}$$

It is more convenient to perform the synthesis of fluorescent polycondensed polymers of the polyamide type (see Chapter 4), based on unsubstituted rhodamine or its analogues, when the dyes are in their basic form (XXXVI).

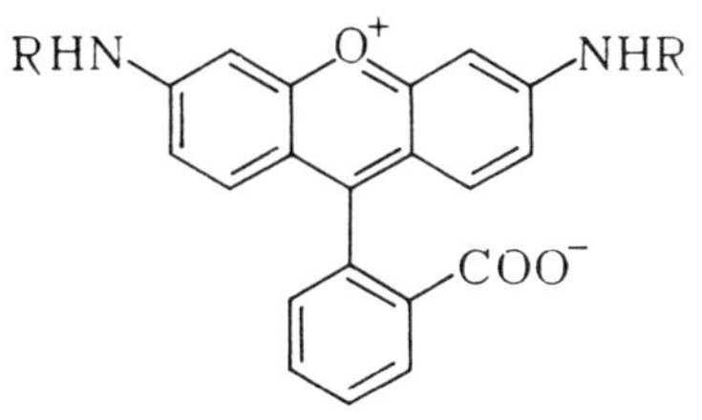

$$\text{XXXVI}$$

where R = —H (XXXI), —C₂H₅

As reported in [91], the unsubstituted rhodamine base is formed from the hydrochloride of the dye by treatment in the 5% alkaline solution. In stronger alkaline solutions, the anion of the carbinol derivative XXXVII is formed.

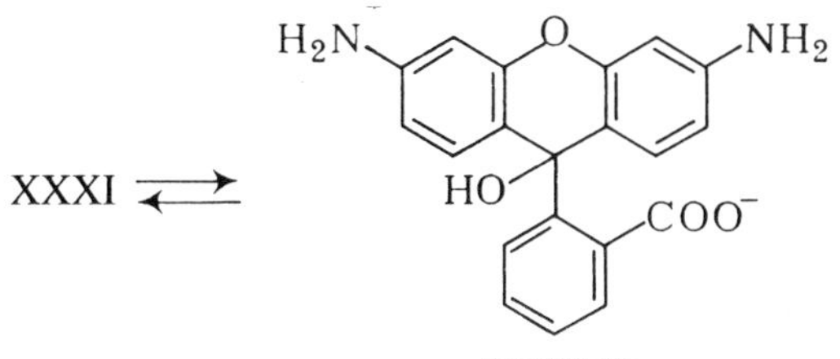

$$\text{XXXI} \rightleftharpoons$$

$$\text{XXXVII}$$

Study of the absorption spectra of rhodamine base XXXI in aqueous solutions of different acidity (pH = 0.1–14) showed that the position of the maximum in the spectrum is independent of pH (λ_{max} = 500 nm). The authors of [85] explain such behaviour by the fact that the base has a bipolar ionic structure (XXXI) in which the negative charge is concentrated on the carboxyl residue, while the positive charge is distributed, as in the case of basic triphenylmethane dyes, among both the amine groups and the atoms of the xanthene group. Under the action of acids, this bipolar ion is liable to add one proton to the carboxyl group residue to form the cation **XXXVIII**. This explains the ability of rhodamines to form relatively stable monobasic salts. Besides, since the proton added to the carboxyl residue does not change the electronic structure determining the colour of the compound, the colour of the dye remains unaltered on changing from the neutral to the acidic medium.

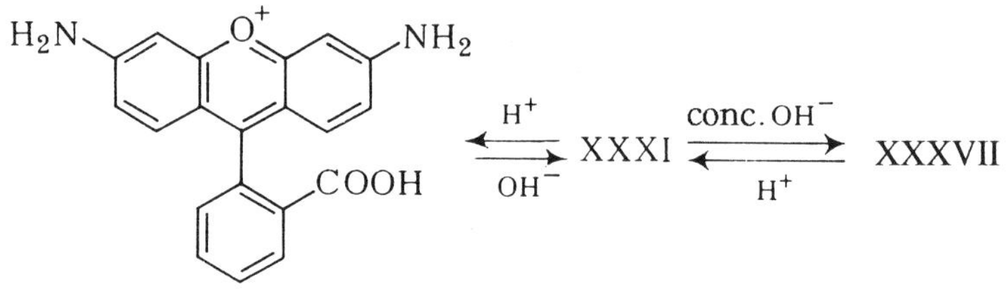

XXXVIII

The tendency of the base (**XXXI**) to add another proton at the expense of the unshared electron pairs of the amine group (nitrogen) or heterocyclic group (oxygen) is weakened owing to the delocalization of the positive charge, as a result of which the ion of the dibasic rhodamine salt can be formed only in concentrated acids. At the same time, according to [85], the electronic structure of the dye changes, which is confirmed by the appearance in the absorption spectrum of highly acidic solutions of rhodamine, of a new band with $\lambda_{max} = 480$ nm and the depression of the intrinsic absorption with $\lambda_{max} = 500$ nm. The stability of the dibasic salt is low, and when diluted with water it hydrolyses to form the cation (**XXXVIII**).

During attack by hydroxyl ions in strongly alkaline solutions, the unsubstituted rhodamine undergoes transformations leading to bleaching of the solution [85]. Since on acidification of the decolorized alkaline solution, the dye quickly regains colour, such decolorization is probably caused by the addition of the hydroxyl groups and formation of the anion of the corresponding carbinol derivative (**XXXVII**) found only in highly alkaline solutions. The rhodamine base used in the synthesis of fluorescent polyamides does not fluoresce in solution. However, acylation of the amine groups by isophthaloyl chloride causes protonation of the carboxyl residue by the hydrogen chloride released during the reaction, [8] p. 1198.

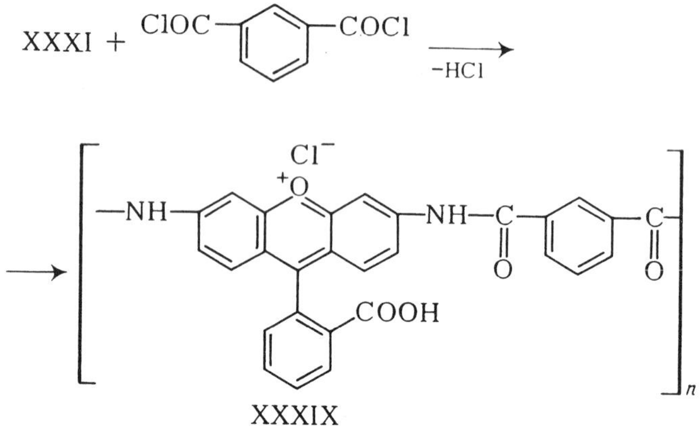

XXXIX

The absence of fluorescence in rhodamine base solutions [8] correlates with the results of study into the spectral and luminescent properties of various forms of rhodamines C and 3B [92]. It has been discovered that the basic forms of rhodamines C and 3B do not fluoresce in benzene solution. However, addition to them of the polar 1-propanol or acidification of the solutions, leads to the development of intense fluorescence.

In [93] the suggestion is made that the bond between the protons and the nitrogen atoms of the amine groups of the dye, formed as a result of acidification, is electrostatic rather than covalent in nature. Consequently, the effect of the proton consists of exclusion from the conjugated bond of the electron pair of the nitrogen atom. At the same time, it can be expected that the hydroxyl group of basic rhodamine 3B establishes a covalent bond with the (hydrocarbon) atom adjacent to the amine group. In this case the electron pair of the atom will not be involved in the conjugated bond either.

So, both in the protonated molecules of rhodamine 3B and in the rhodamine base, the electron pair of the nitrogen atom is excluded from the conjugated chain, which makes them closely similar in their electron structure. It also explains the similarity between the absorption spectra of the protonated molecules of rhodamine 3B to that of the basic form (λ_{max} = 495 nm). The authors of [92] express the opinion that the absence of fluorescence in basic solutions is associated with the nonradiative exchange of the excitation energy caused by the establishment of a covalent bond of the proton and the hydroxyl group. After the addition of polar solvents, or acidification of the rhodamine base solvents, this bond ruptures and the fluorescence of the rhodamine molecules is restored.

The above discussions have been centred on the spectral and luminescent properties of the low molecular luminophors used in the synthesis of fluorescent polymers. There is a wide range of other compounds which may be of interest from this point of view, including, for example, the luminophor diamines such as benzoflavine (XL), safranine (XLI), and cresyl violet (XLII).

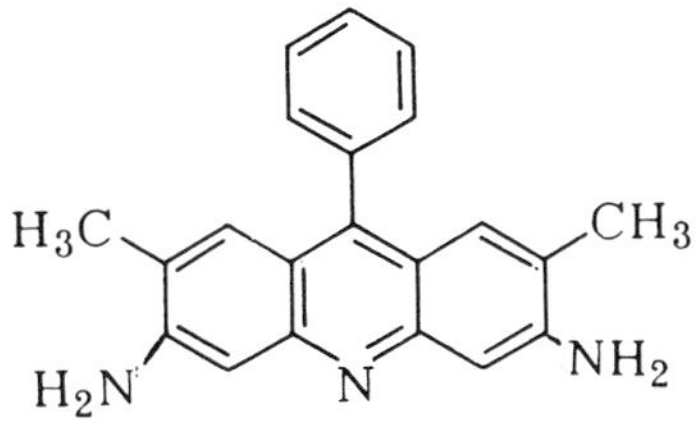

XL

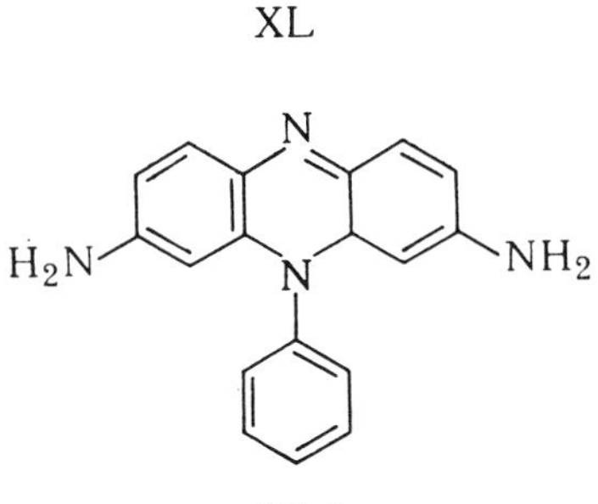

XLI

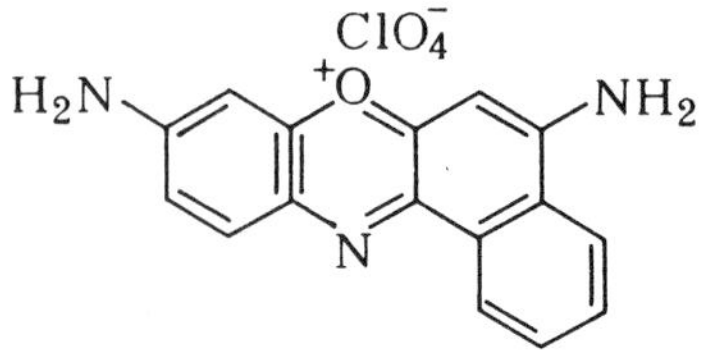

XLII

However, no attempts have been made so far to use these as monomers to obtain poly-condensed fluorescent polymers.

3

Polymers obtained by polymerization methods

When studying the spectral properties of polymers it is expedient to distinguish between fluorescent macromolecules with saturated C—C bonds in the main chain, and polymers with a system of conjugated bonds.

Those polymers in whose molecules the single bond is arranged between two double bonds, or between a multiple bond and atoms with an unshared p-electron pair, differ noticeably in properties from polymers with isolated bonds [94]. The composition of poly-conjugated polymers cannot be described by the structures formed by two-centre bond combinations, since the p- and π-electrons are delocalized over many molecular centres. The delocalization of π-electrons is caused by the overlap of the wavefunctions or orbitals of the π-electrons in the polyconjugated macromolecule. Since the overlap of the wave functions or orbitals of π-electrons is maximal when the arrangement of all the atoms and bonds is planar, the main condition for the maximum degree of delocalization of π-electrons, and the best efficiency of conjugation, is the coplanarity of the components participating in the conjugation. For this reason, any factors weakening the exchange interaction of π-electrons, and preventing the coplanar arrangement of the polymer chain, reduce the efficiency of conjugation, thus affecting the whole complex of properties of these polymers.

3.1 POLYMERS WITH SATURATED C—C BONDS IN THE MAIN CHAIN

Polymers with saturated C—C bonds in the main chain exhibit fluorescent properties if a π-conjugated system is present in their side groups (aromatic or heterocyclic substituents).

Analysis of [95, 96], devoted to the spectral properties of polymers with no π-conjugation both in the main and in side chain (polyethylene, polypropylene, poly(vinyl chloride)) enables us to conclude that the emission of these polymers at $\lambda > 400$ nm is caused by the presence of impurities.

3.1.1 Polystyrene and polystyrene copolymers

The spectral and luminescent properties of polystyrene have often been investigated [33, 34, 97–103]. It was in polystyrene that the excimer fluorescence of polymers was first detected [97]. Another effect of prime importance is the competition between excimer formation and phosphorescence. This was also discovered for the first time in polystyrene [98]. The author of [98] suggested that since the excimer emission is at a longer wavelength than the monomer emission, i.e. $E_E < E_M$, evidently excimers function as traps of singlet excited states. Accordingly, excimer formation competes with the S–T conversion processes preceding phosphorescence ($T_1 \rightarrow S_0$).

This conclusion is confirmed by a study of the temperature dependence of the luminescence of polystyrene film [98] (Fig. 3.1). When the temperature drops from 295 to 4.2 K, it is possible to observe the phosphorescence band of the polymer with $\lambda_{max} = 410$ nm, intensification of the monomer fluorescence band (280–300 nm) and decay of the excimer fluorescence with $\lambda_{max} = 330$ nm. Clearly, at low temperatures at which molecular movement is impeded, the formation of excimers is sterically hindered.

In close agreement with the above are the deductions made in reference [99] devoted to the fluorescence decay kinetics observed during the formation of intramolecular excimers in styrene–methyl methacrylate copolymers. Sufficient evidence points to the reversible dissociation of the excimer into the monomer in the excited singlet state.

The specific features of excimer formation in polystyrene and mixtures of polystyrene with poly(vinyl methyl ether) are covered in [100, 101].

Study of the fluorescence spectra of polystyrene films at different excitation wavelengths (270 and 350 nm) revealed the presence of bands with maxima at 440, 460, and 490 nm which are believed to be due to dimers [102]. The formation of the latter depends on the submicroscopic structure of the polymer and is connected with the nonplanar isomerization of the phenyl rings. It has been found that when initially exposed to UV radiation, polystyrene undergoes profound changes in its macromolecular structure owing to the transannular interaction, and the isomerization and photodimerization of phenyl rings involving the formation of intermediates.

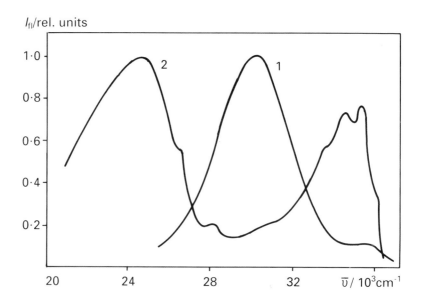

Fig. 3.1. Luminescence spectra of polystyrene films at 295 K (1) and 4.2 K (2).

The methyl groups present in the aromatic ring of homologues of polystyrene (polyvinyltoluene, polyvinylxylene) are responsible for the negligible red-shift of the fluorescence spectrum and enhancement of the emission intensity. In poly(2,4-dimethyl styrene) films, along with the excimer emission band, is the very distinct maximum of the fluorescence band of the monomer unit ($\lambda = 290$ nm). The most significant difference is noted in the fluorescence spectra of solutions. Poly(2,4-dimethyl styrene) is characterized by the sole presence of the monomer emission band. The efficiency of excimer emission of methyl-substituted polystyrenes depends not only on the number of methyl groups but on their ring position.

Many works are devoted to the synthesis of fluorescent polymers based on polystyrene. One of the first publications [104] deals with the polymerization of styrene in the presence of dimethylbenzacridine, with the formation of a polymer with fluorescent units in the main chain. It has been found that with the addition of dimethylbenzacridine, the thermal polymerization rate of styrene increases 2.5 to 3 times. Spectrophotometric analysis indicates that the polymer formed contains 1.4 to 2 luminophor moieties per polystyrene macromolecule with RMM 450 000.

A more devailed study has been made of the polymerization of styrene in the presence of anthracene derivatives with [105–107] and without [105, 108, 109] a vinyl group. As shown in [108], in the process of styrene polymerization, the anthracene additive transforms, being partially converted to compounds exhibiting absorption and fluorescence at longer wavelengths. These compounds are believed to be products of substitution of hydrogen atoms in the meso-positions of the anthracene ring by styryl and polystyryl radicals. It was later discovered [105, 109] that under polymerization conditions, both the unsubstituted anthracene and many of its derivatives [106] are also capable of interacting with styrene via the diene synthesis reaction.

The possibility for interaction with styrene increases with the introduction into the anthracene derivative of a vinyl group. During an examination of the spectral and luminescent properties of polymerization products of styrene solutions of 9-vinylanthracene, 1-vinylanthracene, 2-vinylanthracene and 2-vinyl-9,10-diphenylanthracene, it was established [105–107] that these compounds interact with styrene in the following ways:

(i)   addition of styryl or polystyryl radicals to the meso-position of the anthracene ring,

(ii)   addition of one or several molecules of styrene to the vinyl group of vinylanthracenes to form low molecular products,

(iii) addition of polystyryl radicals to the vinyl group of vinylanthracenes (copolymerization),

(iv)   substitution of hydrogen atoms in the meso-position of the anthracene ring by styryl or polystyryl radicals.

As a result of reaction (i), the molar concentration of anthracene compounds in the polymerization products drops to 50–55% of the initial level in the presence of 9-vinylanthracene and to 20–25% in the presence of 1-vinylanthracene and 2-vinylanthracene. The interaction of 2-vinylanthracene and 2-vinyl-9,10-diphenylanthracene with styrene proceeds mainly via reaction (iii) and practically all the anthracene compounds not taking part in reaction (i) enter into the polymer composition. In the case of 9-vinylanthracene, the anthracene derivatives present in the polymer account for only 5% of the initial molar content, and in the case of 1-vinylanthracene, for some 10%. Reaction (iv) is followed only by 1-vinylanthracene, with which this reaction proceeds so easily that in the fully polymerized samples, 1,9,10-substituted anthracenes constitute the greater part of the anthracene compounds contained in the products of polymerization. The capacity of 2-vinylanthracene molecules to interact via two pathways results in the formation, after polymerization, of an insoluble, non-melting crosslinked polymer when the initial 2-vinylanthracene concentration in styrene exceeds 0.15% by mass (Table 3.1).

Table 3.1. Thermal polymerization of styrene in the presence of anthracene compounds [110]

Compound	Content of polymerization product /% of initial	Compound /% of initial	Content of polymerization product /% of initial
Anthracene (A)	28	9,10-Diallyl-A	6
9-Methyl-A	2	9-Vinyl-A	50
9-Ethyl-A	14	2-Vinyl-A	26
9-*n*-Propyl-A	31	9,10-Dibrom-A	12
9,10-Dimethyl-A	1	9,10-Dichlor-A	19
9,10-Diethyl-A	7	9-Phenyl-A	94
9,10-Di-*n*-propyl-A	28	9,10-Diphenyl-A	105

Polymerization conditions: temperature, 150°C; time, 48 h, initial concentration of anthracene compounds, 2 mass % (in the case of 2-vinylanthracene, 1 mass %).

As shown in Table 3.1, the introduction of a methyl group in the meso-position of the anthracene ring has a strong activation effect. If during the polymerization of styrene solutions of anthracene, its content drops to 28% of the initial level, 9-methylanthracene and 9,10-dimethylanthracene drop to 2 and 1% respectively. At the same time, the meso-phenyl derivatives of anthracene barely interact with styrene via the diene synthesis pathway.

The copolymerization of styrene with the vinyl derivatives of 9,10-diphenylanthracene has been studied more thoroughly. For this purpose, use was made of the vinyl derivatives substituted either in the anthracene ring [107] or in one of the phenyl rings [111]. In the first case, copolymerization was effected by heating the styrene-2-vinyl-9,10-diphenylanthracene mixture in a ratio from 99.95:0.05 to 99.0:1.0% (molar) to 120–150°C. In the second case, the styrene–*p*-vinylphenyl-9-phenyl-10-anthracene mixture (taken in a ratio from 99.8:0.2 to 97.6:2.4% (molar)) was subjected to radical copolymerization in benzene solution at a temperature of 70°C or anionic copolymerization in tetrahydrofuran in the presence of phenylisopropyl potassium. As a result, a copolymer with RMM from 43 000 to 285 000 was obtained.

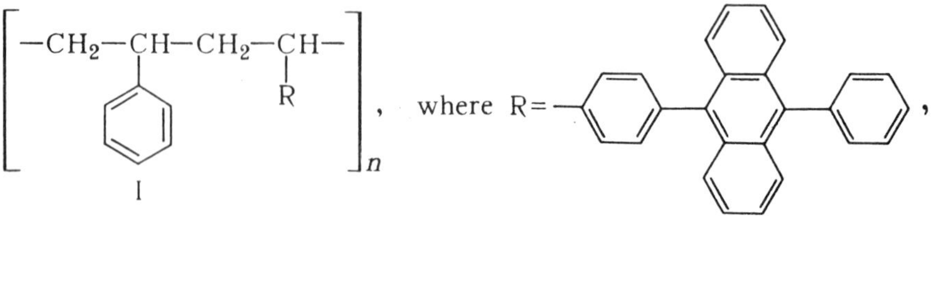

Ref. [107] covers the absorption and fluorescent properties of the copolymers obtained, as well as of the initial 2-vinyl-9,10-diphenylanthracene and model 2-methyl-9,10-diphenyl-anthracene (Fig. 3.2). As seen from Fig. 3.2, the absorption spectra of the polymer and monomer are similar at 340–400 nm. However, at $\lambda < 300$ nm (not shown in Fig. 3.2), the spectrra of these compounds display significant differences associated with the absence in the polymer of an alkene group (hypsochromic shift of the short wavelength band).

The most striking feature evident when comparing the fluorescence spectra of the monomer and 2-methyl-9,10-diphenylanthracene is the intense, resolved maximum to long wavelength of the main maximum. The authors of [93] attribute this to the existence of excited monomer molecules in the form of a mixture of two isomers (*S-cis-trans*-isomers) with their respective, different fluorescence spectra. As a result of polymerization, the fluorescence spectrum undergoes profound changes (Fig. 3.2) and becomes similar to the spectra of 2-methyl-9,10-diphenylanthracene, and other meso-diaryl derivatives with saturated substituents in the β-position.

The absorption spectra of the copolymers obtained in [111] are similar to those shown in Fig. 3.2. However, in the emission spectra of these compounds only one broad structureless band can be observed with a maximum at 440 nm instead of two fluorescence bands with maxima at 415 and 435 nm.

It has been shown [112] that solution of styrene–4-vinyldiphenyl diluted with dichloroethane exhibit the emission of only the monomer unit ($\lambda_{max} \sim 320$ nm). The excimer fluorescence of diphenyl units ($\lambda_{max} \sim 380$ nm) is apparent only at a

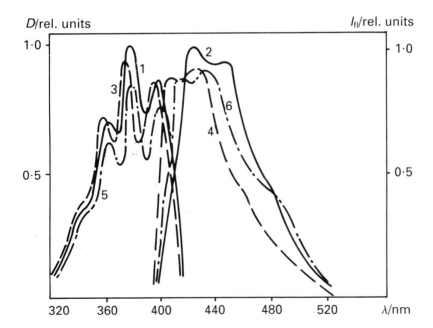

Fig. 3.2. Absorption (1, 3, 5) and fluorescence (2, 4, 6) spectra of benzene solutions of 2-vinyl-9,10-diphenylanthracene (1, 2), model 2-methyl-9,10-diphenylanthracene (3, 4) and styrene–2-vinyl-9,10-diphenyl anthracene copolymer (6, 5).

4-vinyldiphenyl concentration exceeding 20%, and reaching the emission intensity of the monomer unit at a concentration of 50%. In the fluorescence spectrum of the bulk copolymer containing over 50% of diphenyl units, in contrast to the solution spectrum, the monomer band is absent and only the excimer emission is apparent (Fig. 3.3).

The established relationship between the fluorescence intensity of different emission centres and the composition of copolymers enabled the authors of [113] to provide explanations for the high fluorescence quantum yield of styrene–vinyldiphenyl copolymers (up to 20%) and the sharp reduction of its level with increase in the concentration of vinyldiphenyl, which is associated with the appearance of excimer emission (Table 3.2).

Table 3.2. Relative luminescence quantum yields[a] of styrene–4-vinyldiphenyl copolymers [113]

4-Vinyldiphenyl content/% (by mass)	Relative quantum yield/%	4-Vinyldiphenyl content/% (by mass)	Relative quantum yield/%
—	0.09	20	0.15
5	0.20	30	0.13
10	0.20	50	0.12
15	0.21		

[a]Absolute quantum yield of polystyrene given in ref. [114].

During study of the electronic energy transfer in the copolymer–luminophor system (1,3,5-triphenylpyrazoline), it was established that, with an increase up to 30% in the concentration of 4-vinyldiphenyl, the efficiency of energy transfer also improves.

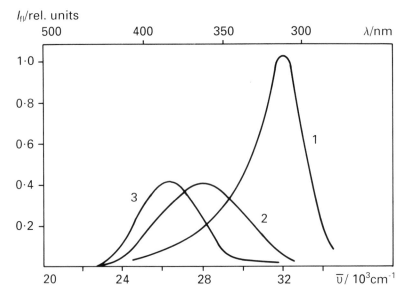

Fig. 3.3. Fluorescence spectra of poly(4-vinyldiphenyl) (3) and styrene–4-vinyldiphenyl (VDP) copolymer films: with 10% VDP (1) and 50% VDP (2).

To elucidate the influence of the luminophor to polymer bond on the efficiency of electronic energy transfer, styrene was copolymerized with vinylterphenyl (v-TP) and with the vinyl derivatives of 2,5-diphenyloxadiazole-1,3,3 (v-PPD) as well as 2,5-diaryl-oxazoles-1,3 (v-PPO).

Refs [115, 116] compare the absorption and fluorescence spectra of the above copolymers with solid luminophor solutions corresponding in structure to the elementary unit of ethyl-TP, ethyl-PPO and ethyl-PPD incorporated in the copolymer.

The spectrum of the elementary unit of ethyl-TP in the copolymer is similar to that of ethyl-TP in polystyrene, the only difference being that the maxima of the former are red shifted by 5 nm. This shift is explained by the chemical interaction of the monomer unit of v-TP with the polymer chain.

The absorption and luminescence spectra of the elementary unit of ethyl-PPD are also red shifted in the styrene copolymer by several nm in contrast to the spectra of ethyl-PPD and PPD forming a solid solution in the polystyrene.

On increasing the concentration in the copolymer of the elementary units of ethyl-TP and ethyl-PPD (5×10^{-3}–2×10^{-1} mol dm^{-3}), their spectra are virtually unchanged. Study of the luminescence spectra of the vinyl-TP and vinyl-PPD polymers revealed the presence in them of a diffuse band with a maximum in the range 430–450 nm. Such a long wavelength shift, compared with the emission band of the isolated elementary unit, can be explained by a strong intermolecular interaction of the chromophore groups. The absence of luminescence in the range 430–450 nm of the copolymer spectra rules out the presence in the polymer chain of bonded elementary units of ethyl-TP and ethyl-PPD. On the basis of a spectral study of repeatedly reprecipitated copolymer samples, it has been found that the main portion of the vinyl-containing luminescent additive (LA) forms with styrene, macromolecules with a high degree of polymerization and only a small portion of the additive (*ca* 10%) forms low molecular compounds.

A different situation arises with the styrene–v-PPO copolymer. An increase in concentration of the elementary unit of ethyl–PPO gives rise to a luminescence band in the region 440–450 nm typical of polyvinyl–PPO. This indicates that the elementary units of ethyl–PPO are directly bonded, rather than separated from each other in the polymer chain.

It follows from the data presented in Table 3.3 that the quantum yield of the activators

Table 3.3. Relative luminescence quantum yields (%) of copolymers with different additives

Luminescent probe	Concentration, $C/10^{-3}$ mol dm^{-3}		
	5.0	7.0	15.0
PPD	0.66[a]	0.67±0.01	0.66±0.02
PPDCH=CH$_2$	0.75±0.01	0.75±0.01	0.68±0.02
PPO	0.80±0.01	0.78±0.02	0.58±0.02
PPOCH=CH$_2$	0.72±0.02	0.54±0.02	0.44±0.03
TP	0.75±0.01	0.75±0.02	0.70±0.02
TPCH=CH$_2$	0.83±0.02	0.81±0.01	0.75±0.02

[a] Absolute quantum yield given in ref. [114].

increases (approximately 1.1 times) when their molecules enter the polymer chain, which can be explained by the more rigid attachment of the molecules to the matrix and, consequently, a decrease in the expenditure of excitation energy by the nonradiative mechanism.

In all the investigations of luminophors, the quantum yield is virtually independent of concentration, except for v-PPO. With increase in concentration of the elementary unit of ethyl-PPO, the quantum luminescence yield of the v-PPO–styrene copolymer decreases owing to the presence of the homopolymer v-PPO.

The spectral data are confirmed by the results of a study of the kinetics of copolymerization of vinyl-containing luminophors and styrene [117]. Thus, the product of relative reactivities of v-PPO and styrene is 1.7, that is, $r_1 r_2 > 1$. This means that the monomer molecules add to the polymer chain randomly to form separate sites of bonded elementary units of ethyl-PPO. During the copolymerization of styrene and vinyl-PPD, $r_1 r_2$ is equal to 0.7. If $r_1 r_2 < 1$ the copolymer tends to have an ordered rather than random structure.

The most suitable candidate systems for studying the process of electronically excited energy transfer are those with a uniform distribution of the elementary units of the luminophor in the polymer chain, for example the copolymer of vinyl-TP and vinyl-PPD with styrene.

Calculations show that the introduction of the luminophor into the polymer chain has no bearing on the efficiency of electronic energy transfer. The theoretical calculations are in full agreement with the experimental data, which confirm the three-dimensional migration of excitation energy in polymers.

1,3,5-Triphenyl-2-pyrazoline can be used not only as an additive in the study of excitation energy transfer in polymers [113] but also in the preparation of fluorescent styrene copolymers (II) and the corresponding homopolymers (III) [118].

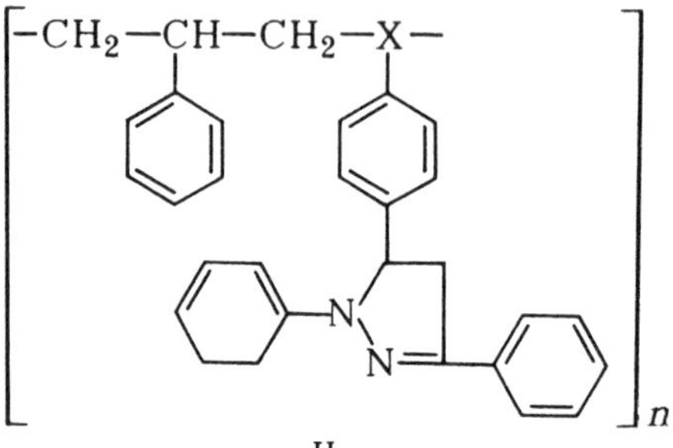

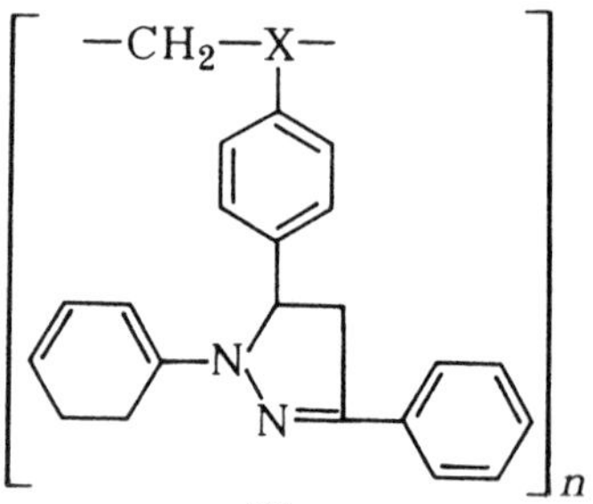

where $X = \!\!>\!CH-$(IIa, IIIa); $\;>\!C(CH_3)OCO-$(IIb, IIIc).

A comparison of the absorption and fluorescence of polymers II and III with the model 1,3,5-triphenyl-2-pyrazoline (Fig. 3.4) reveals that, irrespective of the nature of the solvent, they are spectrally similar. Benzene solutions of polymers IIa and IIIa have high quantum fluorescence yields (0.76 and 0.73 respectively), approaching the quantum yield of the model compound (0.90). The excimer fluorescence of these polymers has been found to occur neither at 293 nor at 77 K. The spectral properties of polymers II and III, when compared with those of the model 1,3,5-triphenyl-2-pyrazoline, are characterized by a larger Stokes shift and low sensitivity of the fluorescence band to the polarity of the solvent.

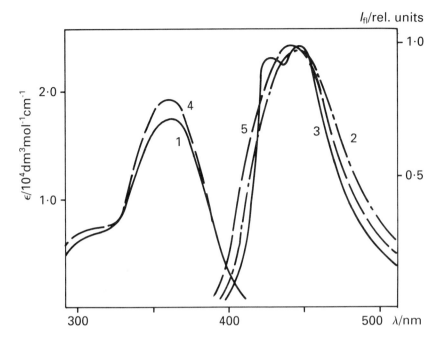

Fig. 3.4. Absorption (1, 4) and fluorescence (2, 3, 5) spectra of benzene solutions of polymer IIa (1, 2), model 1,3,5-triphenyl-2-pyrazoline (4, 5) at room temperature and fluorescence spectrum of methyltetrahydrofuran solution of polymer IIa at 77 K (3).

With an increase in the content of styrene moieties, the copolymers of 1,3-diphenyl-5- *p*-(methacryloyloxy) phenyl-2-pyrazoline (IIb) and styrene show a short wavelength shift of the fluorescence band which is particularly striking in the polymer solutions at 77 K [118].

The copolymer (IV) containing 8.1% (mol) pyrenyl units has been obtained by copolymerization of styrene with 1-pyrenylmethyl methacrylate in the presence of azobisisobutyronitrile [119].

IV

A solution of polymer IV taken in a dimethylformamide–water mixture exhibits excimer fluorescence with λ_{max} = 465 nm. Interesting results were obtained during a study of the photoconversions of the leuco dye of crystal violet in mixed solutions with polymer IV [119].

Refs [18, 120] deal with the synthesis and fluorescent properties of polystyrene with anthryl end groups (V) prepared in accordance with the reaction of 9-chloromethyl-anthracene and the 'living' polystyrene dianion, which is the product of the anionic polymerization of styrene, using sodium naphthalenide in a tetrahydrofuran medium.

$$-CH_2-\left[-CH_2-CH-\right]_n-CH_2- \quad , \quad n = 100-3000$$

V

The process of triplet–triplet annihilation has been studied, and the intramolecular reaction rate constants of anthryl groups have been measured for benzene, butanone, and cyclohexane solutions of this polymer with varying RMM. The author of [120] suggests a kinetic diagram to calculate the rate constants of the diffusion limited reactions. It has been established that the dynamic diffusion of segments occurs at a higher rate than expected for polymer coils with high values of the DP.

The fluorescence behaviour patterns of polystyrene with naphthalene and anthracene groups incorporated in the chain yield valuable information [121, 122] about intra- and intermolecular interactions in polymer solutions. Of interest are data on the monomer and excimer fluorescence of the copolymer of styrene and 2(2')-hydroxy-5'-vinylphenyl-2H-benzotriazole presented in [123]. Ref. [124] describes binary and ternary copolymers of the following type: styrene–acrylic acid, styrene–1-vinylnaphthalene–acrylic acid, styrene–1-vinylanthracene–acrylic acid (containing the rare earth metal ions Dy^{3+}, Er^{3+}, Eu^{3+}, and Sm^{3+}). The fluorescence of the styrene–acrylic acid copolymer containing Eu^{3+} ions has been studied extensively. It has been shown that the intensity of fluores-cence with a maximum at 660 nm, arising from the transition from the $^5D_0$-levels to the multiplet ^7F-levels, increases with increase in the ion concentration to 4–5% and then diminishes as a result of concentration quenching. The conclusion of the authors of [124] is that the aromatic groups of these systems are arranged too far from the zones of ion aggregation and, for this reason, intermolecular transfer of energy from these groups to Eu^{3+} ions is unlikely.

Interest in this type of polymer system is connected with the potential of their monochromatic fluorescence for laser systems.

3.1.2 Poly(methyl methacrylate) copolymers

In the preparation of fluorescent polymers, methyl methacrylate (MMA) is used as often as styrene for the copolymerization of vinyl-containing luminophors. In one of the first studies devoted to the interaction of MMA with anthracene derivatives [110], the anthracene compounds were shown as being capable of entering, as in the case of styrene, into the diene synthesis reaction whether or not they contain a vinyl group. Under the conditions of thermal polymerization of MMA (120°C, 48 h), anthracene has been shown to form no polymer but to yield a low molecular product (VI):

VI

The absorption spectrum of this compound is blue shifted relative to that of anthracene.

In the copolymerization of MMA with 2-vinylanthracene under the above conditions, along with the formation of high-molecular linear products, compounds of type VI are obtained, as is confirmed by the appearance of the absorption and fluorescence spectra (Figs 3.5 and 3.6). In their spectral properties, the reprecipitated polymers appear to be similar to the corresponding alkyl derivatives of anthracene. The formation of compounds of type VI is confirmed by the blue shift in the absorption and fluorescence spectra of the ethanol extracts remaining in solution after separation of the reprecipitated polymer by evaporation (curves 3 in Figs 3.5 and 3.6). Unlike styrene, MMA does not form crosslinked polymers with 2-vinylanthracene which is evidently associated with the greater difficulties in the addition of poly(methyl methacrylate) radicals to the meso-positions of the anthracene ring compared with polystyryl radicals.

A number of anthracene-containing poly(methyl methacrylates) are described in [67, 125, 126]. Thus, ternary copolymers containing anthracene groups in the side chain have been obtained by the free radical copolymerization of 0.1% (mol) 9-anthrylmethyl

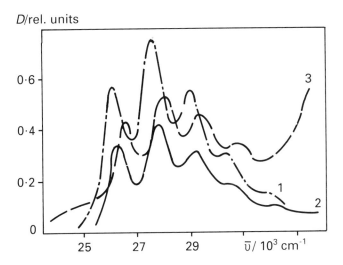

Fig. 3.5. Absorption spectra of benzene solutions of 2-vinylanthracene (1), copolymer of 2-vinylan-thracene and MMA (2) and ethanol extract from the evaporated solution after separation of reprecipi-tated polymer (3).

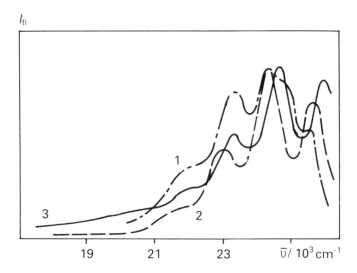

Fig. 3.6. Fluorescence spectra of 2-vinylanthracene (1), benzene solution of 2-vinylanthracene–MMA copolymer (2) and ethanol extract from the evaporated solution after sepation of reprecipitated polymer (3).

methacrylate with a mixture of MMA and styrene in the presence of azobisisobutyronitrile.

The copolymers of MMA with styrene and 4-vinyldiphenyl have a practical interest for the preparation of plastic scintillators (Chapter 7) [113].

Copolymers with a low content of aromatic chromophores display only the monomer fluorescence, whose intensity increases with increasing concentration of the latter. The monomer emission occurs owing to the isolated arrangement of the chromophoric groups in the MMA copolymers. Increase in the concentration of elementary units with aromatic groups to 30% leads to the appearance of the excimer fluorescence, which suggests the possibility of formation of excimer centres.

The copolymers of MMA and 9-vinylanthracene or 4'-vinyl-2,5-diphenyloxadiazole have been studied by fluorescence spectroscopy at a 0.05–2.0% concentration (mass) of the latter [127]. From comparison of the spectral properties of these copolymers with those of the corresponding monomers of the chromophore, it follows that the introduction in the chain of the latter results in a hypsochromic shift (by 10–20 nm) of the long wavelength absorption and fluorescence bands. Table 3.4 compares the relative quantum yields (φ_{rel}) and fluorescence lifetime (τ_{fl}) of the copolymers with a 0.1% content of luminophoric groups and solid solutions, corresponding to the low molecular luminophors, taken at the same concentration, with PMMA.

As seen from the table, the φ_{rel} of the former is 1.4–1.7 times higher than that of the latter, the correlation between the quantum yield values and τ_{fl} being satisfactory. In the case of chemical fixation of the emission centres observed in the copolymers, the possibilities for the nonradiative deactivation of electronic excitation energy must be reduced.

Table 3.4. φ_{rel} and τ_{fl} of MMA copolymers and solid PMMA solutions of lumino-phors measured in the solid phase and in chloroform solution [127]

Sample[a]	φ_{rel}		τ_{fl}/nm	
	solution	glass	solution	glass
MMA–VA copolymer	0.42	0.53	9.7	12.2
PMMA + anthracene	0.15	0.29	3	7.2
MMA–VDPD copolymer	0.13	0.68	3	4.2
PMMA + 2,5-diphenyloxadiazole	0.10	0.42	3	3

[a] VA–9-vinylanthracene, VDPD–4'-vinyl-2,5-diphenyloxadiazole.

An interesting result was obtained in the measurement of τ_{fl} and φ_{rel} for the above solutions in chloroform solvent, which shows the famous 'heavy-atom' effect. Both in the free luminophors and in those bound by the polymer chain, a notable reduction in φ_{rel} and τ_{fl} is evidenced during the transition from the glassy state to solution. However, in the case of the chemically fixed luminophor, the heavy-atom effect is weakened, evidently because of the diffusion limitations arising when molecules of the chloroform quencher enter the polymer coils.

The authors of [128] suggested that in the MMA–styrene copolymer (with tri-phenylbutadiene used as acceptor), the energy transfer efficiency should be lower than in polystyrene owing to the lower migration coefficient of the copolymer. However, this suggestion was not confirmed by experiment: the efficiency of excitation energy transfer proved to be higher in the copolymer than in polystyrene. This disagreement between the-ory and experimental data obtained in [128] can be explained by the fact that the contribu-tion of excimers to the excitation energy transfer was not taken into account. Similar results were obtained in [54] with the reprecipitated polystyrene samples and the styrene–MMA copolymer (74:26) taken with TPB. At room temperature, the efficiency of excitation energy transfer of this copolymer was higher than that of the pure polystyrene.

The fluorescent properties of the copolymers of MMA and acenaphthylene (VII) [129] and of polyacenaphthylene and its copolymers with maleic anhydride and methylacrylate have been thoroughly investigated. Fig. 3.7 illustrates the fluorescence spectra of tetrahydro-furan solutions of polyacenaphthylene and its copolymer with maleic anhydride. The short wavelength band, λ_{max} 325 nm, occurs during monomer emission, while the long wave-length band with λ_{max} 400 nm appears to be associated with excimer emission. The similar-ity of the long wavelength bands of the two copolymers indicates the independence of the excimer interaction energy of the environment of the luminophoric naphthylene groups.

$$\left[\begin{array}{c} -C(CH_3)-CH_2-CH-CH- \\ | \\ COOCH_3 \end{array} \right]_n$$

VII

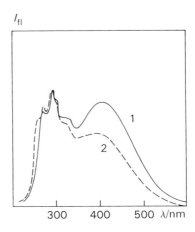

Fig. 3.7. Fluorescence spectra of polyacenaphthylene solutions in tetrahydrofuran (1) and of copolymer of acenaphthylene and maleic anhydride (2).

As seen in Fig. 3.7, the I_E/I_M ratio is higher in the homopolymer than in the copolymer. The authors of [129] explain this by the fact that the distance associated with the
migration of electronic energy required for the formation of excimers is larger in the
copolymer than in the pure polyacenaphthylene.

The fluorescence spectra of the copolymers of acenaphthylene and MMA or methacrylate are similar to those illustrated in Fig. 3.7. In [130] the intramolecular migration
of electronic energy was studied by the depolarization method. As expected, the formation of excimers in polymers of this type is promoted by an increased mole fraction of
naphthyl units in the copolymer. Comparison of the fluorescence properties of the acenaphthylene–methacrylate (MA) and acenaphthylene–MMA copolymers made it clear
that the process of energy migration is more effective in the latter.

The fluorescent properties of the copolymers of MMA and 1-naphthyl methacrylate or
1-vinylnaphthalene, poly(1-naphthyl methacrylate) and poly(2-naphthyl methacrylate)
were investigated in [131–133]. Investigation of emission spectra in solvents of varied
viscosity and at different temperatures provides evidence that with the increasing
viscosity in the sequence of solvents (i) ethyl acetate, (ii) toluene, (iii) a mixture of
$CHCl_3$+cyclohexane (1:1), (iv) CH_2Cl_2, (v) $CHCl_3$, the intensity ratio of the excimer
fluorescence (λ_{max} = 385 nm) and monomer fluorescence (λ_{max} = 335 nm) of poly(1-
naphthyl methacrylate) falls from 3.67 to 1.44. A further decrease in the I_E/I_M ratio is
observed when the temperature of the polymer solution is increased from 6 to 82°C.

The variations in the intensity of excimer emission of this polymer with its RMM
(Fig. 3.8) are discussed in greater detail in Chapter 6.

A study of the fluorescence features of poly(2-naphthyl methacrylate) in a variety of
solvents revealed a dependence of the I_E/I_M ratio on the nature of the solvent varying
from 0.69 (chloroform, tetrahydrofuran) to 0.79 (acetonitrile, methanol) and increasing
with the dielectric permittivity of the solvent [132, 133].

A detailed study of the excimer formation of MMA–1-naphthyl methacrylate (NMA)
copolymers is described in [134, 135]. Copolymers containing from 3 to 100% NMA

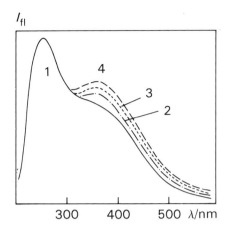

Fig. 3.8. Fluorescence spectra of methylene chloride solutions of poly(l-naphthyl methacrylate) of different RMM: 1, 40 000; 2, 88 000; 3, 250 000; 4, 360 000.

units were subjected to synthesis in a toluene solution at 65°C in the presence of benzoyl peroxide until they underwent minor conversion (from 5 to 10%) enabling samples to be obtained which were uniform in structure and composition. The copolymerization constants calculated according to the Mayo equation were 0.97 ± 0.03 (r_{MMA}) and 0.92 ± 0.03 (r_{NMA}) [134].

The fluorescence spectra of dioxane solutions of the homopolymer NMA and its copolymers with various contents of NMA units are given in Fig. 3.9. Alongside the monomer fluorescence band (λ_{max} 340 nm), those copolymers with a NMA content exceeding 25% show the excimer fluorescence band ($\lambda_{max} = 396$ nm).

Fig. 3.10 features the intensity dependence of the excimer fluorescence on the luminophor group content (curve 1). The high intensity of the excimer fluorescence, maintained until the luminophor group concentration becomes low, indicates an efficient transfer of electronic energy from the isolated chromophores in the S_1 state to the dyads (or higher associates) of the NMA units, whose mutual orientation is favourable to excimer formation. If energy migration did not occur in the macromolecular chains, and excimer states developed only at the expense of direct excitation of one of the dyed groups potentially capable of excimer formation, I_E would diminish more drastically with decreased NMA content (curve 2) in comparison with what can be observed in practice [134]. At low concentrations of NMA units (less than 10%), when the distance between the chromophores is large and the probability of energy transfer is low, I_E seems to be proportional to the number of excimer-forming sites.

As reported in [134], the intensity of singlet fluorescence increases with the MMA content of the copolymer, which is evidently caused by screening of the luminophor naphthyl groups by MMA units incapable of quenching the electronically excited states of isolated naphthyl methacrylate units. At a 10^{-2}–10^{-4} mol dm^{-3} concentration of the NMA homopolymer in solution, excimer formation occurs solely owing to intramolecular interactions, which is evidenced by I_M/I_E being unchanged (Fig. 3.11). The increase in the value of this ratio at a concentration of 10^{-5} mol dm^{-3} was not explained [135].

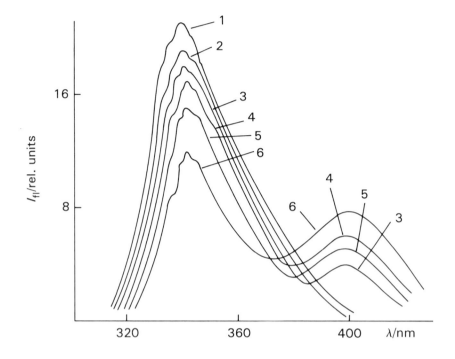

Fig. 3.9. Fluorescence spectra of dioxane solutions of copolymers of MMA with 3% (1), 10% (2), 25% (3), 50% (4), 57% (5) NMA and homopolymer NMA (6).

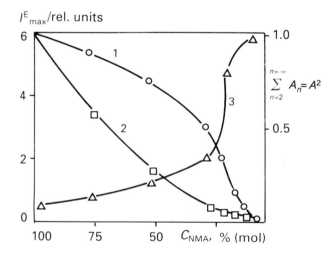

Fig. 3.10. Excimer fluorescence intensity of MMA–NMA copolymers in solution (1, 2) and fraction of excimer-forming sites of dioxane dyads (3) vs. content of luminophoric groups: 1, experimental curve; 2, theoretical curve ($C_{chrom} = 10^{-3}$ mol dm^{-3}).

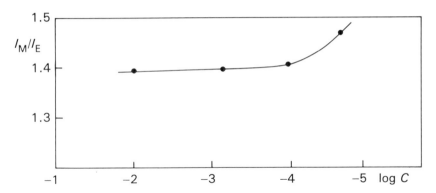

Fig. 3.11. I_M/I_E vs logarithm of NMA concentration in dioxane solution.

Ref. [136] deals with radical copolymerization of methyl methacrylate and 1-pyrenyl methyl methacrylate in benzene solution in the presence of azobisisobutyronitrile as initiator and also the polymerization of methyl methacrylate in chloroform solution containing small amounts of 1,3-bis(1-pyrene)propane and 1,10-bis(1-pyrene)decane. Copolymerization is shown as being controllable by the variation in the I_M/I_E ratio with time (where I_M and I_E are the fluorescence intensity of non-associated pyrene groups (λ_{max} = 377 nm) and of the intramolecular excimer of pyrene (λ_{max} = 480 nm), respectively). On copolymerization, ($10 \times I_M/I_E$) increases gradually from 0.75 to 0.9 within the first 3 hours of the reaction, then rises rapidly up to 3.0 over the next 1 to 1.5 hours because of the sharp increase in the viscosity of the reaction medium. A similar picture was observed on polymerization of methyl methacrylate doped with low molecular pyrene derivatives [136].

The temperature dependence of the I_M/I_E ratio for the copolymers of MMA and 1-vinylpyrene can be used in the development of thermal indicators changing, in their colour of emission, from greenish-blue to blue-violet over the rather narrow temperature range from 90 to 95°C [137, 138].

3.1.3 Poly(vinylcarbazoles) and other polymers

Poly(*N*-vinylcarbazole) and its analogues are the most thoroughly studied fluorescent polymers. The results of the earlier investigations devoted to these polymers are summarized in [9]. For this reason, we shall restrict ourselves to a brief review of the latest publications.

All the carbazole-containing polymers obtained by polymerization can be divided arbitrarily into three types: (i) poly(*N*-vinylcarbazole), (ii) copolymers of *N*-vinylcarbazole, and (iii) polymerization products of *N*-alkylcarbazoles with an unsaturated group on one of the benzene rings.

The authors of [26, 139–141] investigated the excimer fluorescence of solid poly(*N*-vinylcarbazole) and its solutions. At the same time, particular attention was given to the photo-oxidation processes in this polymer. It was shown, for instance in [139], that under the joint action of atmospheric oxygen and UV radiation, the intensity of excimer fluorescence with λ_{max} = 418 nm markedly decreased.

From a comparison of the intensity of the sensitized impurity fluorescence of anthracene and its excimers, it becomes clear that the concentration of excimer formation sites is factors of ten lower than that of the photo-oxidized monomer moieties found in the absorption spectra. This accounts for the high probability of exciton quenching by the photo-oxidized structural units.

The emission referred to in [139, 142] as simply excimer, is regarded by the authors of [143] as complex by its nature and proving the existence of three excimer states. To confirm the above suggestion, they investigated the fluorescence quenching kinetics of a tetrahydrofuran solution of poly(N-vinylcarbazole) at a temperature ranging from 200 to 295 K. No monomer emission of carbazole fragments (λ_{max}= 350 nm) was detected in the fluorescence spectra of these solutions, though they exhibited two fluorescence bands peaking at 370 and 420 nm. This emission is attributed to the low energy sandwich excimers (E_2) formed by the neighbouring carbazole groups. The excimer with a fluorescence maximum at 370 nm (E_3) is a dimer structure showing a marked deviation from coplanarity of two carbazole rings. The emission spectrum of the excimer state E_1 is thought by the authors to coincide with the fluorescence spectrum of E_3. The fluorescence intensity in the 370 nm region has been established as decreasing significantly with increasing temperature. At the same time, the emission is practically invariant at *ca* 420 nm.

As shown in [143], the fluorescence quantum yield of the excimer state E_1 decreases, and that of E_3 increases, with decreasing temperature. The variations in the concentration of high energy excimers (E_1 and E_3) due to the fall in temperature confirms that the formation of excimers E_3 is enhanced at low temperatures. It is suggested that the probability of formation of the excimer state E_2 is highest for polymers of superior isotacticity generally achieved by low temperature cationic polymerization.

Fluorescence quenching was used in [144], dealing with the specific features of energy migration in poly(N-vinylcarbazole). The quencher CCl_4 was added to a dichloromethane solution of the polymer to investigate the effect of its concentration on the quenching efficiency. The quenching of the excimer emission was found to be more efficient than that of the monomer, owing to the difference in fluorescence lifetime. From the Stern–Volmer quenching constants reported in [144], it follows that the simple excimer–monomer scheme suggested in [46] for small molecules cannot be extended to polymeric systems. The calculated migration rate of the triplet exciton of poly(N-vinylcarbazole) exceeds 10^{-5} cm^2 s^{-1} for an energy migration path of 4 nm.

The partially nitrated poly(N-vinylcarbazole) [139, 145] and its copolymers with octyl methacrylate [139] and dimethyl fumarate [146], can be distinguished from N-vinylcarbazole copolymers as regards their spectral luminescent properties. Thus, Ref. [139] covers the fluorescence properties of the first two copolymers and the energy transfer characteristics in these systems. Fig. 3.12 illustrates the fluorescence spectra of poly(N-vinylcarbazole) and its copolymer with octyl methacrylate, as well as of the partially nitrated poly(N-vinylcarbazole) at 77 K and room temperature. Comparison of these spectra reveals the absence of the excimer emission with λ_{max} 415 nm in the case of the copolymer. The emission spectrum of the partially nitrated polymer shows a new band, λ_{max} 500 nm, whose intensity increases sharply with decreasing temperature. When comparing the quenching of the main fluorescence of poly(N-vinylcarbazole) as the fraction of nitrated units increased in the polymer, with an increasing band intensity at

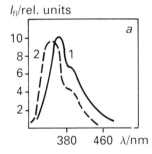
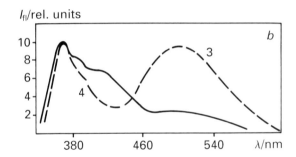

Fig. 3.12. Fluorescence spectra of dioxane solutions: a) poly(*N*-vinylcarbazole) at 77 K (1), vinyl carbazole-octyl methacrylate copolymer at room temperature (2); b) partially nitrated poly(*N*-vinylcarbazole) at room temperature (3) and at 77 K (4).

500 nm, the authors of [139] obtained convincing evidence for the new model suggested by them of the non-coherent one-dimensional energy transfer in these systems.

The spectral luminescent properties of poly(*N*-alkylcarbazoles) are dealt with in [147, 148]. Specifically, [147] covers the fluorescence properties of the poly(methyl methacrylates) based on monomers VIII, IX, and X.

VIII

IX

X

where $R = -CH_2-O-CO-C(CH_3) = CH_2$, $-(CH_2)_2-O-CO-$
$-C(CH_3) = CH_2$, $-CH_2-CO-O-(CH_2)_2-O-CO-O-C(CH_3) =$
$= CH_2$.

Compounds VIII to X were polymerized in benzene solution in the presence of azobis-isobutyronitrile as initiator at 60°C. Fig. 3.13 illustrates the absorption and fluorescence spectra of solutions of the initial monomeric IX- and IX-based poly(methyl methacrylates) in dichloromethane. The emission spectrum of the polymer shows both the monomer bands at 362 and 377 nm and the excimer fluorescence band at 440 nm. The position of the excimer emission band and the ratio between the intensities of the excimer and monomer fluorescence are dependent on solvent polarity (Table 3.5).

Table 3.5. Excimer fluorescence of polymethacrylate based on monomer IX [147]

Solvent	λ^{fl}_{max}/nm, of excimer	I_E/I_M
C_6H_6	418	3.3
CH_2Cl_2	440	2.7
$CHCl_3$	435	2.0
Film	440	1.0
CH_3CN	425	1.0
$CH_3COOC_2H_5$	425	100.0

The polymethacrylates based on monomers VIII and X exhibit solely monomer fluorescence with λ_{max} *ca* 336–374 nm. The authors of [147] explain this in the basis of the geometry of the macromolecules. Measurements of the luminescence quenching kinetics made with and without the anthracene additive do not suggest any actual energy migration in the chains of the polymers based on monomers VIII and X. The chain conformation in these polymers is such that no excimer state is formed during the lifetime of the excited state of the monomer. The flexibility of the macrochains seems to destroy the parallel arrangement of the planes of the chromophore molecules which is a necessary pre-condition for excimer formation and the resonance transfer of energy. However, the steric hindrances existing in the chains of the monomer IX-based polymer make possible the spatial arrangement of the chromophores yielding the excimer state with minimal conformation change.

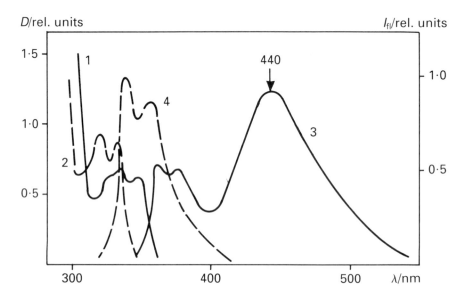

Fig. 3.13. Absorption (1, 2) and fluorescence (3, 4) spectra of solutions of inital monomer IX (2, 4) and IX-based polymer (1, 3).

The carbazole-containing polyacrylates and polymethacrylates of type XI are featured in [148]:

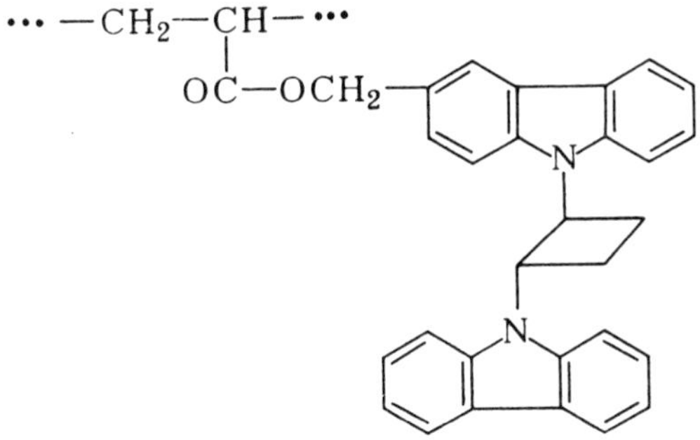

$$\cdots-CH_2-\underset{\underset{\displaystyle\text{carbazole group}}{|}}{CH}-\cdots$$

XI

These polymers were synthesized by radical polymerization of the corresponding derivatives of 1,2-*trans*-dicarbazolylcyclobutane in dimethylformamide at 74°C in the presence of azobisisobutyronitrile. The primary photoprocesses occurring in the solutions of these polymers were investigated by laser induced photolysis. It was established that the intensity of the triplet–triplet absorption of the chromophore groups decreases during the photochemical processes.

Of great interest are the vinyl polymers with anthracene [150], acridine [73, 151], and pyrenyl moieties [151], with phenanthrene and *N*, *N*-dimethylaniline groups [152], and also polyacrylates and polymethacrylates with anthracene [153] and stilbene [154] groups. In particular, it has been suggested in [150] that the excimer fluorescence of poly(9-vinylanthracene) be used to study the electrochemical polymerization of the corresponding monomer.

The fluorescence spectra of solutions of poly(9-vinylacridine) prepared by polymerization of 9-vinylacridine, using sodium naphthalenide in tetrahydrofuran solution [73], show only the excimer emission band with $\lambda_{max} = 520$ nm in such solvents as dioxane, chloroform, and acetonitrile. A methanol solution of this polymer displays a fluorescence spectrum containing an additional band with $\lambda_{max} = 425$ nm emitted by the non-associated acridine moieties. The model 9-ethylacridine is characterized, under the same conditions, solely by a short wavelength fluorescence with $\lambda_{max} = 435$ nm [73].

Poly(1-vinylpyrene) (XII) was synthesized by the polymerization of 1-vinylpyrene in benzene or toluene solution in the presence of benzoyl peroxide or di-*tert*-butyl peroxide as initiator [69, 151].

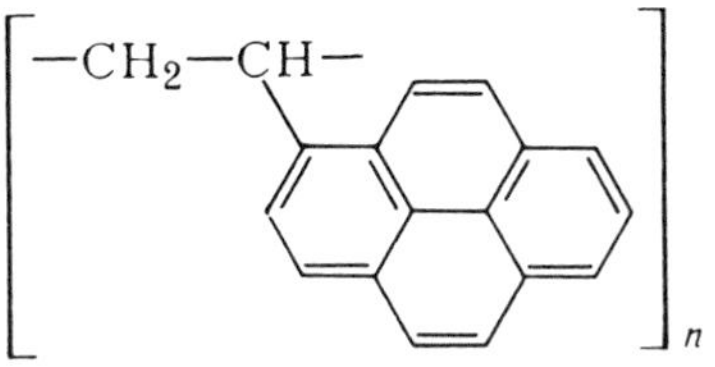

$$\left[-CH_2-CH-\right]_n$$

XII

The fluorescence properties of poly(1-vinylpyrene) solutions were compared with those of the poly(1-pyrenylmethylvinyl ester) solutions (XIII).

$$\left[\begin{array}{c} -CH_2-CH- \\ | \\ O \\ | \\ CH_2 \end{array}\right]_n$$

XIII

Measurements of the fluorescence spectra of tetrahydrofuran-methyltetrahydrofuran solutions of both these polymers show [151] that, at room or dry ice temperature, the emission of the former polymer is characterized solely by an excimer band with λ_{max} = 480 nm. Under the same conditions, the fluorescence spectrum of the latter polymer solutions shows bands of both the monomer and excimer fluorescence. At 77 K both these polymers feature only monomer fluorescence which, in the opinion of the authors of [151], is associated with the freezing of molecular motion necessary for excimer formation.

The specific features of the exciplex fluorescence of 9-vinylphenanthrene and N,N-dimethylaminostyrene copolymers prepared by radical polymerization in benzene solution in the presence of azobisisobutyronitrile are covered in [152]. The fluorescence spectra of dioxane solutions of copolymers containing 23 to 75% phenanthrene units (mol) show an exciplex emission band with λ_{max} = 437 nm and a less intense monomer band with λ_{max} = 356 nm, the intensity of the exciplex fluorescence and its band position being independent of the copolymer composition and slightly dependent on solvent polarity.

The existence of nonfluorescent complexes of ion radical type was discovered on quenching the fluorescence of poly(9-vinylphenanthrene) and the copolymer with phenanthrene and dimethylaniline groups by dicyanobenzene (DCB). Lifetime measurements by flash photolysis showed that the anion radicals of DCB are formed in greater quantity in the poly-9(vinylphenanthrene)-DCB system than in the copolymer–DCB system [152]. This happens because of the opposite direction of electron transfer from the phenanthrene anion radical to the cation radical of the dimethylaniline moiety in the copolymer. Additionally, the electron transfer from the excited phenanthrene molecule to DCB proceeds rather efficiently owing to the migration of energy along the poly(9-vinylphenanthrene) chain.

Ref. [153] deals with the polymerization of the anthracene-containing methacryl ethers yielding poly[(9-anthryl)methyl methacrylate] (XIV) and poly[(10-phenyl-9-anthryl)methyl methacrylate] (XV) (see next page).

The polymerization was performed in benzene at 60°C in the presence of azobisisobutyronitrile as initiator. Measurements of the fluorescence spectra of polymer films and dichloromethane solutions of the polymers at room temperature and at 77 K showed that, unlike the model (9-anthryl)methyl-2-methyl propanoate, a broad structureless excimer band with λ_{max} = 450 nm is present in the spectra of the former polymer. Conversely, the fluorescence spectra of the latter polymer in dichloromethane solution are closely similar to those of the corresponding model compound, and present the characteristic vibrational structure peculiar to monomeric anthracene moieties.

$$-\left[\begin{array}{c} -\overset{\underset{\displaystyle C}{|}}{C}(CH_3)-CH_2- \\ CO \\ | \\ O \end{array}\right]_n$$

XIV

$$-\left[\begin{array}{c} -\overset{\underset{\displaystyle C}{|}}{C}(CH_3)-CH_2- \\ CO \\ | \\ O \\ | \\ CH_2 \end{array}\right]_n$$

XV

The polymers obtained were studied in terms of the migration of electronic energy and singlet–singlet annihilation, in particular, as well as photo-oxidation and photodimerization [153].

Of great interest are two other types of fluorescent polymer, namely poly(vinylnaphthalenes) [155–160] and polymethacrylates with phenyl groups in the side chain [161].

Thus, the fluorescence properties of a tetrahydrofuran solution of poly(1-vinylnaphthalene) [155] suggests the presence of two bands with maxima at 330 and 475 nm, corresponding to emission from the monomer and excimer units. Study of the fluorescence decay curves and decay times enabled the authors to conclude as to the occurrence of effective intermolecular energy transfer in this polymer.

A more complex pattern of the emission spectra of solutions of this polymer is featured in [156]. Specifically, three types of centre were identified: a monomer emission centre peaking at 325 nm and two excimer centres. One of these fluoresces in the region 325–470 nm (λ_{max} = 375 nm) with a lifetime of 14 ns, while the other does so in the 400–540-nm region (λ_{max} = 470 nm), having a lifetime of 22 ns.

The excimer with λ_{max} = 375 nm is conceived as originating from units with a parallel configuration (XVI) of the alkylnaphthyl sites, while the excimer with λ_{max} = 470 nm is associated with the 'antiparallel' configuration (XVII).

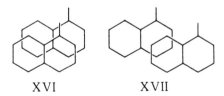

XVI   XVII

The fluorescence quantum yields are 0.0015 and 0.078 for the monomer and excimer emission respectively. Experiments on quenching the excimer emissions by anthracene show that the decrease in their fluorescence quantum yields is not associated with quenching of the monomer units. The opinion of the authors of [156] is that the concentrations of units with parallel and 'antiparallel' configuration, capable of forming the respective excimers of both types, is rather high, and during the lifetime of the monomer singlet state, no energy transfer over large distances occurs.

Study of the luminescence properties of poly(1-vinylnaphthalene) films at temperatures from 77 to 260 K showed [157] that the luminescence spectra possess two bands, that is, a short wavelength band with λ_{max} 410 nm which is a temperature independent slow fluorescence band, and a band lying at longer wavelengths with λ_{max} 590 nm, which is brought about by phosphorescence and undergoes a bathochromic shift with increasing temperature. A triplet state model is suggested to explain the red shift of the phosphorescence band and the origin of the slow fluorescence.

The fluorescence properties of dilute solutions of poly(2-vinylnaphthalene) were investigated at a pressure up to 450 MPa [158]. Valuable information was gained on the influence of the free volume in the polymer and of diffusion processes on excimer formation. It was found that the intensity of the monomer bands in the 310–370 nm region of the polymer spectra increases with pressure, while the intensity of the excimer band peaking at 416 nm remains practically unchanged.

Of great interest are data [158] on the influence of the solution viscosity on the ratio of the monomer and excimer emission intensities. It is evident from these data that the diffusion controlled intramolecular formation of the excimer in the polymer depends on the polymer free volume rather than follows any pattern arising from the Stokes–Einstein theory. As is apparent from the results obtained, the red shift of the excimer band occurs with increasing pressure, which indicates the stabilizing effect of pressure on the excimer complex.

The characteristics of electronic energy transfer and excimer formation due to the macromolecular conformation of the copolymers of methacrylic acid and its esters with 2-vinylnaphthalene are discussed in [159, 160].

Polymethacrylates with phenyl groups in the side chain [161] belong to those polymers whose monomer and excimer emission is exhibited in the shortest wavelength region as compared with all the abovementioned chromophore-containing polymers. Fig. 3.14 illustrates the fluorescence spectra of tetrahydrofuran solutions of poly(benzyl methacrylate) (XVIII), poly(benzhydryl methacrylate) (XIX), and poly(trityl methacrylate) (XX).

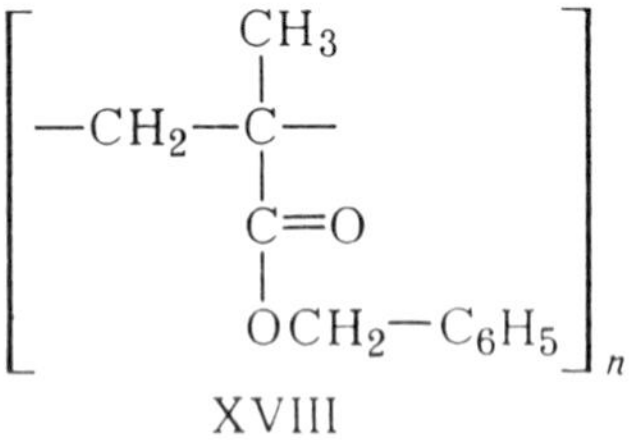

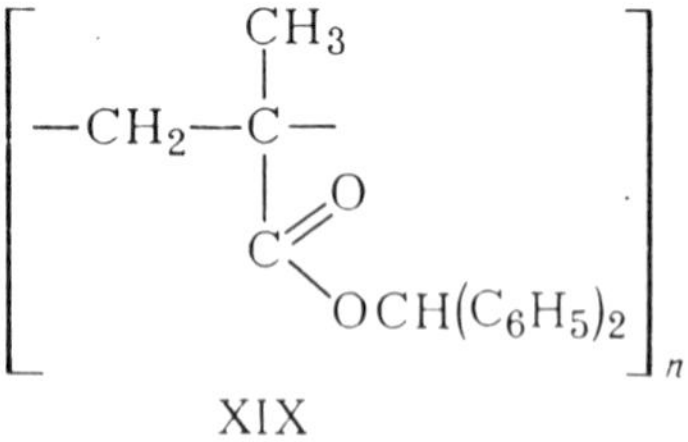

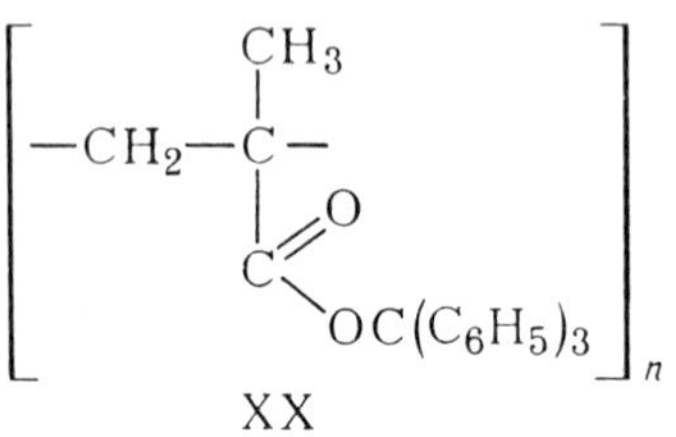

I'_{fl}/rel. units

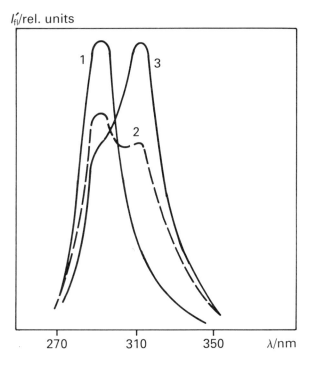

Fig. 3.14. Fluorescence spectra of dilute solutions of polymers XVIII (1), XIX (2), XX (3) in tetrahydrofuran.

It is seen that the polymer emission is characterized by two bands—a short-wavelength band with λ_{max} 285 nm corresponding to emission of the isolated benzene rings and a broad band λ_{max} 315 nm related to the radiative deactivation of excimers formed by the chromophoric moieties. It has been found that the intensity of the monomer fluorescence is markedly higher for polymer XVIII than for polymers XIX and XX, while the intensity of the excimer fluorescence is, by contrast, lower. A study of fluorescence quenching with increasing temperature and in the presence of added CCl_4 enables one to conclude about the intramolecular nature of the excimer emission.

As is evident from the spectral luminescent properties of the chromophore-containing polymerization systems, most of them show more or less effective processes of energy migration and transfer from the sites of absorption to the sites where the proximity and parallel arrangement of the planes of the chromophoric moieties promote excimer formation. These processes give rise to one or several excimer emission bands in the fluorescence spectra of the polymers along with the monomer emission band, the nature and position of which are determined mainly by the nature of the chromophore. In rare cases, when the absence of excimer emission can be regarded as proven [90, 147], it can be said that no electronic energy migration occurs along the polymer chain. One of the possible reasons for the low excimer-forming capacity in such polymers is the increased flexibility of the macrochains which destroy the necessary planarity in the arrangement of the chromophores.

3.2 POLYMERS WITH CONJUGATED BONDS IN THE MAIN CHAIN

Polyconjugated polymers attract general attention primarily owing to their unusual electrophysical properties. This class of properties has been covered in sufficiently great detail in [94, 162]; however, the luminescence properties of such polymers has received undeservedly little attention.

The fluorescence properties of polyconjugated systems such as polyhexyne, poly(*p*-diethynylbenzene), and polyphenylacetylene were first investigated in [163]. It was discovered that the short wavelength fluorescence band with λ_{max} *ca* 450 nm present in the emission spectra of these polymers weakens with increase in their RMM. Simultaneously the long wavelength band undergoes a bathochromic shift in the region 540–650 nm, the total intensity of the emission being sharply decreased. The appearance of the maximum at 540 nm evidences formation of the conjugated system on the thermal treatment of polyacrylonitrile [163].

It should be noted that in [163] use was made of the weakly sensitive photographic method for spectrofluorimetry, and some of the data presented have not been confirmed in later studies [164, 165, 166]. Specifically, [165] deals with the fluorescence spectra of polyphenylacetylene available as a suspension in a polystyrene film and in dichloromethane solution (Fig. 3.15). It can be seen that the emission spectrum of the polymer in the polystyrene matrix is the superposition of two bands, while the dichloromethane solution shows only one short wavelength structureless band. The fluorescence excitation spectra indicate the presence in polyphenylacetylene of two emission centres: one centre responsible for the short wavelength fluorescence with λ_{max} 420 nm and the other centre related to the long wavelength band with λ_{max} 490 nm. The opinion of the authors of [165] is that the luminescence properties of these systems depend on the mobility of the chromophore as determined by the conformational limitations of the polymer chain itself and the physical condition of the polymer. One pos-

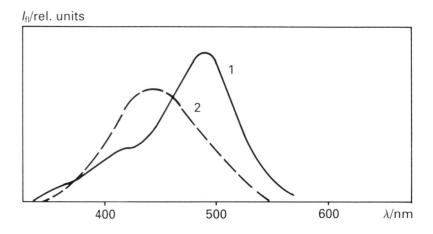

Fig. 3.15. Fluorescence spectra of polyphenylacetylene in polystyrene matrix (1) and methylene chloride solution (2).

sible explanation for these results might be the presence in the system of two singlet states, in which the relative positions of the energy levels is determined by solvent polarity.

Some unique considerations about the nature of the luminescent properties of poly-conjugated polymers are put forward in [167–170]. For instance, the luminescence spectra of polyacenaphthylene partially dehydrated by chloranil showed [166–169] that with an increasing level of dehydration up to 28%, the conjugated blocks steadily increase in length, as evidenced by the red shift of the luminescence maxima (Fig. 3.16). As seen from a comparison of the luminescence spectra of polyacenaphthylenes dehydrated to 28 and 42%, a further increase in the level of dehydration causes only the accumulation of the conjugated blocks, the length of the blocks being unchanged. The position of the maximum in the 530–550 nm region is thought by the authors of [167] to correspond to a length of the conjugated block of 6 or 7 monomer units, which is supported by a comparison of the absorption and luminescence spectra.

The Stern–Volmer dependences of the fluorescence quenching efficiency of polyace-naphthylene, partially dehydrated by chloranil in benzene solution, indicated the following [169]: if the emission spectrum of the polymer is overlapped by the absorption spectrum of chloranil, then quenching occurs owing to dipole–dipole energy transfer. If such an overlap does not occur, the fluorescence is quenched by the migration of energy over short distances or by the formation of donor–acceptor complexes.

Refs [168–170] cover in detail the luminescent characteristics of poly(propiolic acid) (PPA) samples of varied spatial structure prepared under different conditions. In particular,

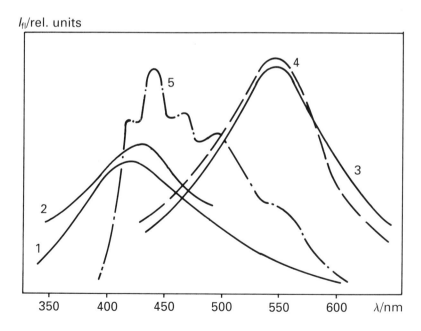

Fig. 3.16. Fluorescence spectra of polyacenaphthylene: 1, initial powder; 2, 3, 4, powder with degrees of dehydration of 9, 28 and 42%, respectively; 5, benzene solution of partially dehydrated polymer.

the effect was investigated of conformation changes in the PPA chain prepared by liquid state and solid state polymerization (*cis–trans*-configuration XXI and *trans–trans*-configuration XXII respectively) on the luminescence spectra (Figs 3.17 and 3.18).

$$
\begin{array}{c}
\overset{\displaystyle H}{|}\quad \overset{\displaystyle H}{|}\quad \overset{\displaystyle H}{|}\quad \overset{\displaystyle H}{|}\\
\sim C{=}C{-}C{=}C{-}C{=}C{-}C{=}C\sim\\
|\qquad |\qquad |\qquad |\\
COOH\ \ COOH\ \ COOH\ \ COOH
\end{array}
$$

XXI

XXII

It was suggested that the spectral maximum shift towards longer wavelengths corresponds to the increase in size of the effective conjugated block. At the same time interesting results were obtained with regard to the luminescence behaviour of PPA solutions of various acidities. In aqueous and alcoholic solutions of PPA, an increase in the pH of the medium from 1 to 9 is accompanied by straightening of the macromolecule and an increase in size of the conjugated block as a result of the increasing dissociation of the carboxyl groups. A further increase in pH causes the screening of charges and coiling of the macromolecule, which corresponds to the shift of the band maximum to shorter wavelengths (Fig. 3.19).

The transition of PPA from solution to the solid state is accompanied by extension of the conjugated blocks. The luminescence spectra of the polymer with *cis*-transoid and *trans*-transoid configuration and the conjugated sequences of different lengths suggests

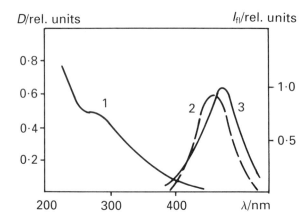

Fig. 3.17. Absorption (1) and fluorescence (2, 3) spectra of aqueous solutions of PPA of *cis*-transoid (2) and *trans*-transoid (3) configuration.

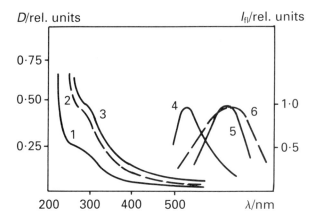

Fig. 3.18. Absorption (1, 2, 3) and fluorescence (4, 5, 6) spectra of solid PPA samples with RMM 2100 (1, 5), 1540 (2), 1400 (3, 4) and mixture of samples with RMM 2100 and 1400 (6).

the intermolecular transfer of energy from shorter to longer sequences.

This latter idea is confirmed by a comparison of the luminescence spectra of solid PPA samples of different RMMs (Fig. 3.18) [170]. The authors of [170] draw our atten-

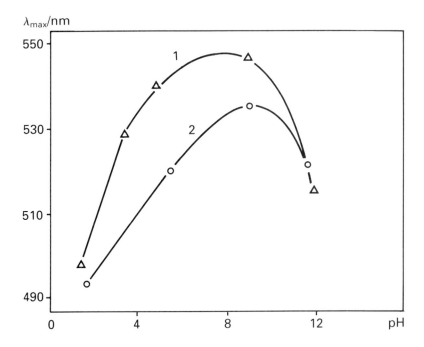

Fig. 3.19. Position of λ_{max} in luminescence spectra of aqueous solutions of PPA of *trans–trans* (1) and *cis–trans* (2) configuration vs pH of medium.

tion to the absence of mirror symmetry and other differences in the absorption and luminescence spectra. However, they do not deny the possibility of emission emanating from
impurities. To interpret their results, they resort to the scheme suggested in [171] to
describe dipole–dipole energy transfer from the excited state of one molecule to less
excited states of other molecules by the Förster mechanism.

As a result of the redistribution of energy by this mechanism, the luminescence spectrum shifts to longer wavelengths, the absorption spectrum being unchanged, since the
luminescence spectrum is given by the transition from a low excited state to the ground
state of the molecules in which the difference in energy transfer is the smallest, that is, in
similar polydisperse systems, the position of the luminescence band depends solely on
the arrangement of the energy levels in those molecules with the longest sequence of
effective conjugation. According to the authors of [170], the coincidence of the luminescence band maximum of the PPA sample with mixed RMM (Fig. 3.18, curve 6) with that
of the highest polymer supports this view.

The spectral and luminescent properties of polyphenylacetylene and polydiphenylacetylene were studied in [172, 173]. It was shown, in particular, that the luminescence
spectra of polyphenylacetylene feature a broad structureless band λ_{max} 580–660 nm
whose position is determined by the temperature and duration of the thermal polymerization of the initial monomer. The suggestion was made in [173] of assessing the effective
value of the conjugation n_c by the position of the maximum in the luminescence spectrum. On assuming that the positions of maxima in the luminescence spectra of
polyphenylacetylene and polyenes coincide, the authors of [173] concluded that the polymer prepared at 150°C contains conjugation sequences with $n_c = 5$ to 6. This is explained
by the fact that the mid-position of the polyphenylacetylene spectrum maximum is at
about 620 nm, that is, in the range between the maxima of diphenylhexatriene (530 nm)
and diphenyloctatetraene (670 nm).

The luminescence spectra of polydiphenylacetylene measured in [172, 173] were
closely similar to those of polyphenylacetylene. For the position of the maximum given,
$n_c = 5$ to 7. For the polymer prepared by heating diphenylacetylene at 300°C for 6 hours,
the benzene rings are involved in continuous conjugated chain over the entire length of
the system. The quantum luminescence yield of this polymer was not measured, but the
luminescence intensity is reported to be high and comparable to that of special technical
luminophors [173].

Although the luminescent properties of polymers with alternating single and double
bonds are covered adequately in the literature, macromolecules with acetylene units in
the chain have not in fact been studied by this method. The absorption and fluorescent
properties of water soluble polydiacetylenes (XXIII) prepared by 1,4-addition of crystalline monomers $RC{\equiv}C{-}C{\equiv}CR$ are discussed in adequate detail in only one reference [174].

$$[{=}C(R){-}C{\equiv}C{-}C(R){=}]_n,$$

where $R = -(CH_2)_3-OCONHCH_2COOK$ (XXIIIa); $-(CH_2)_3-COOK$
(XXIIIb),

The nature of the solution absorption spectra of these polymers is established to be
critically dependent on the pH of the medium: the absorption spectral maximum of
polydiacetylene solution (XXIIIa) shifts from 426 nm at pH = 9.5 to 535 nm at pH = 3.

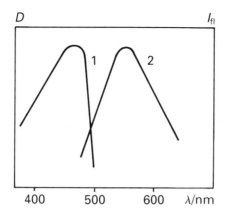

Fig. 3.20. Absorption (1) and fluorescence (2) spectra of 0.001% aqueous solution of polymer XXIIIa
at pH = 10.15.

At pH < 3 a new absorption band with λ_{max} 575 nm appears, owing to the conformational transition from the nonplanar to the planar form reflecting the increased delocalization of electrons in the main polymer chain. The repulsive forces between the COO^--groups, in interfering with the formation of the planar conformation in polymer molecules, are assumed to weaken as a result of the formation of carboxyl groups on neutralization. Fig. 3.20 shows the absorption and fluorescence spectra of a 0.001% aqueous solution of polymer XXIIIa at pH = 10.15. The fluorescence band peaking at 545 nm shifts to longer wavelengths and the emission intensity falls when the solution is doped with electrolytes such as KCl, LiCl or NaCl.

The results of an investigation into the absorption and fluorescence spectra of hexadiyne-2,4-dicarboxylic-1,6-acid suggest that, on heating or illuminating, the solid state polymerization of the monomer occurs via conversion of the parallel diyne molecules into the polyacetylene macromolecule [175].

Thus, just as the tendency to form excimers enables interpretation of the fluorescence characteristics of chromophore-containg polymers with a saturated main chain, so can the characteristics of the spectral luminescent properties of polyconjugated systems be determined in full by the size of the effective conjugated sequences and the conformational structure of the macromolecule.

4

Polycondensed polymers and products of polymer-analogous conversions

Polycondensed fluorescent polymers have been extensively studied in recent years. Polycondensed polymers are distinguished by the presence in their main chain of polar functional groups such as ether, amide, urethane, etc.; their susceptibility to form hydrogen bonds has a significant effect on their nature and the character of the absorption and fluorescence of these polymers.

4.1 ALIPHATIC POLYMERS

Aliphatic polycondensed polymers containing chromophoric groups in the chain are characterized by bright colouring and apparent fluorescence. They were first prepared early in the 1960s [176, 177]. The most comprehensively investigated are stained polycaproamides with diaminoanthraquinone moieties [176] and poly(ethylene terephthalates) with the chemically bonded luminescent dye rhodamine C [177]. Most of the subject matter of these investigations has been summarized in review [178].

Later investigations describe polycondensed polymers beginning to fluoresce on the introduction into them of luminophoric moieties such as carbazole, anthracene, or any other condensed aromatic rings. Such polymers are covered in greater detail below.

4.1.1 Carbazole-containing polymers

The spectral luminescent properties of a small number of carbazole-containing polymers have been studied [179]. The synthesis of luminescent polyesters is covered in [180, 181], of polyurethanes with ω-9-carbazolylbutyl groups in [182], and of polyamides based on 3,6-diamino-9-ethylcarbazole in [183].

The polyesters of general formula I were prepared by polycondensation of 6-carbazyl-2,2-dihydroxymethylhexane with the diethyl esters of dicarboxylic acids at 150 to 160°C in a nitrogen atmosphere for 3 h, then *in vacuo* over 10 h [180, 181].

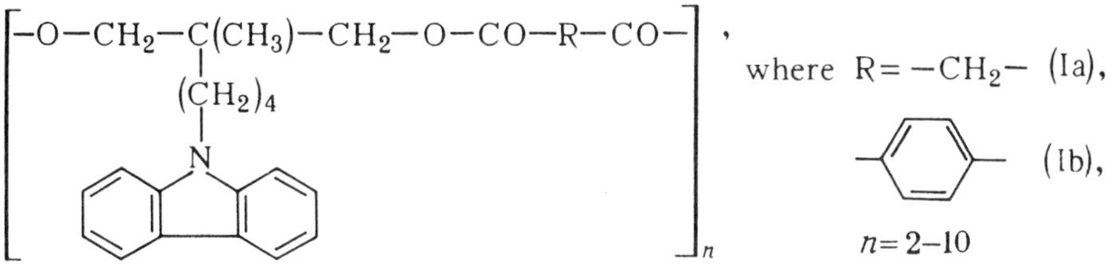

$$\left[-O-CH_2-\underset{\underset{\displaystyle\overset{|}{N}}{(\overset{|}{C}H_2)_4}}{\overset{|}{C}}(CH_3)-CH_2-O-CO-R-CO-\right]_n$$

I

where $R=-CH_2-$ (Ia),

(Ib),

$n=2-10$

The absorption and fluorescence properties of polymers Ia and Ib were investigated along with the properties of model compound II (Figs 4.1 and 4.2).

As Fig. 4.1 suggests, the absorption properties of the polymers are similar in the region with $\lambda_{max} > 270$ nm where the carbazyl group is the only light absorbing chromophoric group. The differences in the region with $\lambda_{max} < 270$ nm are assigned to the absorption of the terephthalate unit. The fluorescence spectra of aliphatic polyester Ia and model compound II

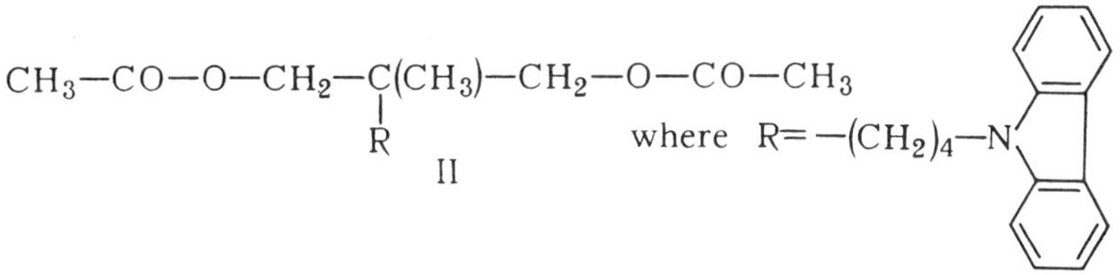

$$CH_3-CO-O-CH_2-\underset{\underset{\displaystyle II}{R}}{\overset{|}{C}}(CH_3)-CH_2-O-CO-CH_3$$

where $R=-(CH_2)_4-N$

are identical and show only the monomer fluorescence band with $\lambda_{max} = 368$ nm. Conversely, the fluorescence spectrum of polyester Ib (see Fig. 4.2) shows an additional broad structureless band with $\lambda_{max} = 490$ nm caused by the emission of the exciplex formed during interaction of the terephthalate group of the polymer with the singlet excited state of the carbazyl group. The ratio between the intensities of the exciplex and monomer fluorescence depends on the degree of polycondensation n [181] and decreases when the solution is diluted. The latter point indicates the intermolecular mechanism of exciplex formation even at a concentration of 10^{-4} mol dm^{-3} [180]. From the quantum yields of the total fluorescence of polyester Ia and model II (0.3) and polyester Ib (less than 0.1), it follows that the probability of fluorescence is lower in the exciplex formed during the interaction of the excited carbazyl group of model II with the terephthalate group than in the carbazyl group itself. The fluorescence lifetimes determined for compounds Ia, Ib, and II suggest that the reduced fluorescence intensity of the carbazyl groups of polyester Ib results from dynamic quenching by the terephthalate groups.

Ref. [182] describes the synthesis of carbazole-containing polyurethanes III by polycondensation of 9-(6-hydroxy-5-hydroxymethylhexyl)carbazole with diisocyanates in a refluxing anisole medium for 6 h.

$$\left[-O-CH_2-\underset{\underset{\displaystyle\overset{|}{N}}{(\overset{|}{C}H_2)_4}}{\overset{|}{C}H}-CH_2-O-CONH-(CH_2)_m-NHCO-\right]_n$$

$m = 2-6.$

III

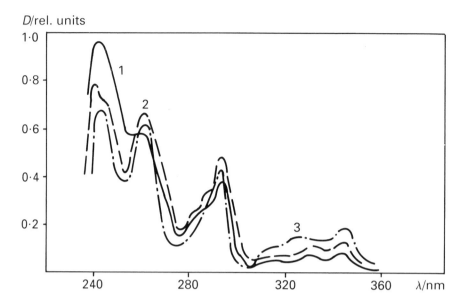

Fig. 4.1 Absorption spectra of dioxane solutions of polymers Ib (1), Ia (2) and of model compound II (3).

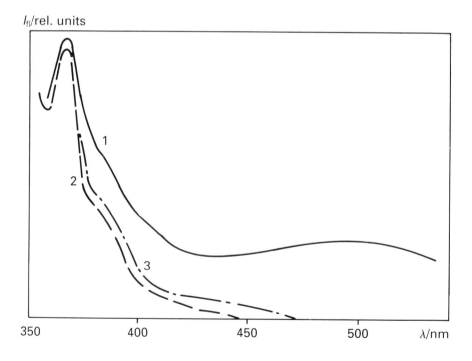

Fig. 4.2. Fluorescence spectra of dioxane solutions of polymers Ib (1), Ia (2) and of model compound II (3).

For polymer III with $m = 6$, the degree of polycondensation determined by gel permeation chromatography is about 35. Polyurethanes of this type are soluble in dimethylformamide, chloroform, and anisole.

The spectral properties of polymers III and carbazole-containing polyesters Ia are very similar, being determined mainly by the carbazole chromophore free from exciplex states.

Polyamides V containing the carbazole residue in the main, rather than the side, chain are prepared by polycondensation of 3,6-diamino-9-ethylcarbazole (IV) with dichloroanhydrides of dicarboxylic acids in 9-ethylcarbazole with triethylamine as catalyst at 45 to 50°C [183].

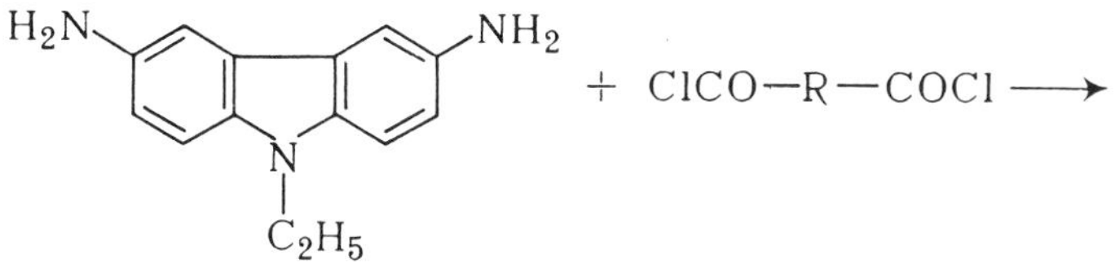

$$R = (-CH_2-)_m, \ m = 2\text{--}8.$$

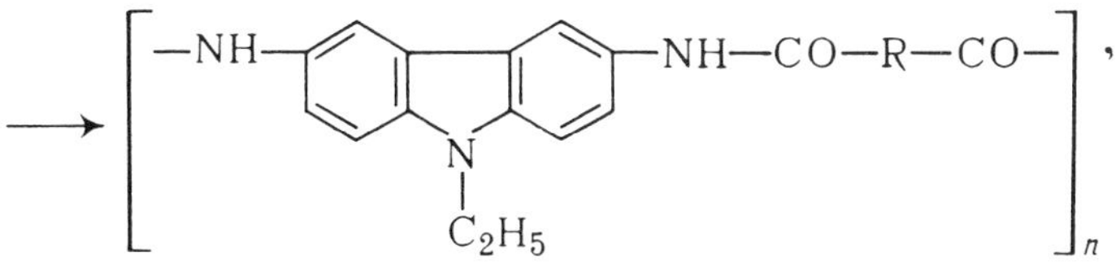

Polyamides V are a white or pale yellow powder with $[\eta] = 0.25\text{--}0.60 \ cm^3 \ g^{-1}$ (m-cresol, 45°C). Special experiments showed that any deviation of the proportion of reactants from the stoichiometric value markedly reduces $[\eta]$ of the polymer formed.

The fluorescent properties of the polymers and the model 9-ethyl-3,6-bis(propionyl-amino)carbazole were investigated in solutions of hexane and m-cresol (Fig. 4.3). The solutions of all the polymers and the model compound were characterized by similar monomer fluorescence free from any detectable excimer emission. The significant red shift of the fluorescence band observed in m-cresol solution is explained by the formation of a hydrogen bond between the OH-group of the solvent and the amide linkage of the compounds in question. Measurements of the fluorescence spectrum of the polyamide solutions in m-cresol could not, however, be fully correct, and the spectral curve presented in Fig. 4.3 with a maximum found in the region 360 to 400 nm is apparently a reflection of the emission emanated from impurities, since for the fluorescence excitation of the solutions the authors of [183] used the region with $\lambda_{max} = 360$ nm in which the polymer shows virtually no natural absorption.

4.1.2 Anthracene-containing polymers

Practically all polycondensed polymers of this type contain anthracene in the side chain. Of primary interest are polyesters [78, 80, 82, 184–186], polyurethanes [80, 185], and polymeric quaternary ammonium salts (polyionenes) [187, 188]. Anthracene-containing

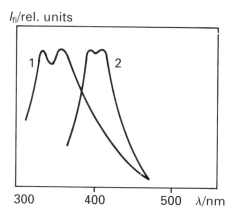

I_{fl}/rel. units

Fig. 4.3. Fluorescence spectra of carbazole-containing polyamide V in hexane with λ_{excit} = 300 nm (1) and in *m*-cresol with λ_{excit} = 360 nm (2).

polyesters VI were prepared by polycondensation of 2-(9-anthryl)methyl-1,3-propanediol with diethyl esters and the dichloroanhydrides of dicarboxylic acids [78].

$$\left[-O-CH_2-CH-CH_2-O-CO-R-CO- \right]_n$$

VI

where R = $-CH_2-$ (VIa), $(-CH_2-)_2$ (VIb), $(-CH_2-)_4$ (VIc),

$(-CH_2-)_8$ (VId), $-CH-CH_2$ (VIe),   (VIf).

The reaction of propanediol with diethyl esters was conducted at 150–160°C in the presence of catalytic amounts of calcium acetate and antimony trioxide in a nitrogen atmosphere for 3 h and then *in vacuo* for 10 h. When dichloroanhydrides were used instead of diethyl esters, the reaction was run at room temperature for 4 to 6 h, the HCl formed in the reaction being removed by the nitrogen.

The polymers obtained are soluble in tetrahydrofuran, and, as seen from the results gained by gel permeation chromatography, have a low RMM (n = 2–7).

I_{fl}/rel. units

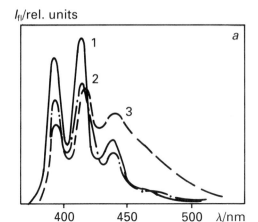

I_{fl}/rel. units

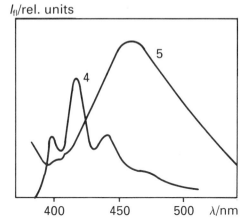

Fig. 4.4. Fluorescence spectra of model compound (a) and polyester (b) in tetrahydrofuran: 1, model compound VII; 2, polyester VId; 3, polyester VIa; 4, polyester VIc; 5, polyester film VIb.

Polyesters VIb, VId, VIf were synthesized by interaction of propanediol with the diethyl esters of the corresponding dicarboxylic acids at 150–160°C for 2 to 3 h in a nitrogen atmosphere and then for 13 to 18 h *in vacuo* to remove thiophenol [184]. This method was used to prepare polyesters of a higher RMM ($n = 4$–25).

Study of the spectral luminescent properties of the polymer in solution and film as well as of solutions of the model 1,3-diacetoxy-2-(9-anthryl)methylpropane (VII) indicates that they are similar as regards absorption properties. The fluorescence spectra of the polymer solutions show both monomer bands with λ_{max} = 392, 415, 440 nm and excimer bands with λ_{max} = 460 nm (Fig. 4.4). It has been established that for the polymer solutions, I_E/I_M varies from 0.431 to 0.024 in the polymer sequence VIa > VIe > VIb > VIc > VId > VIf. In the polymer films, excimer fluorescence dominates over monomer fluorescence (Fig. 4.4b). The highest excimer fluorescence intensity of polyester VIa appears to be caused by the high efficiency of energy transfer in this polymer.

As was revealed in a study of the dependence of I_E/I_M on the polymer (VIa) content of the solution, the intramolecular interaction of anthryl units prevails (I_E/I_M = const) at concentrations of up to 10^{-3} mol dm^{-3}, while at higher concentrations the contribution of interamolecular interactions increases (I_E/I_M increases with concentration) [184]. It was confirmed with polyester VIe that luminophor units take part in the process of energy transfer: on excitation of fluorescence with light (λ_{max} = 280 nm), which is absorbed by the naphthyl groups, it was possible to observe emission typical only of the anthryl units (Fig. 4.4b).

Refs [70, 185] suggest the synthesis of polyesters (VIII) and polyesterurethanes (IX) by polycondensation of 9,10-bis-4-(hydroxybutyleneoxycarbonyl)dianthracene and the dichloroanhydrides of acids or diisocyanates.

$$[-O-(CH_2)_4-OCO-C(Ar)_2-COO-(CH_2)_4-OCORCO-]_n$$
VIII

$$[-O-(CH_2)_4-OCO-C(Ar)_2-COO-(CH_2)_4-OCO-NHR'NHCO-]_n,$$
IX

where R= $(-CH_2-)_6$ (VIIIa), $(-CH_2-)_{10}$ (VIIIb), Ar = ![anthracene structure]

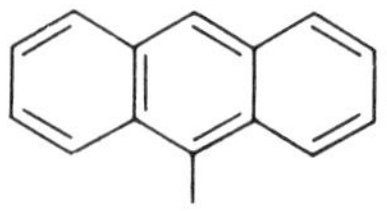

R'= $(-CH_2-)_m$, m= 4, 8 and 12 (IXa), ![structure with CH₂] (IXb)

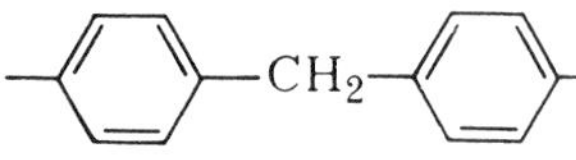

The polymers obtained are highly soluble in chloroform and dichloromethane, their values of [η] being 0.37 to 1.68 (chloroform, $C = 0.8–1\%$ (mass), 25°C).

It has been shown in [185] that, on irradiation at 196 K of polymers VIII and IX with light of wavelength 300 nm, a small fraction (up to 5%) of dianthracene units undergoes photolysis to give anthroate groups (X).

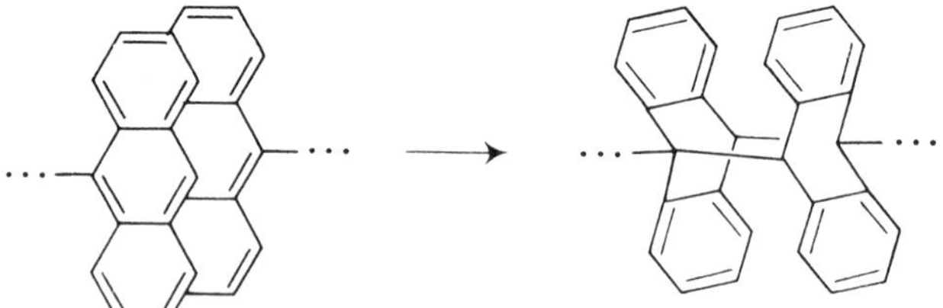

X

The terminal anthroate groups appeared in the form of sandwich dimers which, on heating, gradually dissociated into monomer moieties. The formation of dimers was evidenced by a broad structureless band peaking at 500 nm in the fluorescence spectrum of the polymers subjected to photolysis at 196 K. In the samples of polyesterurethanenes XI, the sandwich dimer structure is retained at room temperature, and in polyesters VIII these complexes are stable only at low temperatures.

It has been established from the fluorescence spectra of isolated terminal anthroate groups that the rotational mobility of the luminophors is less in polymers IX than in polymers VIII. The limited mobility of these groups affects the formation of new dianthracene groups on irradiation of the anthroate groups with light in the absorption band ^1L_a.

Anthracene-containing polyurethanes XI were prepared in [186] by the interaction of bis(2-hydroxyethyl)-9-anthrylmethyl malonate and the corresponding diisocyanates in anisole for 6 h. Polymers XIa and XIb are soluble in tetrahydrofuran, and, according to gel chromatographic measurements, have RMM from 4 100 to 11 000.

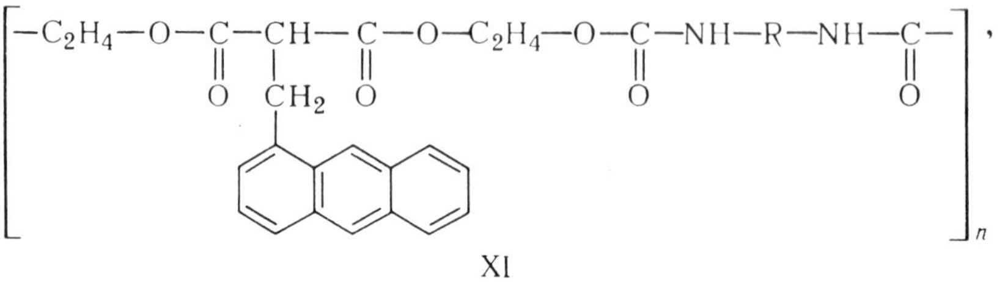

XI

where R= $(-CH_2-)_6$ (XIa), ![structure with H₂C and CH₂] (XIb).

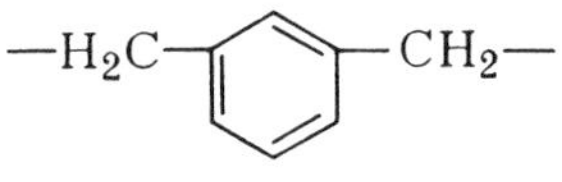

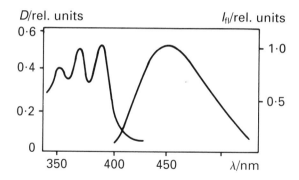

Fig. 4.5. Absorption (1) and fluorescence (2) spectra of polyurethane XIa films.

Fig. 4.5 shows the absorption and fluorescence spectra of polyurethane (XIa) films. The opinion of the authors is that the broad absorption bands of this polymer at 360 and 380 nm appear to be due to the sandwich dimers which are formed as a result of the close parallel arrangement of the anthryl groups. The fluorescence spectra of tetrahydrofuran solutions of polymers XIa and XIb are due to transitions caused by emission of the monomer units and excimers. The authors of [186] showed that the fluorescence excitation spectra of the polymer films are broader than the analogous spectra of the solutions, which confirms the significant role of interaction of the macromolecules in the solid state.

Refs [187, 188] deal with the absorption and fluorescence properties of polyionenes (XII) prepared by the Menschutkin reaction between 1,3-bis(haloacetoxy)-2-(9-anthryl)methylpropane or 2-(9-anthryl)methyl trimethylene halides and ternary diamines.

$$XCH_2-R--CH_2X + (CH_3)_2N-(CH_2)_6-N(CH_3)_2 \longrightarrow$$

where R $=$—CO—O—CH$_2$—CH—CH$_2$—O—CO—(XIIa),

—CH$_2$—CH—CH$_2$— (XIIb); X $=$ Cl, Br.

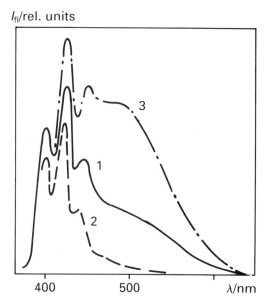

Fig. 4.6. Fluorescence spectra of polyionene XIIa in aqueous solution (1), in a mixture of equal volumes of $H_2O - MeOH$ (2) and aqueous HCl with ionic strength $\mu = 1.0$ (3).

The reaction was conducted in dimethylformamide, dimethylacetamide, dimethyl sulfoxide or sulfolane at 40–60°C over 20–126 h. Polymers XIIa and XIIb are readily soluble in water and in the solvents used for their synthesis. $[\eta] = 0.5$ to 0.26 (dimethylformamide, $C = 0.25$–0.33% (mass)).

Study of the fluorescence properties of these polymers disclosed that in aqueous solutions they form hydrophobic domains from the anthrylmethyl groups which give rise to the broad structureless excimer emission band with $\lambda_{max} = 500$ nm (Fig. 4.6). As is seen from Fig. 4.6, the addition of methanol facilitates the destruction of the hydrophobic domains as a result of which the intensity of the excimer fluorescence decreases. Conversely, the excimer emission intensifies on the addition of electrolytes: the higher the ionic strength of the solution, the stronger is this effect.

Study of the photochemical properties of polymers XIIa and XIIb shows that irradiation of their solutions with light at a maximum ≥ 340 nm leads to photodimerization of the anthryl groups.

4.1.3 Pyrene-containing and other polycondensed polymers

The pyrene-containing polyesters (XIII) and (XIV) based on 3-pyrenyl-1,3-propanediol [184, 189] or 2-(4-N,N-dimethylaminobenzyl)-2-(1-pyrenylmethyl)-propane-1,3-diol [190] are synthesized by the interaction of the diol with the dithiol esters of dicarboxylic acids at 150–170°C, first in an inert atmosphere and then *in vacuo*. As a result, the RMM of the polymers obtained is 2–5 times higher than that of the polymers obtained by the conventional polycondensation of diols with acid dichloroanhydrides [184, 190].

Table 4.1 features the composition and some properties of the pyrene-containing polyesters (XIII) and of the model compound (XV).

$$[-O-CH_2-CH(R)CH_2-OCOR'CO-]_n \; ;$$
XIII

$$CH_3-COO-CH_2-CH(R)-CH_2-OCO-CH_3,$$
XV

where R= $-CH_2$ (pyrene) , R'= $(-CH_2-)_2$ (XIIIa), $(-CH_2-)_4$ (XIIIb),

$(-CH_2-)_8$ (XIIIc), $(-CH_2-)_{10}$ (XIIId), $-\langle phenyl \rangle-$ (XIIIe)

As follows from Table 4.1, the absorption properties of polyesters XIII are virtually independent of the nature of the dicarboxylic acid residue and are similar to the properties of the model compound (XV). Conversely, the fluorescence properties of these compounds reveal significant differences (Fig. 4.7) assigned to the appearance, in the case of the polymers, of the intense excimer emission band with λ_{max} = 480–490 nm.

In their emission spectra, polyesters XIII were found to be similar to other pyrene-containing polyesters (XIV) having dimethylaniline groups in the side substituents [189]:

$$\left[-O-CH_2-\underset{R}{CH}-CH_2-O-CO-\underset{R'}{CH}-CO-\right]_n ,$$

XIV

where R= $-CH_2$ (pyrene) , R'= $-CH_2-\langle phenyl \rangle-N(CH_3)_2$

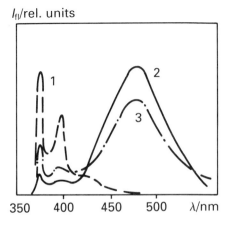

I_{fl}/rel. units

350   400   450   500   λ/nm

Fig. 4.7. Fluorescence spectra of model compound XV (1), polymers XIIa (2) and XIIId in tetrahydro-furan solutions (3).

Table 4.1. Absorption properties of polyesters XIII with pyrene groups in side chain and of model compound XV [184]

Compound	$M_n{}^a$	Absorption	
		λ_{max}/nm	ε_{max}/dm^3 mol^{-1} cm^{-1}
XIIIa	5 100	344	31 700
XIIIb	13 000	344	31 100
XIIIc	46 000	344	33 200
XIIId	11 000	344	32 100
XIIIe	1 600	344	31 500
XV		343.8	36 700

^{a} Measured by gel permeation chromatography.

Polymers XIV were synthesized by bulk polycondensation of the respective propane-diol with diethyl-4-(N,N-dimethylamino)benzyl malonate in the presence of calcium acetate as catalyst at 160–210°C *in vacuo*.

Fig. 4.8 enables a comparison of the absorption and fluorescence spectra of solutions of the model compound XV and polyesters XIV, which suggests a weak interaction between the chromophores in the ground state. It has been established that the intensity ratio of the exciplex fluorescence of the solution spectrum of polymer XIV (with λ_{max} 498 nm) and the monomer fluorescence (with λ_{max} 377 nm) increases with the solution concentration. This result confirms, in the opinion of the authors of [189], intramolecular association in the concentration region below 10^{-5} mol dm^{-3} due to the weak interactions in the ground state between the electron-donating N,N-dimethylaniline and the electron-deficient pyrenyl groups. Such an explanation is considered as authentic: no exciplex fluorescence was observed at the same concentrations in solutions of the model compound containing a pyrene and a N,N-dimethylaniline group.

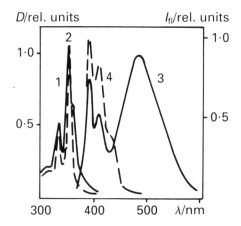

Fig. 4.8. Absorption spectra (1, 2) of 1,2-dichloroethane solutions and fluorescence spectra (3, 4) of model compound XV (2, 4) and polymer XIV (1, 3) in dioxane solutions.

Polyesters XVI also containing pyrene and dimethylaniline groups but, unlike polymers XIV, attached to one carbon atom, were investigated in [190]:

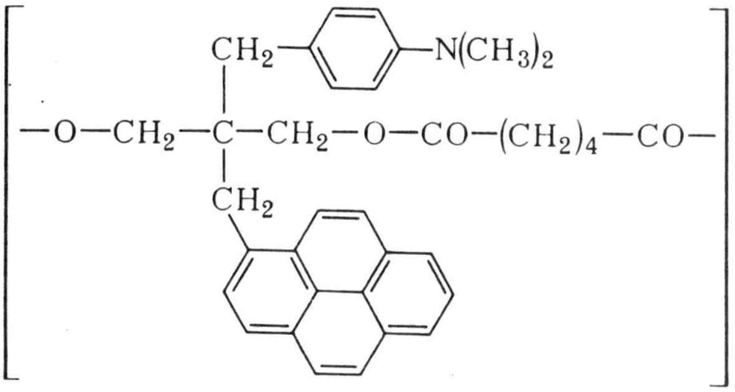

XVI

Measurements of the fluorescence spectra of tetrahydrofuran solutions of the polyesters and model compound XVII showed the presence of a weak band with λ_{max} = 380 nm in the region of emission of the monomer pyrene units, and a strong band of the excimer fluorescence with λ_{max} = 520 nm. Comparison of the emission spectra of polymers XIV and XVI discloses that the mutual approach of the electron-donating and -deficient moieties gives rise to a bathochromatic shift of the exciplex fluorescence. As is emphasized in [190], the excimer bands of the fluorescence spectra of solutions of polyesters XVI and model compound XVII are different in shape. This is apparently explained by the fact that solutions of the model compound of concentration below 10^{-5} mol dm^{-3} show only the intramolecular interaction of the donor and acceptor, while both the intra- and intermolecular interactions can be observed in the polymers.

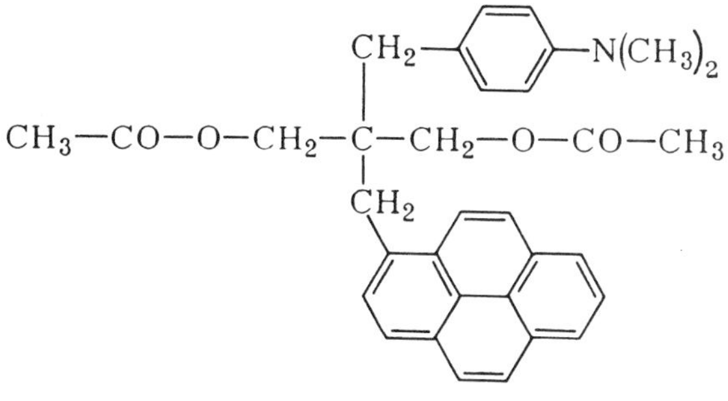

XVII

In addition to the above polymer systems with pyrene moieties, of great interest are the polycondensed polymers with luminophor units such as naphthalene [76, 191–193], p-phenylenediacryl [194], 3-methoxybenzanthrone [195], and naphthalimide [196].

Polyesters with side naphthylmethyl groups (XVIII) were synthesized by the polycondensation of 2-[(1-naphthyl)methyl]-1,3-propane with diethyl esters of dicarboxylic acids at 150–160°C for 3 h in an inert atmosphere and then for 10 h *in vacuo* [76].

Gel permeation chromatography measurements show that the polyesters obtained have a low RMM ($n \sim$ 3–10).

Study of the fluorescence properties of polymers XVIII was preceded by a study of model compounds XIX and XX. It was established that, unlike the spectra of model XIX, the emission spectra of the polymer solutions (XVIII) show, along with the monomer emission band with $\lambda_{máx} = 324$ nm, a broad excimer band with $\lambda_{max} = 400$ nm (Fig. 4.9). The intramolecular formation of excimers takes place even when the naphthyl groups are separated by 16 carbon and 2 oxygen atoms. The I_E/I_M ratio is higher for the polymer system than for the corresponding dimer system (XX). This is explained in [76] by the fact that the local concentration of the naphthyl units in the ground state is higher near a separate excited group of the polymer macromolecule than in the dimer.

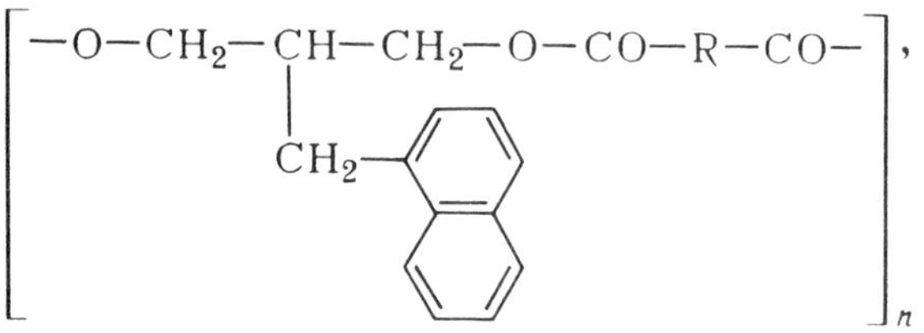

$$\left[-O-CH_2-CH-CH_2-O-CO-R-CO- \right]_n ,$$

XVIII

where R= $(-CH_2-)_2$ (XVIIIa), $(-CH_2-)_8$ (XVIIIb), $-CH-$ (XVIIIc)

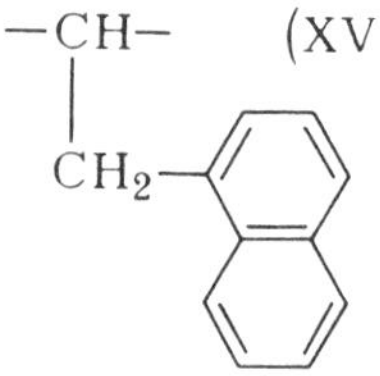

$$CH_3-CO-O-(CH_2)_3-$$

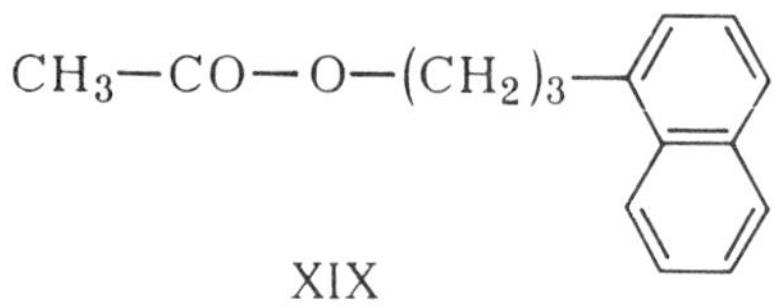

XIX

$$-(CH_2)_3-O-CO-R-CO-O-(CH_2)_3- ,$$

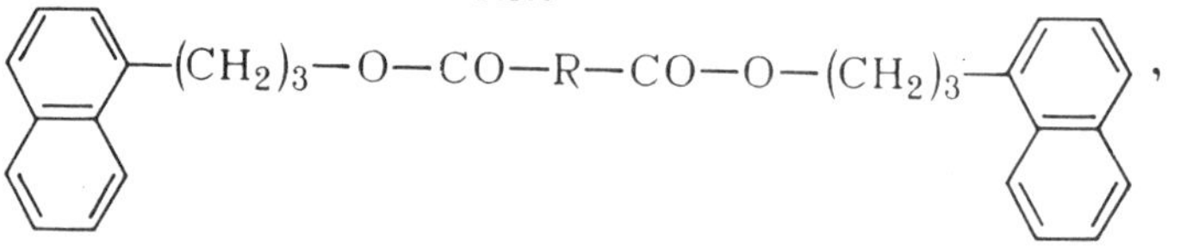

XX

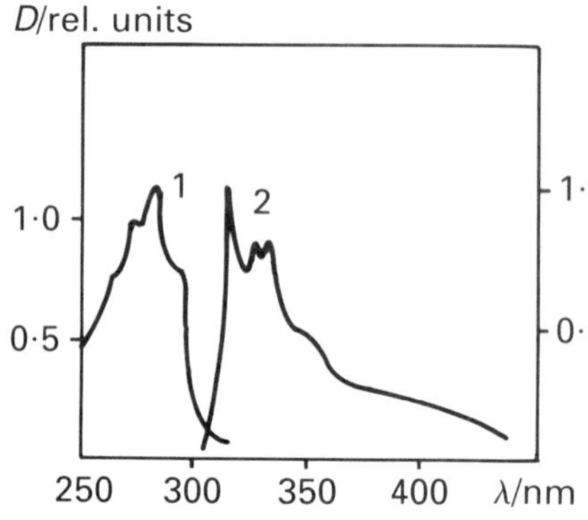

Fig. 4.9. Absorption (1) and fluorescence (2) spectra of polymer XVIII in tetrahydrofuran solutions at –57°C.

The temperature dependence of the fluorescence intensity shows that the temperature at which I_E/I_M is at its maximum is highly dependent on the polymer structure [76]. In contrast to the data given by the same authors in [180] on the excimer fluorescence of polyesters with a ω-carbazylbutyl group, it has been established in [76] that the polymer structure does not create any energetic advantages for excimer formation. For this reason, the higher efficiency of excimer formation in polymers should be assigned to smaller entropy losses sustained during the formation of excimer configurations as a result of steric hindrances in the orientation of naphthyl units.

Ref. [191] describes the fluorescence properties of naphthalene-containing polymers of another type represented by polyamides with naphthalene rings incorporated directly into the main polymer chain:

$$\left[\begin{array}{c} -N-CH_2 \\ | \\ CH_3 \end{array} \quad CH_2-N-CO-(CH_2)_m-CO- \\ | \\ CH_3 \end{array}\right]_n ,$$

XXI

where $m = 2$ (XXIa), 4 (XXIb), 8 (XXIc).

The formation of fluorescent intermolecular complexes with charge transfer is discussed below, using the example of polymers XXI. Investigations of such complexes have been reviewed in some detail in [192].

Fig. 4.10 illustrates the fluorescence spectra of solutions of polyamide XXI diluted with dichloromethane in the presence and absence of triethylamine as donor. It is seen that on adding triethylamine to the polyamide, the intensity of the natural fluorescence of the polymer peaking at 360 nm diminishes, and a broad exciplex fluorescence band with $\lambda_{max} = 500$ nm is formed. There is a linear relationship between the concentration of triethylamine and I_E/I_M. It has been shown that the susceptibility to exciplex formation

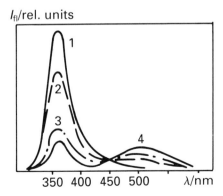

Fig. 4.10. Fluorescence spectra of polyamide XXIb in methylene chloride solutions with 0.01 (2), 0.06 (3), 0.19 mol (4) of triethylamine and without it (1).

falls in the polyamide series with a reduction in the value of m in the acid residue, or in accord with increased steric hindrances to the mobility of the naphthyl group.

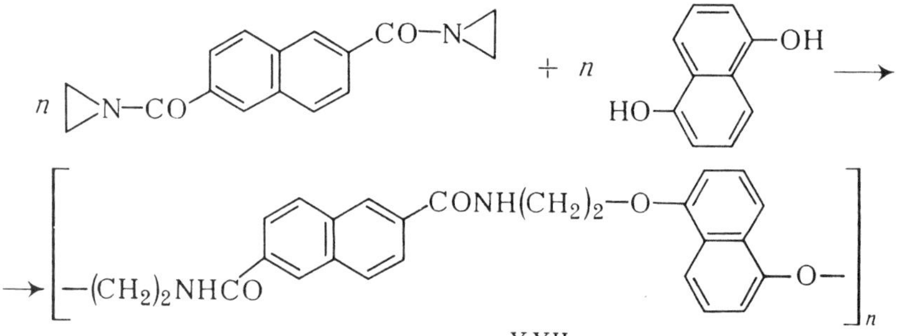

XXII

Polymers (XXII) containing naphthalene moieties in their chain were prepared in [193] by the photoinitiated polyaddition of 1,1'-(2,6-naphthalenedicarbonyl)diaziridine and 1,5-dihydroxynaphthalene in tetrahydrofuran solution exposed to UV radiation with > 290 nm.

On irradiation of diaziridine in the absence of 1,5-dihydroxynaphthalene, the former yields an oligomeric product with side 1-naphthoylaziridine groups.

The fluorescence spectra of tetrahydrofuran solutions of this oligomer display a broad structureless band with λ_{max} 460 nm caused by the formation of intramolecular excimers between the side naphthalene chromophores. The spectrum of the polymer chain with naphthalene moieties shows another broad excimer fluorescence band with λ_{max} 390 nm [193].

Polymers of another type containing naphthalene groups in the chain and prepared by the polycondensation of 1,5-naphthalene diisocyanate and the oligomer poly(tetramethyleneglycoloxide) are described in [194]. It follows from a comparison of the fluorescence spectra of the polyurethane solutions and the model bis-propylcarbamate 1,5-naphthalene diisocyanate` that, along with the band at 350 nm, it is possible to observe, in the former case, a decay of the long wavelength fluorescence with λ_{max} > 400 nm which fails to be described (as for the model) solely by an exponential function. For the long wavelength region, the emission curve can be represented as three exponents with lifetimes of 1.3, 2.4, and 22.3 ns. According to the authors of [194], the presence of the long-lived component is associated with the formation of naphthalene chromophore excimers, the exciter fluorescence being predominant in the polyurethane films.

The fluorescence properties of the commercial photoconductive polymer (XXIII) based on p-phenylenediacrylic acid and 1,4-bis(2-hydroxyethoxy)cyclohexane are covered in [195].

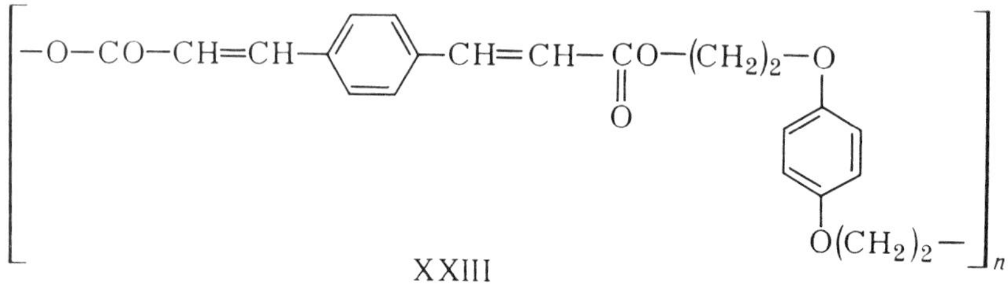

XXIII

The fluorescence characteristics of the model diethyl ester of p-phenylenediacrylic acid (XXIV) were studied at the same time.

$$C_2H_5O_2C-CH=CH-\underset{\text{(benzene ring)}}{\bigcirc}-CH=CH-CO_2C_2H_5$$

XXIV

The emission spectra of solutions of the model compound and polymer XXIII illustrated in Fig. 4.11 show the monomer fluorescence short wavelength bands with λ_{max} 370 nm and a broad excimer emission band lying at longer wavelengths, the intensity ratio of these bands being dependent on the solution concentration. The polymer film exhibits only excimer fluorescence with λ_{max} 470 nm.

As shown by [195], the emission decay kinetics at 470 nm are approximated by the sum of two exponentials: a short-lived component assigned to monomer emission and a long-lived component assigned to excimer emission. The lifetime of the long-lived fluorescence shortens on increasing the concentration of phenylenediacrylic groups, which points to the bimolecular nature of the quenching of the excimer emission. Excimer formation is supposed to occur as a result of excitation of pairs of chromophores present in solution before excitation. The absence of a long-lived component in the monomer fluorescence decay curve at 370 nm does not support any dissociation of the excimer into excited and ground state monomers.

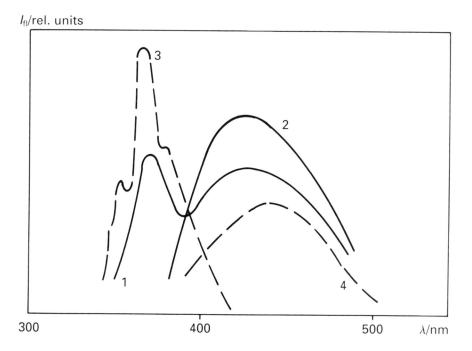

I_{fl}/rel. units

300   400   500 λ/nm

Fig. 4.11. Fluorescence spectra of dilute solutions of polymer XXIII (1) and model compound XXIV (3) in dichloroethane, film of polymer XXIII (2) and concentrated solution of compound XXIV (4).

Melamino-*p*-toluenesulfamideformaldehyde resins (MTSF), which are frequently used as the polymeric substrate in the preparation of daytime fluorescent pigments, can be classed as an aliphatic polycondensed polymer [5] p. 173. If during the synthesis of MTSF, or on heating it first in an inert atmosphere and then *in vacuo*, the *p*-toluenesul-famide-melamine-formalin mixture be doped with a luminophor containing a reactive group, then the polymer formed will be chemically bonded to the luminophor molecule. Chloromethyl, hydroxymethyl, carboxyl, anhydride, or amine groups can be used as the reactive group. Thus the suggestion is made in [196] of using the derivatives of 3-meth-oxybenzanthrone (XXV) for bonding to MTSF.

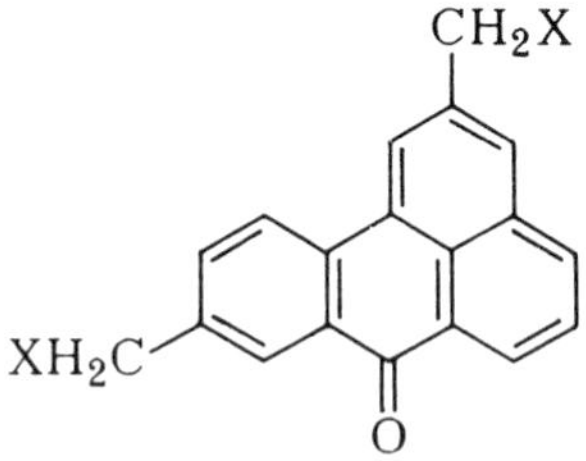

$$CH_2X$$

$$, \quad X = -Cl, \ -OH.$$

$$XH_2C$$

XXV

The polymers obtained contain the luminophor units (XXV) in their chains and show intense fluorescence at 520–540 nm.

The derivatives of 1,8-naphthalic acids or their anhydrides have found wide application in bonding to MTSF [197]. During the selection of the naphthalic acid anhydrides as reactants in the preparation of MTSF with the amine groups of melamine, a popular choice is the 4-substituted heterocyclic derivatives such as:

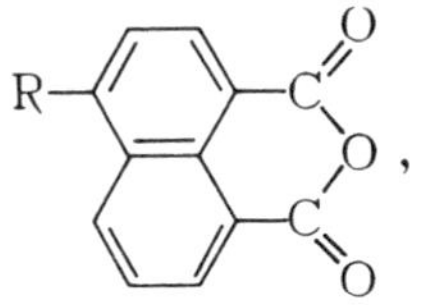

$$R= -C \qquad (XXVIa),$$

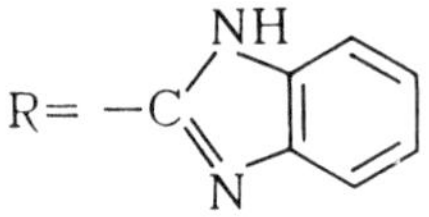

XXVI

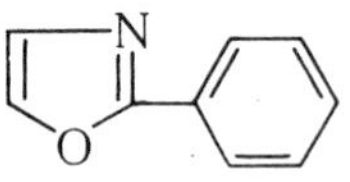

$$(XXVIb)$$

Thus, addition of the anhydride of phenyloxazolylnaphthalic acid (XXVIb) (0.25% mass) to the reaction mixture during the synthesis of MTSF leads to the formation of a lemon yellow polymer fluorescing with λ_{max} 475 nm [197]. The optical properties of the polymers obtained by this process and used as daytime fluorescent paints and varnishes are discussed in greater detail in Chapter 7.

In addition to MTSF, we shall refer to two other types of polycondensed polymer-containing chromophore groups in the chain, namely oxirane polymers [198] and poly-caproamide [199, 200]. The anthracene, carbazole, and naphthalene moieties were incorporated in oxirane polymers by hardening the diglycidyl ethers of 2,2-di(hydroxy-phenyl)propane in the presence of 0.1–5.0% (mass) of the glycidyl ethers of the corresponding luminophors, using conventional hardeners of the polyamine or anhydride type [198]. Depending on the chromophoric group, the polymers obtained were characterized by absorption spectra with λ_{max} = 295–380 nm and fluorescence spectra with λ_{max} = 340–450 nm assigned to the electronic transitions within the respective chromophoric moieties. The fluorescence technique was used to determine the kinetics of hardening epoxy oligomers by diethylenetriamine in the presence of the glycidyl ether of 9-hydroxyanthracene (see section 6.1).

Fluorescent polycaproamides (PCA) were synthesized by copolycondensation of caprolactam with 0.5–5.0% (mass) of chromophoric diamines in the melt at 260–270°C in the presence of 5.0% (mass) aminocaproic acid as activator [199, 200]. Table 4.2 features the composition and some properties of the modified PCAs which in most of cases produce an intense golden yellow (dibenzimidazole units), red (1,5-diaminoanthraquinone units), or blue (diaminodihydroxyanthraquinone units) colour of outstanding homogeneity and lightfastness. As seen from Table 4.2, the introduction of luminophoric diamine units in the PCA chain leads to a 15–20% reduction of the polymer RMM. Despite this fact, PCA fibres containing up to 1% (mass) chromophore compare well with unmodified fibres in their mechanical and physical properties, and, in some cases, they are superior to them in tensile strength and elastic modulus.

The mechanical and physical characteristics of chromophore-containing PCA fibres and the data obtained by X-ray and differential scanning calorimetric methods indicate that no significant change in the polymer structure is introduced by this modification [200].

The fluorescent properties of the solutions and fibres of the luminophor-containing PCAs are determined by the nature of the diamine residue (Table 4.2). Additionally, as shown by the example of diaminostilbene derivatives, the position of the λ^{fl}_{max} of the fibres depends on the concentration of chromophore units. On increasing the concentration from 1–5% (mass), it is possible to observe a bathochromic shift of the emission band and an appreciable decrease in the emission intensity, which must be related to concentration quenching of fluorescence.

It has been found [200] that all luminophor-containing PCAs exhibit higher UV resistance than the initial PCA. This may be accounted for both by the screening effect of the chromophore residues and the electronic energy transfer from the macromolecules to the luminophoric diamine residues. For PCAs containing dibenzimidazole moieties in their chain, the transfer of energy is supported by the almost complete overlap of the fluorescence band of the unmodified PCA (λ_{max} = 445–455 nm), assigned to the impurity keto-imide units (see section 6.1), and of the long wavelength absorption band of the chromophor-containing PCA. Examination of the evolution of the fluorescence spectrum of the latter as the content of chromophoric units is increased shows that, at low concentrations of dibenzimidazole residues, the spectrum displays, along with their intrinsic fluorescence with λ_{max} 518 nm, the emission of the unmodified PCA. With increasing concentration, the intensity of the former band is enhanced while that of the latter is diminished. For a luminophoric moiety content ≥ 0.008 mol dm^{-3}, the donor excited states are completely quenched, which is indicated by the disappearance of the short wavelength fluorescence.

Analysis of the available data on the synthesis and spectral luminescent properties of aliphatic fluorescent polycondensed polymers shows that the best studied are those polymers whose luminophoric residues are introduced into the macromolecule via the diamine (diol), rather than the acid component. It is the luminophor groups that are mainly responsible for the long wavelength absorption and fluorescence characteristics of the polymers, and both monomer and excimer (exciplex) fluorescence bands are likely to

Table 4.2. Composition and some properties of modified fluorescent polycaproamides

Chromophore comonomer	%mass	Solution in HCOOH			Properties of fibres			
		Conc. $/10^{-4}$ mol dm^{-3}	λ_{max}^{abs} /nm	λ_{max}^{fl} /nm	Strength SN/tex	Elongation /%	λ_{max}^{fl} /nm	Light[b] resistance /%
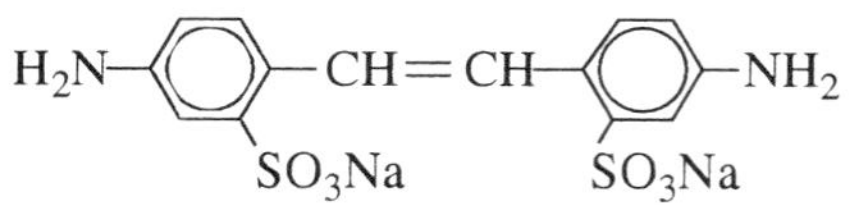 H_2N—⬡—CH=CH—⬡—NH_2, SO_3Na, SO_3Na	1	1.72	340	390, 410	37.6	33.1	435	73
	3	1.47	340	390, 410	31.5	24.3	438	75
	5	1.21	340	390, 410	24.8	26.2	455	70
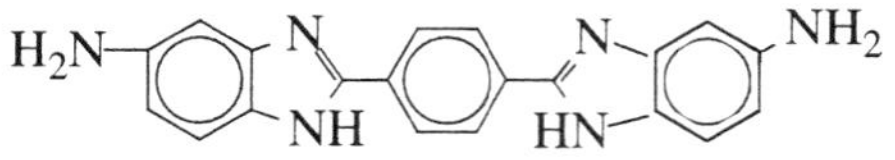 H_2N—⬡—N=NH— ... —N=NH— ... NH_2 (amidine/benzimidazole type)	1	1.80	415	509	38.0	24.0	518	91
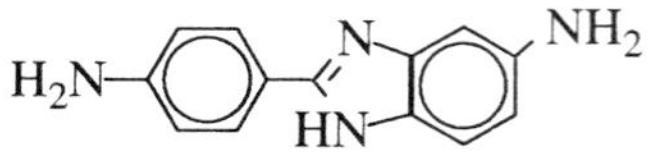 H_2N—⬡—N=HN— ... NH_2 (benzimidazole type)	1	1.74	347	405	42.0	15.0	426	73
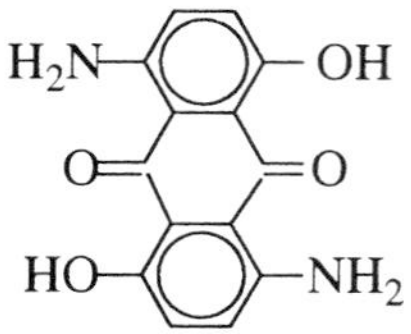 H_2N—anthraquinone—NH_2, with OH and HO groups	1	1.55	580, 625	633	25.3	20.0	—	65
Unmodified polycaproamide	1	1.94	300	—	31.1	23.0	445[a]	48

[a] Fluorescence is attributed to presence of impurities.
[b] Ratio of fibre strength after UV irradiation to initial strength.

appear in the emission spectra. The tendency to form exciplexes increases markedly when the macromolecules possess moieties capable of intra- and intermolecular donor–acceptor interaction.

4.2 AROMATIC POLYMERS

4.2.1 Carbon chain polymers

Aromatic carbon chain polymers of the polyphenylene or polyxylylidene type are well known as regards their absorption properties. Their fluorescence characteristics, however, are much less characterized. The absorption spectra of such polymers follow the same general rules as other conjugated polymers (see section 3.2).

Ref. [1] states the following empirical rule derived for sequences of linear polyenes and poly(p-phenylenes): the longer the conjugated chain, the smaller the energy of the light quantum required to excite the π-electron system. As seen in Fig. 4.12 [201], this rule holds good for poly(p-xylylidenes) with the following general formula (XXVII):

$$\left[-\!\!\left\langle\bigcirc\right\rangle\!\!-CH=CH- \right]_n \qquad (n=1\text{–}7)$$

XXVII

Fig. 4.12 shows another distinct generality characteristic of polycondensed systems: any further bathochromic shift of the long wavelength absorption band disappears once the conjugated chain reaches a certain length. From this fact, the authors of [94] inferred that the macromolecular moiety functioning as a chromophore corresponds to a model

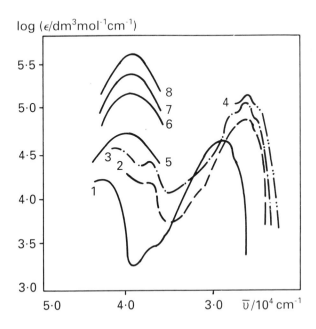

Fig. 4.12. Absorption spectra of poly(p-xylylidenes) (1–4) and poly(m-phenylene) (5–8): 1, in hexane (2); 2, in chloroform (3); 3, in dioxane (4); 4, in methylnaphthalene (7); 5, in chloroform (2); 6, in chloroform (3); 7, in chloroform (4); 8, in chloroform (5).

with the conjugated π-bonds smaller in number rather than corresponding to the length of the polymer chain. The chromophore group of the polymer is approximately equivalent to three or four repeat units of the macromolecule [201].

The opposite picture is observed in the absorption spectra of poly(*m*-phenylene) solutions (Fig. 4.12, curves 5 to 8). It is seen [3] that with an increase in the number of phenyl groups, the position of the long wavelength absorption band does not change. What can be observed is an enhancement of the band intensity, suggesting an *apparent* increased solute concentration. In [94] the independence of λ_{max} of the absorption band of the number of phenyl groups is interpreted as being due to a 'conjugation interruption effect' on the *meta*-phenylene units.

The fluorescence properties of poly(*p*-phenylene) and polymer XXVII are described in [164, 169]. Fig. 4.13 illustrates the emission spectra of a benzene solution of poly(*p*-phenylene) at a concentration of 5×10^{-5} mol dm^{-3} and of powdered polymer XXVII. The fluorescence band of poly(*p*-phenylene) is found to undergo a long wavelength shift on increase of concentration of its solutions [169]. Investigation of the specific features of the fluorescence of this polymer in solutions of different concentration revealed that in dilute solution, the nature of the spectrum is determined by the contribution to the emission of each of the transitions corresponding to particular polycondensed sequences. With increase in the concentration of the solutions, the processes of energy transfer to the lower levels are enhanced. As a result of these processes, a shift of the fluorescence band toward longer wavelengths occurs without any notable change in the absorption spectra. On increasing the concentration up to 10^{-3} mol dm^{-3}, energy transfer becomes so efficient that fluorescence quenching occurs.

In addition to the concentration quenching of poly(*p*-phenylene) fluorescence, [169] details the effect produced on the emission of the polymer solutions by such quenchers as tetracyanoethylene, chloranil, trinitrobenzene, dimethylamine, aminostyrene, and triethylamine. It is shown that if the absorption spectrum of the quencher overlaps the

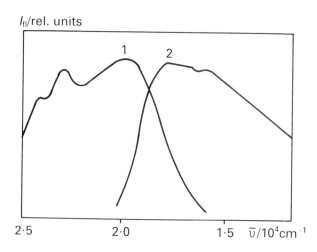

Fig. 4.13. Fluorescence spectra of benzene solution of poly(*p*-phenylene) (1) and powdered poly(*p*-xylylidene) XXVII (2).

emission spectrum of the polymer, then fluorescence quenching proceeds by the Förster mechanism of dipole–dipole energy transfer (for example, with chloranil). If the absorption spectrum of the quencher and the emission spectrum of the polymer do not overlap, quenching occurs via short distance energy migration or by the formation of donor–acceptor complexes (e.g. with aminostyrene, dimethylamine, or triethylamine).

The absorption and luminescence properties of polymers containing one or two methylene groups between the aromatic rings were investigated [202–205]. The electronic absorption spectra of poly(p-xylylene) [202] show weak bands at 260–280 nm. Of interest are the optical properties of such polymers as poly(9,10-dimethyleneanthracene) (XXVIII) examined in [203].

Polymer XXVIII was synthesized by the pyrolysis of 9,10-dichloromethylanthracene through the formation of 9,10-anthraquinodimethane as an intermediate compound.

$$\left[-CH_2-\text{(anthracene)}-CH_2- \right]_n \qquad (XXVIII)$$

Fig. 4.14 shows the electronic absorption spectra of the initial 9,10-dichloromethylanthracene at 20°C, 9,10-anthraquinodimethane at −195°C and polymer XXVIII at 20°C. It is seen that the bands at 330–420 nm typical of the anthracene ring disappear almost completely during the transformation from 9,10-dichloromethylanthracene to 9,10-anthraquinodimethane. The spectrum of the latter displays only an intense band at 270–290 nm assigned to the absorption of the anthraquinodimethane chromophore. On heating this compound from −195°C to room temperature, the intense absorption band with λ_{max} 291 nm disappears, giving rise to a band with λ_{max} 260 nm and a few bands at 330–430 nm, thus confirming the poly(9,10-dimethyleneanthracene) structure of the polymer XXVIII formed.

As shown in [203], polymer XXVIII does not fluoresce in the solid state yet shows intense fluorescence in dimethylformamide solution (Fig. 4.15). Fig. 4.15 illustrates the absorption and fluorescence spectra of polymer (XXVIII) solutions as well as the fluorescence excitation spectra peaking at 432 and 482 nm. The dependence of the nature of the solution emission spectrum on the wavelength of the exciting light implies the existence of two emission centres, which is supported by the data on the fluorescence excitation spectra. It seems likely that a short wavelength emission centre with λ_{max} 432 nm is typical of polymer XXVIII [203]. The long wavelength emission centre ($\lambda_{max} = 482$ nm) is evidently related with the presence in the solution in question of polymer associates or low molecular impurities such as dianthrylethylene derivatives [5] p. 39.

The absence of marked fluorescence of the polymer in the solid state is most readily explained by the presence of dianthrylethylene impurities. The specific structural feature of dianthrylethylene molecules is that they are open chains offering steric hindrance for the conjugation of π-electron systems of 'overlap zones'. The presence of such zones markedly reduces the fluorescence intensity of some dianthrylethylenes compared with other anthracene derivatives [5] p. 40.

Polymers containing two methylene groups between phenylene nuclei were studied

D/rel. units

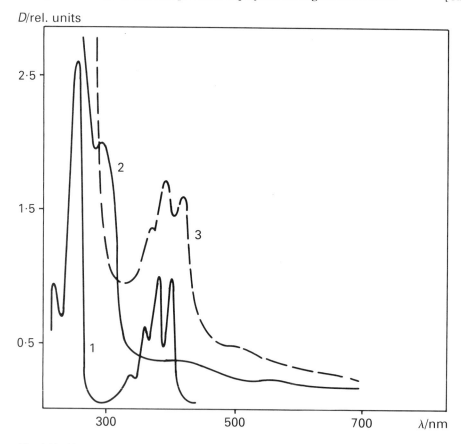

Fig. 4.14. Absorption spectra of 9,10-dichloromethylanthracene in *n*-heptane solution at 20°C (1), of 9,10-anthraquinodimethane precipitated on the substrate at −195°C (2) and of polymer **XXVIII** on the substrate at 20°C (3).

via absorption and luminescence spectroscopy in [204, 205]. It was established that, along with the band λ_{max} 260 nm which corresponds to the $\pi \rightarrow \pi^*$ transition in the benzene ring, the absorption spectra of polyarylenealkyls **XXIX** show bands at 300–400 nm reflecting vibronic interaction between the ring and the substituent and also bands at 450–500 nm caused by the intermolecular mechanism of interaction. As disclosed in [204], the $\pi \rightarrow \pi^*$ transition in this polymer is determined by the excitation vector whose value depends on the nature of the substituent R in the monomer unit, and the intensity of this transition varies linearly with the increase in its transition moment.

$$\left[-CH_2-CH_2-\!\!\!\bigcirc\!\!\!-CH_2-CH_2-\!\!\!\bigcirc\!\!\!-\right]_n , \quad \text{where } R = H, CH_3,$$

$$\qquad\qquad\qquad R \qquad\qquad\qquad R \qquad\qquad Cl, \text{ etc.}$$

XXIX

On analysing data on the electronic absorption spectra and vibrational spectra of polyarylenealkyls, the author of [206] concluded that a multi-centre type of bonding is real-

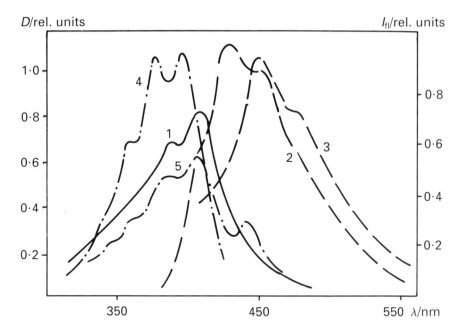

Fig. 4.15. Absorption and fluorescence spectra of polymer **XXVIII** in dimethylformamide solution: 1, absorption spectrum; 2, 3, fluorescence spectra with λ_{excit} = 365 and 412 nm, respectively; 4, 5, fluorescence excitation spectra with maxima at 432 and 482 nm, respectively.

ized in polymers **XXIX** as a result of intra- and intermolecular interactions. To confirm this view, the author used the EPR method to detect the appropriate signal for the polymer with R = CH_3 but failed to detect it for the polymer with R = H. From the temperature dependence of the EPR-absorption signal, the author determined the number of rings over which the electron is delocalized. Based on current concepts [207], these results cannot be accepted as correct, and the EPR signal is apparently caused by paramagnetic impurities present in the polymer. On the basis of the above, we find as incorrect the conclusion of the author of [206] that the structure of polymer **XXIX** is intermediate between a conjugated and nonconjugated structure.

Ref. [265] is concerned with the low temperature luminescence of polymers **XXIX** in the solid state and in solution. These compounds are noted as possessing fluorescence, the decay in emission intensity immediately on removal of the light source being described by a hyperbolic curve, and then an exponential curve. Based on these data which, to our mind, are not wholly reliable, the author of [205] classes polymers **XXIX**, according to their emissive character, as crystallophosphors and suggests the existence of such polymers at low temperatures in the meso-state (thermotropic liquid crystals).

Thus, the data available on the spectral properties of polyarylenealkyls are not very reliable and need elucidation. On the whole, it should be noted that the literature on the spectral and luminescent properties of carbon chain aromatic polymers is scrappy, and further investigation is required.

4.2.2 Heterochain aromatic polymers

The optical properties of heterochain aromatic polymers have been studied in greater detail than those of carbon chain polymers. Data on the spectral luminescence characteristics of aromatic polyethers [208], poly-Schiff bases [209, 210], polyamides [211–218], and polyureas [211] can be found in the literature.

Polyoxy(2,6-dimethyl)-1,4-phenylene was examined as solution and film by absorption and luminescence spectroscopy [208]. The absorption and fluorescence spectra of the polymer were compared with those of the initial 2,6-dimethylphenol (Fig. 4.16). As is evident from Fig. 4.16, the fluorescence spectra of the monomer and polymer solutions are close in appearance, whereas the polymer film shows an abnormal long wavelength fluorescence. It was suggested by the authors that the fluorescence of the films emanated from the chromophore impurities added to the polymer during its preparation and are impossible to remove on its reprecipitation. During continuous photolysis of the films with a xenon lamp, the impurity fluorescence disappears completely, and, in parallel, the optical density falls in the visible spectral region. The phosphorescence spectra of the polymer films and crystals of 2,6-dimethylphenol at 77 K are almost identical, which suggests a single type of emissive centre. Unlike poly(phenylene oxide), 2,6-dimethylphenol displays structured phosphorescence spectra, a difference attributed to differences in the rigidity of the polymer and monomer.

As regards heterochain aromatic polymers, primary attention is given to the poly-Schiff bases, for example the derivatives of iso- and terephthalic acid XXX:

$$\left[=\!N\!-\!Ar\!-\!N\!=\!CH\!-\!\!\!\bigcirc\!\!\!-\!CH\!= \right]_n ,$$

XXX

where Ar is an aromatic group.

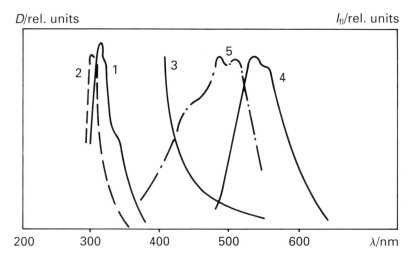

Fig. 4.16. Fluorescence spectra of polyoxy(2,6-dimethyl)-1,4-phenylene (1) and 2,6-dimethylphenol (2) in chloroform solution; absorption spectrum of polymer film (3), fluorescence spectrum (4) and fluorescence excitation spectrum (5).

These polymers are synthesized by the polycondensation of aromatic diamines with iso- or terephthalic aldehyde. Ref. [209] reviews data on the absorption properties of poly(p-xylyl-p-phenylene diamine). The authors mention a rapid and asymptotic reduction in the bathocromic shift of the long wavelength absorption band from 290 to 365 nm in a dimethylacetamide solution of oligomeric p-phenyleneazomethines with increase in the length of the conjugated chain.

The spectral and luminescence properties of poly-Schiff bases and model compounds are studied in detail in [210]. When examining the absorption and fluorescence spectra, the authors discovered that protonation affords a planar molecular configuration thus reducing the possibility of nonradiative deactivation of the S_1 state. As shown in [210], the fluorescence of free aryl azomethine bases is suppressed by nonradiative intersystem crossing whose rate constant is estimated for N-benzalaniline as 10^{12} s^{-1}. The high value of k_{ST} is attributed to the large spin–orbit coupling facilitated by the nonplanar molecular structure and the orbital character of the S_1 state which is $n\pi^*$. When comparing the absorption spectra of hydrocarbon and sulphuric acid solutions of N-benzalaniline, the authors of [210] concluded that the lone pair of electrons on the N-atom of the molecule in its protonated form are highly bonded and excluded from the π-system of aryl azomethine, as a result of which the long wavelength band of the acid solution spectrum undergoes a bathochromic shift of 25 nm.

On examining the spectral and luminescence properties of poly-Schiff bases (XXXI) and a number of model compounds (XXXII) and (XXXIV), the authors concluded that the specific optical behaviour of N-benzalaniline in various media persists even after its conversion into more complex compounds (Fig. 4.17a). It is seen that the absorption spectra of polymer XXXI and model compound XXXII are very similar, the maximum of the long wavelength band of both the compounds being 340 nm whereas λ_{max} is 330 nm for a cyclohexane solution of polymer XXXII. Consequently, the polymers and low molecular bis(arylazomethine) derivatives are N-protonated, like N-benzalaniline, in H$_2$SO$_4$. It is thus possible to observe the activation effect on the fluorescent properties of these materials. While in the solid state, the poly-Schiff bases and corresponding model compounds emit a weak unsteady fluorescence which disappears quickly on UV irradiation of the samples, their sulphuric acid solutions exhibit an intense emission especially on freezing [210].

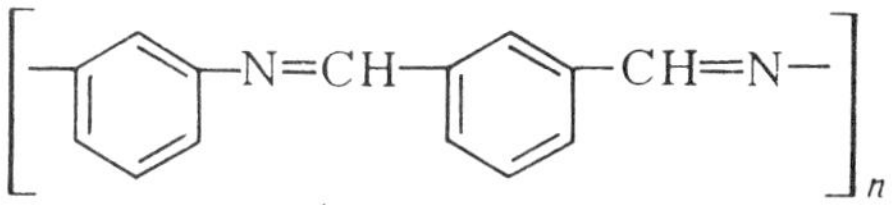

XXXI

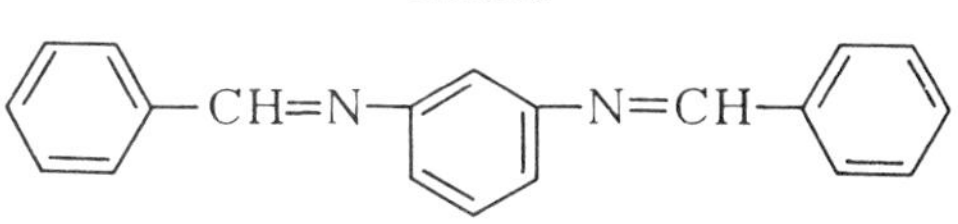

XXXII

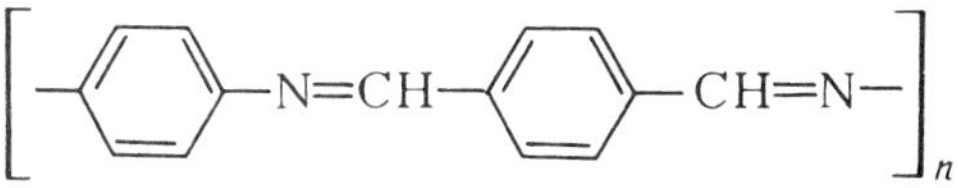

XXXIII

XXXIV

As seen in Fig. 4.17a, the fluorescence spectra of the compounds in question are very similar and have an allowed vibrational structure with the intervals of 1300–1000 cm^{-1} between the bands. Similar conclusions can be reached from an analysis of the absorption and low temperature fluorescence spectra of polymer XXXIII and model XXXIV (Fig. 4.17b).

A comparison of Figs 4.17a and 4.17b shows that during the transition from the *meta*-isomers of polymeric and model arylazomethines to their *para*-isomers, the long wavelength absorption band and all bands of the fluorescence spectrum undergo a bathochromic shift.

Polymer XXXIII powder exhibits a green fluorescence of higher intensity than others of these polymers (Fig. 4.17b). The fluorescence spectrum of this compound is complex and, apparently, comprises the superposition of two types of emission spectra. The long wavelength region of the spectrum with λ_{max} 580 nm is associated with transitions in the chromophore group corresponding to the central moiety of the polymer backbone. The correctness of this inference is confirmed by the similarity of the long wavelength region of the fluorescence spectrum of the polymer powder to the fluorescence spectrum of the model *p*-xylal-di(*p*-dimethylamino)anil powder [210]. The chromophoric groups incorporating the end macromolecular moieties with the amine (XXXV) or aldehyde group (XXXVI) are evidently responsible for the short wavelength region of the fluorescence spectrum of the polymer powder.

XXXV

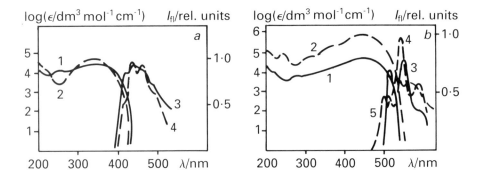

Fig. 4.17. Absorption spectra at 293 K (1, 2) and fluorescence spectra at 77 K (3, 4, 5): (a), polymer XXXI (1, 3) and model compound XXXII (2, 4) in solutions of concentrated H$_2$SO$_4$; (b), polymer XXXIII (1, 3) and model compound XXXIV (2, 4) in concentrated H$_2$SO$_4$ solutions and powdered polymer XXXIII (5).

$$\text{—}\underset{}{\bigcirc}\text{—N=CH—}\bigcirc\text{—C}\overset{\displaystyle O}{\underset{\displaystyle H}{\diagdown}}$$

XXXVI

For the chromophore with an end amine group, this suggestion is confirmed by the similarity of the fluorescence spectrum of N-benzal(p-diethylamino)aniline powder to the short wavelength region of the fluorescence spectrum of the polymer.

The spectral properties of aromatic polyamides are discussed below. It is known [21] p. 95, that polyamides containing no chromophoric groups in the chain, such as poly(m-phenylene isophthalamide) emit blue fluorescence with λ_{max} 445 nm which is attributed to the presence in the polymer of impurities. With the introduction of substituents in the diamine or acid components, the polyamides begin to show natural fluorescence. For example, polyamidoacid XXXVII fluoresces as a film with λ_{max} 470 nm [215] and the poly(m-phenylene isophthalamide) analogue with a nitrile group in the diamine residue (the product of polycondensation of 2,5-diaminobenzonitrile and isophthaloyl chloride [8]) fluorescences with λ_{max} 425 nm.

$$\left[\text{—NH—}\bigcirc\text{—O—}\bigcirc\text{—NH—CO}\underset{\text{HOOC}}{\overset{\text{—COOH}}{\bigcirc}}\underset{\text{CO—}}{}\right]_n$$

XXXVII

The first studies of the optical properties of luminophor-containing aromatic polyamides and polyureas go back to the mid-1970s [8, 211]. The specific subject of investigation was the polarized luminescence of polyamidoacid (PAA) which is close-ly related in structure to polymer XXXVII but contains no luminescent label in the main polymer chain. Unfortunately, the authors failed to furnish any data on the nature and methods of introduction of the luminophoric moieties in the chain. It is indicated that each label is introduced between flexible hinged links C–O–C, with one label per 300 amidoacid units. Data on dimethylformamide solutions and dimethylformamide-glycerine mixtures of the PAA polymer made it possible to show a direct relationship between the inverse luminescence polarization $1/P$ and T/η (where η is the solvent vis-cosity). From the diffusion parameters, the conclusion can be drawn that the kinetic flexibility of the PAA macromolecules is much better than that of carbon chain poly-mers. It was shown that the external viscous friction is the main factor reducing the kinetic flexibility of the polyamidoacid chain. The specific fluorescence behaviour of the polyamide-containing phenylbenzimidazole moieties in the chain (XXXVIII) was investigated in [214].

$$\left[\text{—NH—}\underset{\text{CN}}{\bigcirc}\text{—NH—CO—}\bigcirc\text{—CO—}\right]_n$$

XXXVIII

It was found that N-methylpyrrolidone solutions of polyamide XXXVIII, like the solutions of the model 6,4'-dibenzoylamino-2-phenylbenzimidazole, do not fluoresce at room temperature. The spectral properties of the latter compound were studied as solu-tions in amide solvents and H_2SO_4 at 77 K. Frozen solutions were found to exhibit a long-lived green phosphorescence and blue fluorescence whose intensity sharply increased on going from N-methylpyrrolidone to sulphuric acid solutions. On the basis of the phosphorescence data, the energy of the $T_{\pi\pi*}$ state responsible for this radiative transition is estimated as 19 000 to 20 000 cm^{-1}. Owing to the presence of $>$C$=$O groups, the molecules of the model diamide are characterized by the occurrence of the $T_{n\pi*}$ state whose energy is 28 000 to 29 000 cm^{-1}. The efficiency of the intersystem crossing process in these molecules ($S_{\pi\pi*} \rightsquigarrow T_{n\pi*}$) leading to the decay or quenching of the fluorescence must be attributed to the fact that the energy of the $T_{n\pi*}$ level is sufficiently close to that of the fluorescent $S_{\pi\pi*}$ level (27 000 to 28 000 cm^{-1}). At low temperatures, the effect of the $T_{n\pi*}$ state diminishes. In H$_2$SO$_4$ solutions this effect diminishes to a much greater extent because of the increase in the energy of the $T_{n\pi*}$ state resulting from protonation of the carbonyl groups.

Unlike solutions, films of polyamide (XXXVIII) are characterized by a green fluorescence with λ_{max} 525 nm which is abnormally shifted relative to the long wavelength absorption band [214, 217]. To elucidate the nature of this emission, an investigation was carried out into the fluorescence properties of the initial and protonated molecular forms of 6,4'-diaminophenylbenzimidazole [218], a monomer used in the synthesis of polyamide XXXVIII. The fluorescence in question was found to belong to the dication formed by loss of a proton (by an adiabatic photoprotolytic dissociation reaction) from the ammonium group of the benzimidazole ring, that is, the end groups given in the formula of XXXVIII are responsible for the emission of polyamides.

The synthesis and fluorescence properties of aromatic polyamides and polyureas containing three types of luminophor moiety in the chain are covered in [211] and Japanese Patent No. 74121893, 1973. These polymers were prepared by interphase polycondensation of acridine yellow, thionine or sodium 4,4'-diaminostilbene-2,2'-disulfonate with dichloroanhydrides such as terephthaloyl chloride or sebacyl chloride or with 2,4-toluylene diisocyanate. The synthesized polymers did not fuse at temperatures up to 300°C, dissolved in water (polymers with stilbene units in the chain), or in N,N-dimethylformamide (polymers with acridine and thionine units in the chain). A detailed study of the fluorescence properties of the polymer solutions showed a significant dependence of λ_{max}^{fl} and the emission intensity on the solution concentration. Thus, the polymers with stilbene and acridine groups display a marked blue shift of λ_{max}^{fl} of 10–40 nm, depending on the polymer, when the concentration of the polymer in solution is reduced from 10^{-3} to 10^{-7} mol g^{-1} (in terms of the luminophor group concentration). The fluorescence intensity fell 10–25 times for stilbene-containing polymers and 150–300 times for acridine-containing polymers. A comparison of the fluorescence properties of the initial luminophoric diamines with those of the corresponding polymers showed that the position of λ_{max}^{fl} is determined by the nature of the polymer. For example, at a concentration of 10^{-4} mol dm^{-3} an aqueous solution of sodium 4,4'-diaminostilbene-2,2'-disulphonate shows a maximum at 444 nm, while a polyurea solution with the same concentration of luminophoric groups has λ_{max} at 440 nm, and a solution of the polyamide based on terephthaloyl chloride peaks at 504 nm. This dependence is revealed to a lesser extent in acridine-containing polymers. The significant bathochromic shift of the fluorescence band of polyamide XXXIX, as compared with that of the initial diamine, is thought by the authors of [211] to be connected with the lengthening of its conjugated chain. The shift of λ_{max}^{fl} up to 510 nm in the spectra of polyamide (XL) solutions is explained in a similar way.

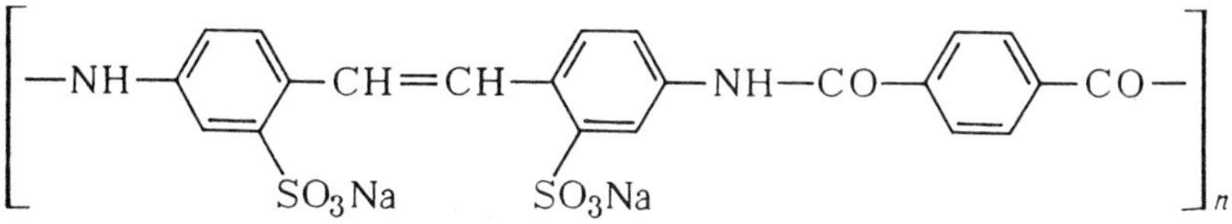

$$\left[-NH-\underset{SO_3Na}{\bigcirc}-CH=CH-\underset{SO_3Na}{\bigcirc}-NH-CO-\bigcirc-CO-\right]_n$$

XXXIX

$$\left[-HN-\underset{N}{\overset{H_3C}{\bigcirc\bigcirc\bigcirc}}\overset{CH_3}{}NH-CO-\bigcirc-CO-\right]_n$$

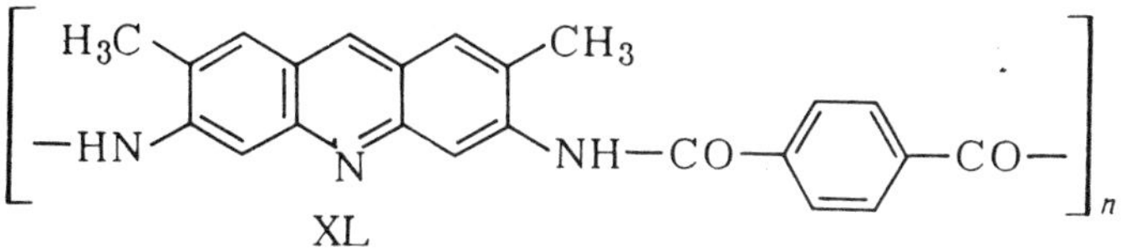

XL

For solutions of acridine yellow λ_{max} is 500 nm.

Ref. [211] gives data on the reduced fluorescence intensity of solutions of polymers XXXIX and XL as compared with that of solutions of the initial luminophor exposed to UV radiation. It was found that after 1.5 h of irradiation, the fluorescence intensity of sodium 4,4'-diaminostilbene-2,2'-disulphonate solutions diminishes by 56%, and of solutions of the corresponding polyamide XXXIX by only 26.7%. Acridine yellow and polyamide XL show a reduction of intensity by 18.4 and 6.5% respectively. Unfortunately, [211] provides no data on the absorption spectra of the fluorescing polyamides and emission spectra of the polymer films.

Polyamides XLI close in structure to those described above were obtained in [213, 214, 216]:

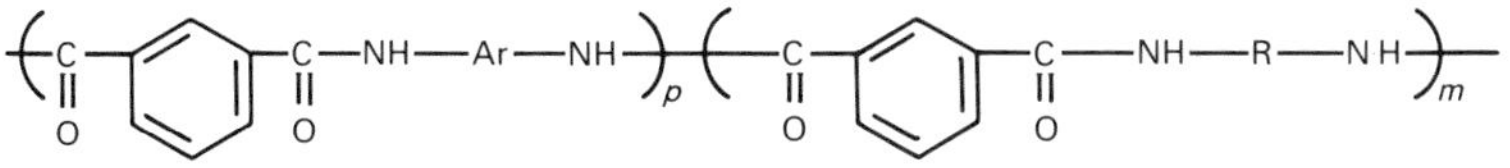

$$\left(\underset{O}{\overset{C}{\bigcirc}}\underset{O}{\overset{C}{}}-NH-Ar-NH\right)_p\left(\underset{O}{\overset{C}{\bigcirc}}\underset{O}{\overset{C}{}}-NH-R-NH\right)_m$$

XLI

$$R = -\bigcirc-O-\bigcirc- \; ; \qquad \bigcirc$$

$$Ar = \underset{N}{\overset{H_3C}{\bigcirc\bigcirc\bigcirc}}\overset{CH_3}{} \; ; \quad -\underset{SO_3Na}{\bigcirc}-CH=CH-\underset{NaO_3S}{\bigcirc}- \; ; \quad -\bigcirc-\overset{N-N}{\underset{O}{\bigcirc}}-\bigcirc-$$

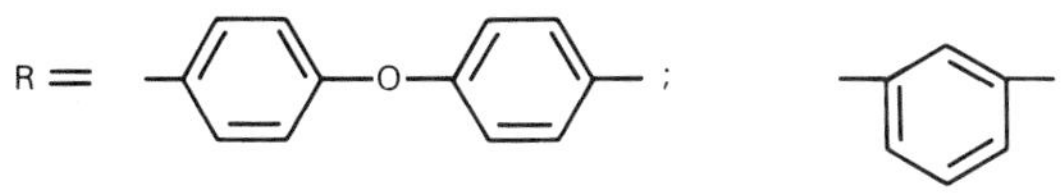

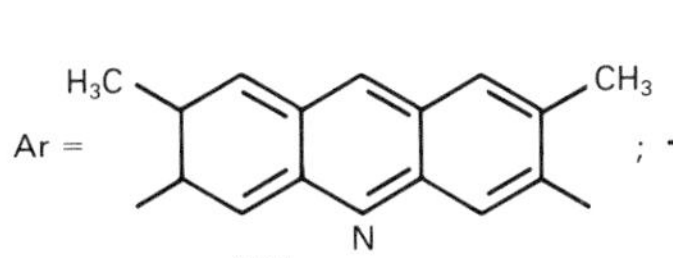

XLIa 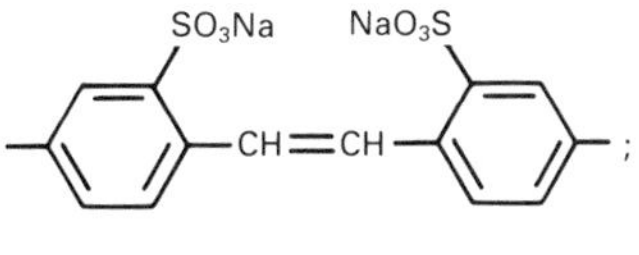   XLIb 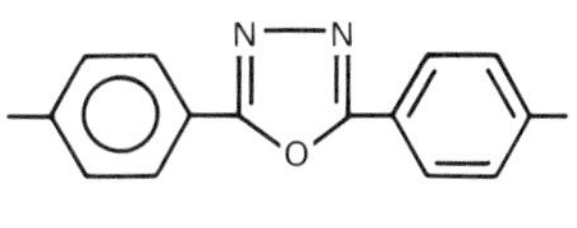   XLIc

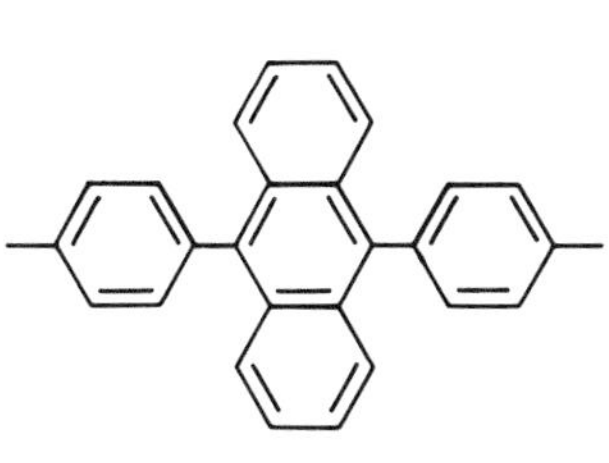

XLId 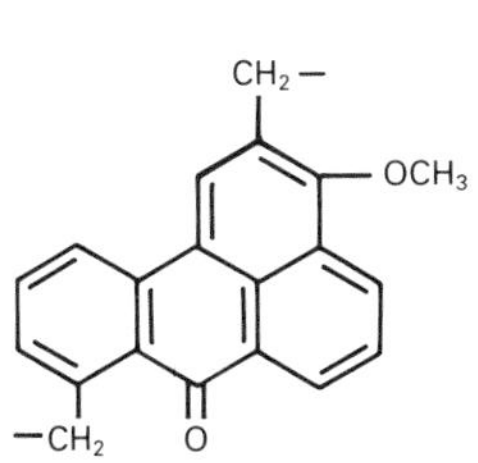   XLIe

Unlike interphase polycondensation, the technique suggested in [211] enabled the preparation of rather high molecular polymers by low temperature polycondensation of the conventional and luminophor/diamine mixture in isophthaloyl chloride solution. Moreover, this technique made it possible to vary over a wide range the concentration of luminophor groups in the chain by changing the proportion of the initial diamine. For the composition and some properties of synthesized polyamides XLI, see Table 4.3.

Table 4.3. Composition and properties of aromatic fluorescent polyamides

Polymer	Proportion of diamine /mol	p/m	η_{log}[a] $/10^2$ cm^3 g^{-1}	$T_{i.d.}$[b] /°C	I_{film}[c] /rel. units
XLIa	1:5	1:90	0.19	470	18
	1:10	1:110	0.24	460	27
	1:50	1:420	0.25	430	60
	1:100	1:640	0.46	490	100
XLIb	1:10	1:20	0.16	—	40
	1:50	1:90	0.31	—	67
	1:100	1:120	0.36	480	100
XLIc	1:10	1:10	0.34	—	45
	1:100	1:200	0.23	400	100

[a] For 0.5% solution in N-methylpyrrolidone at 30°C.
[b] $T_{i.d.}$ is temperature of initial degradation in air.
[c] Natural maximum fluorescence intensity of each of the three types of polymer is taken as 100 units.

The spectrophotometric technique was used to determine the actual proportion of luminophoric and nonluminophoric moieties in the composition of polyamides XLIa–XLIc. For this purpose, use was made of standard solutions of model diamides XLII, XLIII and XLIVa which show intense absorption bands with maxima at 468, 360 and 342 nm respectively [212].

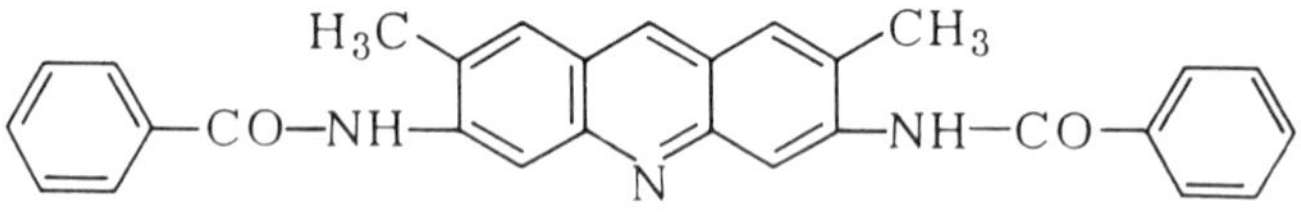

XLII

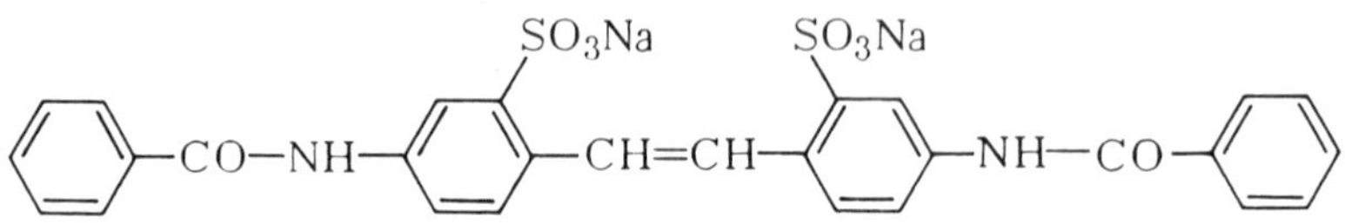

XLIII

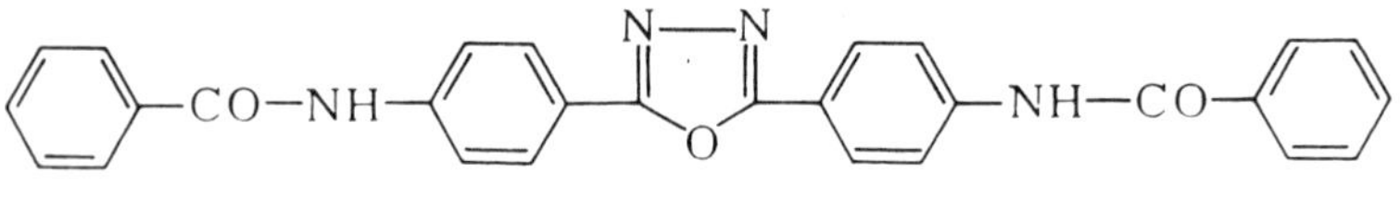

XLIV

XLIVa

XLIVb

XLIVc

As seen from Table 4.3, the concentration of luminophor groups determined spec-trophotometrically is several times below the value calculated on the basis of the initial molar proportions of the diamines used for polycondensation. This discrepancy must be explained by the low reactivity of the initial luminophor diamines, as a result of which a considerable portion of them take no part in the reaction. The values of η_{log} of acridine- and stilbene-containing polyamides are markedly reduced with increasing luminophor content of the chain owing to the low basicity of the amine groups contained in the initial diamines [212].

The thermal resistance of polyamides was found to be critically dependent on the type of luminophor group. Thus, for polyamides prepared with the initial diamines taken in a molar proportion of 1:100, the temperature of initial destruction increases in the series of units diphenyloxadiazole-stilbene-acridine.

Fig. 4.18a shows the absorption and fluorescence spectra of polyamide XLIa, contain-ing acridine units in the chain, as a film, and in N-methylpyrrolidone solution. To identify the long wavelength absorption and fluorescence bands, the absorption and fluorescence properties of low molecular compounds, namely, the initial monomer of acridine yellow and XLII, were investigated. Since acridine dyes are prone to self-association in solution [212], the spectra of these compounds were measured in a variety of solvents. In dimethylformamide solution these compounds were present mainly as associates charac-terized by an absorption maximum lying at shorter wavelengths (Fig. 4.18c). In solution, acridine yellow was almost wholly in the molecular form with λ_{max} 460 nm and an emis-sion maximum at 500 nm. It was established that in ethanol–dimethylformamide mixture (5:1, vol.), the optical density of the monomer solutions at the maxima of the absorption bands of the molecular and associated forms (λ_{max} = 410) was much the same. As seen from Figs 4.18a and 4.18b, the long wavelength absorption and fluorescence bands of the polymer proved to be similar to the corresponding bands of the model compound XLVII, being in the molecular form. These data provide sufficient grounds to relate the absorp-tion and emission bands of this polyamide to the acridine unit [212].

The band with λ_{max} 410 nm observed in the absorption spectra of solutions of polyamides XLIa with acridine units should be assigned to the interchain association of the units. In fact, from measurements of the absorption spectra of polyamides with vary-ing content of the acridine yellow moiety, it was found that with reduction in the concen-tration of such moieties the optical density ratio D_{410}/D_{468} is also decreased.

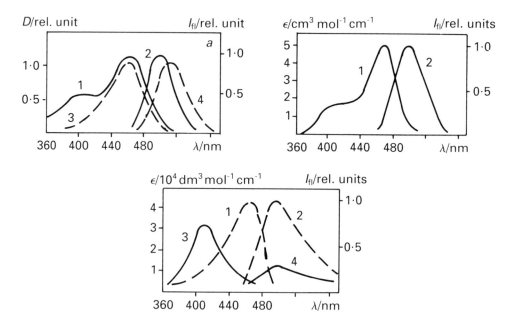

Fig. 4.18. Absorption (1, 3) and fluorescence spectra (2, 4) of acridine-containing compounds: a, polyamide XLIa in N-methylpyrrolidone solution (1, 2) and as film (3, 4); b, model XLII (1, 2) in N-methylpyrrolidone solution; c, acridine yellow in ethanol (1, 2) and dimethylformamide (3, 4) solutions.

Similar results were obtained from a study of the absorption and fluorescence of solutions and films of polyamides, modified with stilbene with and diphenyloxadiazole units.

Investigation of the absorption properties of thin polymer films showed that the pattern of the spectral absorption curves of all structure modified polyamides coincides in the near-UV region with that of the polyamides containing no luminophor units. In addition, the emission intensity of polyamide films turned out to decrease appreciably with an increase in the luminophor group content (Table 4.3), which can be attributed both to the association of such groups and the occurrence of concentration quenching.

Of separate interest is the influence of the luminophor groups on the UV resistance of polyamides; it is known that in a number of cases the organic luminophor additives significantly improve the light resistance of polymers [57, 214].

To understand the mechanism of light stabilization, it is necessary to analyse the reasons for the reduction in the natural fluorescence intensity of the polymer following the introduction into it of chromophore fragments or low molecular chromophore additives. Taking as examples the luminophor-containing polyamides XLId and XLIe, as well as solid solutions of model diamides XLIVb and XLIVc taken in unmodified poly(m-phenylene isophthalamide) (PPhIPhA), such effects as screening, fluorescence reabsorption, and electronic energy transfer from macromolecule to chromophore component [214] have been investigated. Study of the spectral properties of polyamides shows that the introduction of diphenylanthracene moieties (polymer XLId) gives rise to the fluorescence of PPhIPhA with λ_{max} 440 nm virtually coinciding with the natural fluorescence

region of the unmodified polyamide. The introduction of benzanthrone moieties (polymer XLIe) induces a fluorescence with λ_{max} 530 nm. In the latter case, when the chromophore content is low, the spectra of polyamide XLIe show two types of band: a short
wavelength band with λ_{max} 440 nm and a long wavelength band with λ_{max} 520 nm.
When the chromophore content increases, the short wavelength fluorescence becomes
quenched. Similar changes take place in the fluorescence spectra of PPhIPhA on addition
to the polymer of diamide XLIVc at different concentrations.

To assess the contribution of screening and reabsorption (coefficient K_{sr}), the authors
of [214] used the following formula:

$$K_{sr} = \frac{D_n}{D_n + D_s + D} \cdot \frac{1 - 10^{-(D_n + D_s + D)}}{1 - 10^{-D_n}} \qquad (4.1)$$

where D_n and D_s are the values respectively of the optical density of the unmodified
PPhIPhA film and the light stabilizer solution (chromophore additive) in PPhIPhA at

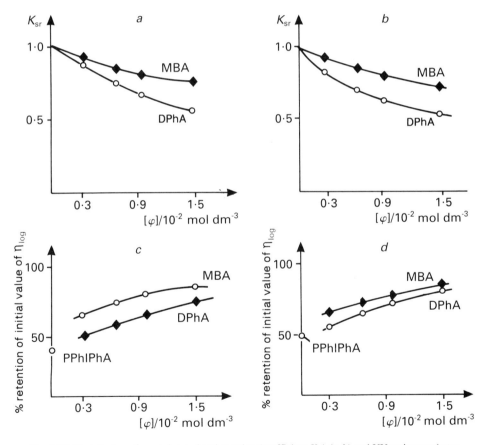

Fig. 4.19. Contribution of screening and reabsorption (coefficient K_{sr}) (a, b) and UV resistance (retention of initial value of η_{log}) (c, d) vs chromophore concentration for chromophore-containing
polyamides (a, c) and solid solutions of model diamides XLIVb (DPhA) and XLIVc (MBA) in
unmodified PPhIPhA.

λ_{excit}; D is the optical density of the light stabilizer at λ_{fl} PPhIPhA = 445 nm.

The UV resistance of these chromophore-containing polyamides and solid solutions estimated by changes in the polymer on exposure to irradiation an unfiltered 500 W mercury lamp over 6 h is shown in Fig. 4.19. Fig. 4.19 also illustrates the values of K_{sr} (λ_{excit} = 340 nm) calculated from eqn (4.1) for various systems. The contribution of screening to the protective effect of the light stabilizer was calculated from the formula:

$$S = \left(1 - K_{\text{s}}\right)100\% \tag{4.2}$$

where K_{s} is the screening coefficient less the reabsorption determined from eqn (4.1) assuming that $D = 0$.

From an analysis of the values of S which, at a chromophore concentration of 0.015 mol dm^{-3}, are 62 and 68% for polyamide XLId in solid solution and 7 and 15% for polyamide XLIe in solid solution respectively, it follows that the contribution of screening to the light-stabilizing effect is dominant in the former case and low in the latter [214].

The quenching of fluorescence emanating from PPhIPhA in the presence of benzan-throne chromophores, and the significant overlap of the emission spectrum of PPhIPhA with the long wavelength absorption band of diamide XLIVc, suggest an efficient elec-tronic energy transfer from the macromolecules to the benzanthrone rings. The Perrin model was used to determine the effective quenching radii for the excited states of the donor, which turned out to equal 45 and 36 Å for polyamide XLIe and a solid solution of diamide XLIVc respectively.

The authors of [213, 214] believe that the high efficiency of energy transfer in polyamides containing chromophore units within the macromolecules is associated with the operation of electronic energy migration along the polymer chains. As seen from Fig. 4.19, this mechanism gives rise to an enhanced light stabilizing action of the benzathrone moieties compared with the protective effect of diamide XLIVc on introduction into the unmodified PPhIPhA at the same concentration.

A range of film-forming aromatic polyamides containing different luminophoric units in the chain, either in the diamine (XLV) or acid component (XLVI), were synthesized in [8]: where $p{:}m = 0{:}1$, $1{:}2$, $1{:}1$, $2{:}1$.

In the majority of cases, the generalities pertaining to the varying properties of polyamides XLIa, b, c, such as the increase in η_{log} with a decreased luminophoric con-tent of the chain, or the reduction in fluorescence intensity of films with an increased concentration of these moieties, also apply to polyamides XLV [8], p. 1199.

Figs 4.20 and 4.21 illustrate the absorption and emission spectra of solutions and films of polyamides XLV and XLVI as well as of solutions of the corresponding monomers—9,10-dianilinoanthracene XLVII, 3,6-diaminophthalimide XLVIII and model compounds IL and L.

XLV

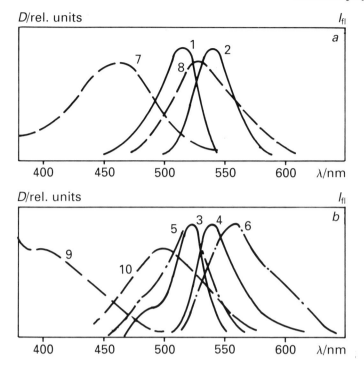

Fig, 4.20. Absorption (1, 3, 5, 7 and 9) and fluorescence (2, 4, 6, 8 and 10) spectra in dimethylacetamide (p:m = 1:10): a, 3,6-diaminophthalimide (1, 2) and model compound IL (7, 8); b, film of
polymer XLVa (5, 6); solutions of polyamides XLVa (3, 4) and XLVb (9, 10) in dimethylacetamide.

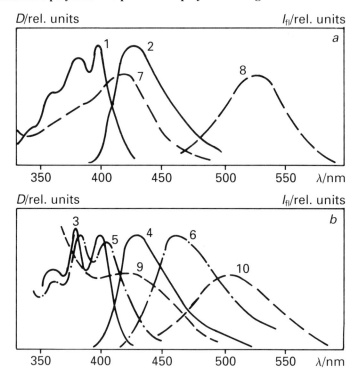

Fig. 4.21. Absorption (1, 3, 5, 7, 9) and fluorescence (2, 4, 6, 8, 10) spectra of model compound and polymer in dimethylacetamide solution: a, model compound L (1, 2) and 9,10-dianilinoanthracene (7, 8); b,
polyamides XLVI (3, 4), polyamide XLVc (9, 10) and film of polyamide XLVI (6, 5); *p:m* =1:10
(polyamide XLVc and *p:m* =1:2 (polyamide XLVI).

Thus, from comparison of the absorption and emission spectra of the monomers or model compounds and the corresponding polymers XLV and XLVI, it becomes clear that luminophoric units make a decisive contribution to the development of the spectral and luminescence properties of these systems.

4.3 AROMATIC HETEROCYCLIC POLYMERS

The aromatic polymers containing heterocyclic moieties in the chain are the least studied polymers from the point of view of absorption and fluorescence properties. There are only a few publications on the spectral properties of polybenzimidazoles [219], polybenzoxazoles [220], polyimides [215, 221–225], polyhexazocyclanes [226], polyisoindoxazenes [227], and dipyridylium polymer salts [228].

It has been established in [219] that solutions of polybenzimidazole LI in dimethylacetamide and formic acid have an intense blue fluorescence.

A comparison of the spectral properties of polymer LI and model bis [p-(2-benzimidazolyl)]phenyl ether (LII) (Fig. 4.22) showed that the fluorescence of model LII is determined by radiative transitions from the vibrational levels 0–0, 0–1, 0–2 of the benzimidazole ring in the excited state $^1L_{excit}$, whereas the fluorescence of a solution of polymer LI is related to the vibrational transitions 0–0 and 0–1 with maxima at 394 and 413 nm. The fluorescence spectra of solutions of polymer LI and model LII in dilute formic acid show, along with the monomer emission band, a band lying at longer wavelengths with λ_{max} 460 nm. This band is referred to the fluorescence of intermolecular associates, and it undergoes a red shift with increase in the solution concentration.

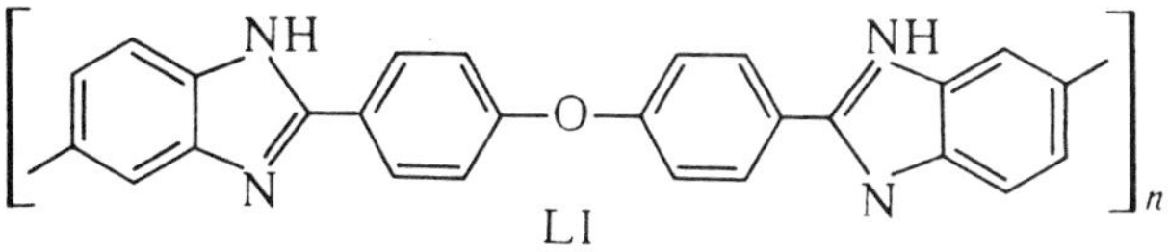

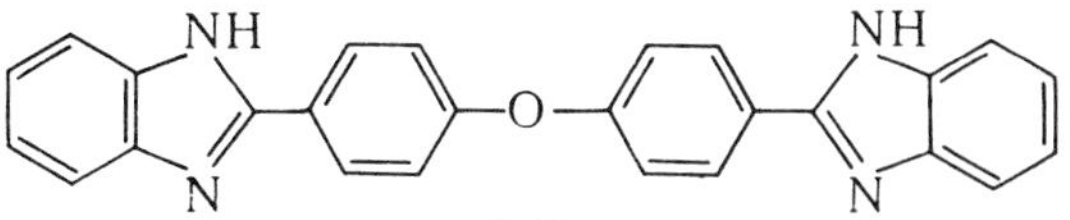

On increasing the concentration of polymer LI in solutions, the emission intensity falls and a dependence of the nature of the fluorescence spectrum on the wavelength of the exciting light is observed. However, the experiments dealt with in [219] were conducted not quite correctly from the spectroscopic point of view: in a number of cases, fluorescence was excited beyond the absorption bands of the solutions of polymer LI. In our opinion, these drawbacks reduce the value of the results reported in [219].

There is even less published work on the fluorescent properties of polybenzoxazole. In particular, Yakubovich *et al.* [220] mentions the presence in films of polymer LIII of a rather intense green fluorescence not found in the model compound LIV. These authors restrict themselves to measurements of the emission spectrum, making no attempts to explain its origin (Fig. 4.23).

The fluorescent properties of polypyromellitimides are open to discussion. In particular, the authors of [215, 221, 222] dismissed the presence in polyimides, such as kapton, of

LIII

LIV

LV

LVI

natural fluorescence with a quantum yield close to that of the corresponding polyamido-acids, which is explained by the nature of the spectral luminescent properties of the diimide moiety of the polymers. In a study of the spectral characteristics of the model compound of N,N'-dimethylpyromellitimide (LV), it was discovered [215] that its solutions do not fluoresce at room temperature but have an intense phosphorescence at 77 K. On the basis of data about the absorption and phosphorescence spectra, it is possible to suggest an energy level diagram for the molecule of LV with a low-level $T_{\pi\pi*}$ state (20 000 cm^{-1}) and an intermediate $S_{n\pi*}$ state arranged between the $S_{\pi\pi*}$ level and $T_{\pi\pi*}$ level. Molecules characterized by such an arrangement of their energy levels are known not to fluoresce [3] p. 25. The lack of fluorescence is associated with the high probability of the intersystem crossing $S_{n\pi*} \Rightarrow T_{\pi\pi*}$.

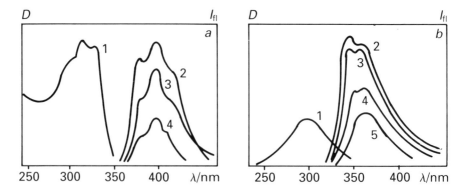

Fig. 4.22. Absorption (1) and fluorescence (2, 4, 3, 5) spectra of model compound and polymers in dimethylacetamide solution: a, concentration of model compound LII is 7.9×10^{-4} (2); 3.2×10^{-3} (3) and 6.3×10^{-3} g cm^{-3} (4); b, concentration of polymer LI is 4.6×10^{-6} (2), 1.5×10^{-4} (3), 4.3×10^{-4} (4) and 5.2×10^{-3} g cm^{-3} (5).

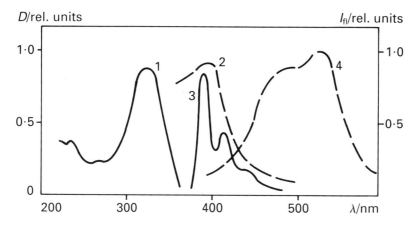

Fig. 4.23. Absorption (1, 2) and fluorescence (3, 4) spectra of model benzoxazole LIV in solutions (1, 3) and polybenzoxazole film (2, 4).

The low-level $S_{n\pi*}$ and $T_{n\pi*}$ states of compound LV are likely to gain more energy on protonation of the pyromellitimide ring. Actually, as shown in [215], the probability of intersystem crossing becomes low in concentrated sulphuric acid solution owing to the increase in the energy of the $S_{n\pi*}$ state, as supported by the occurrence of an intense green fluorescence in compound LV under such conditions.

Thus, the lack of fluorescence in polypyromellitimides is attributed to the presence of low-lying $n\pi*$ states. To obtain fluorescence in aromatic polyimides, a diimide component with higher $S_{n\pi*}$ and $T_{n\pi*}$ levels should be introduced. For example, the polyimides based on the dianhydride of 3,3',4,4'-diphenyl oxide tetracarboxylic acid and 4,4'-diamino-diphenyl oxide exhibit a detectable blue fluorescence [8].

At the same time, some studies [223–225] report the presence of fluorescence in films of kapton polyimide. This emission with λ_{max} = 530, 560 and 595 nm is assigned to exciplex states. It should be noted that some deductions made by the authors of this paper are open to question. Firstly, the most intense fluorescence was detected on light excitation of the weak natural absorption region (λ_{excit} 352 and 386 nm) of the polymers. Secondly, the fluorescence excitation and absorption spectra of kapton films do not coincide [223]. Thirdly, the fluorescence spectra of polyimide films obtained by different authors are all different in nature [223, 224].

These facts suggest that the observed fluorescence refers to impurity centres to which the noticeable absorption of polymer films does not correspond. This suggestion is supported by the dependence of the fluorescence intensity and spectral character on the temperature of polycyclodehydration of the initial polyamidoacids [223].

It is known [229] that at temperatures above 350°C, polyimides are likely to undergo conversions forming amide structures LVI. On UV irradiation of moieties of the LVI type, associated with fluorescence excitation, the possibility exists of forming end groups with primary amine groups [230]. The groups obtained possess electron donating properties and are apparently prone to form charge transfer complexes with the participation of the diimide fragments of polyimide chains. These complexes may be responsible for the fluorescence in question.

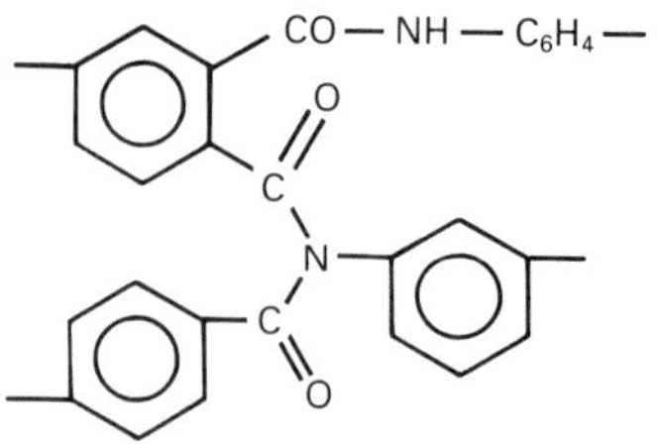

LVI

A comparative study of the spectral properties of polypyromellitimides and poly-imides based on the dianhydride of 3,3',4,4'-diphenyl oxide tetracarboxylic acid with acridine moieties in the chain (LVIIa and LVIIb) is reported in [8, 220].

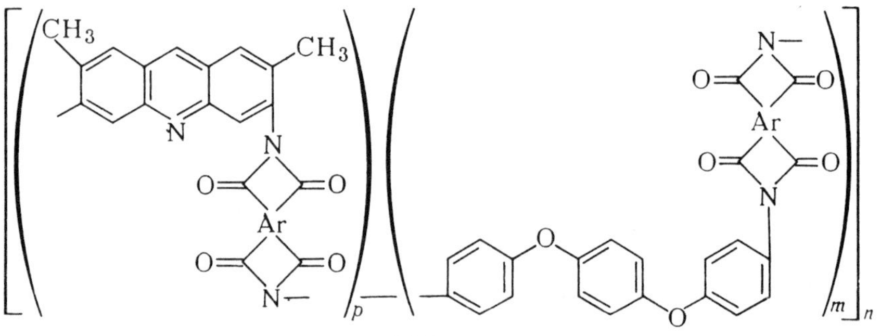

LVII

Ar, LVIIa, 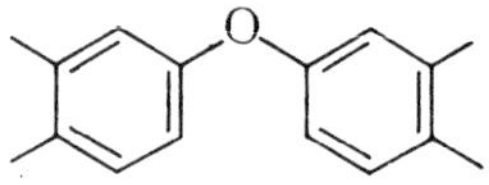 LVIIb

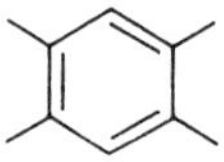

Polyimides LVIIa and LVIIb were prepared by the low temperature polycondensation of a mixture of acridine yellow and the 4,4'-diaminodiphenyl ether of hydroquinone with the anhydrides of the respective tetracarboxylic acids and subsequent thermal polycy-clodehydration of prepolymer (polyamidoacid) films formed.

Acridine-containing prepolymers with η_{log} = 0.25 to 0.88 dl g^{-1} (0.5% solution in dimethylformamide) were yellow coloured and showed an intense green fluorescence with λ_{max} 500 nm (in solution) and a yellow-green fluorescence with λ_{max} 530–540 nm (as films). It is seen from Fig. 4.24a, which illustrates the absorption and emission spectra of polyimide LVIIa and LVIIb films and the corresponding polyamidoacids, that the nature of the dianhydride component has little effect on the spectral properties of the prepolymers. Conversely, the spectral luminescent characteristics of the poly-imides are strongly dependent on the nature of the diimide moiety (Fig. 4.24b). Now the absorption spectrum of polyimide LVIIb shows only one band with λ_{max} 460 nm, yet in the case of polyimide LVIIa, an additional short wavelength absorption band arises with λ_{max} 415 nm. The origin of this band, as shown in [8] p. 706, from the example of model diimide LVIII, is related to aggregation of the acridine fragments (see next page).

Polyimides LVIIb, irrespective of the p:m ratio, showed an intense fluorescence at 515–525 nm, whereas polyimides LVIIa failed to fluoresce.

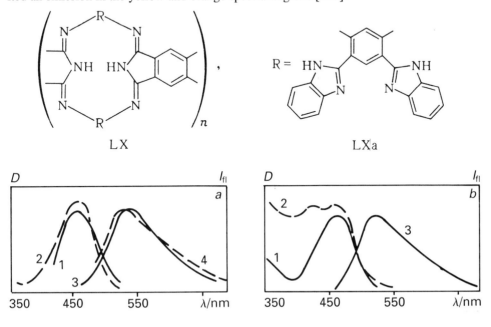

LVIII

Similar results were obtained in [220] for polyimides with stilbene units in the chain (LIX).

LIX

where Ar = LIXa;   LIXb

It was established that polypyromellitimide LIX does not fluoresce, whereas the polymer containing a diphenyl oxide unit in the diimide component fluoresces with λ_{max} 425 nm.

Polyhexazocyclanes LX synthesized by polycondensation of pyromellitonitrile with the diamines of the phenylbenzimidazole and phenylbenzoxazole series in phenol exhibited an emission in the yellow and orange spectral regions [226].

LX   LXa

Fig. 4.24. Absorption (1, 2) and fluorescence (3, 4) spectra (*p:m* =1:3); (a) prepolymer films of polyimides LVIIb (1, 3) and LVIIa (2, 4); (b) films of polyimides LVIIb (1, 3) and LVIIa (2).

It was established that the absorption and fluorescence spectra of polyhexazocyclanes and model hexazocyclanes of similar structure have UV bands related to electronic transitions in the phenylbenzimidazole and phenylbenzoxazole moieties, and also fluorescence at λ_{max} 540–650 nm lying at longer wavelengths and related to charge transfer from the electron-donating chromophoric fragment to the electron-accepting isoindole moiety [226].

Polyisoindoxazenes LXI [227] belong to another type of polymer prepared from the monomer pair used in polyhexazocyclanes, with naphthol or chlorophenol acting as solvent:

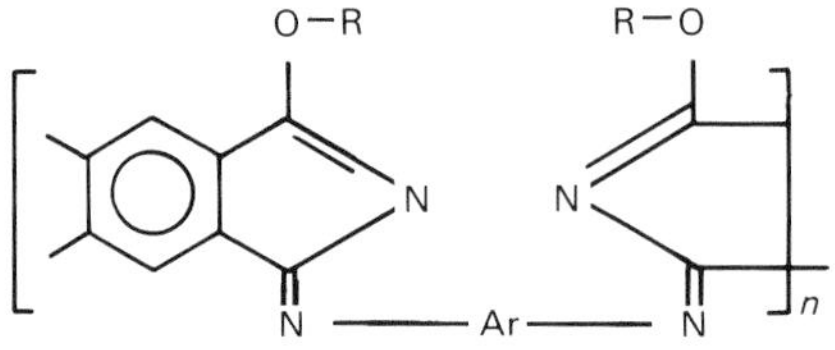

LXI

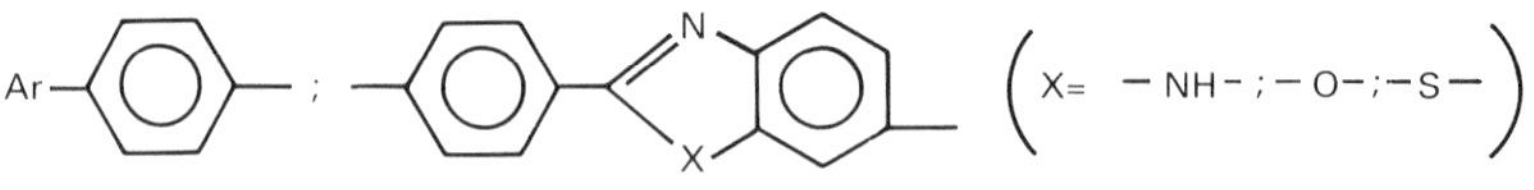

It has been found that the fluorescence properties of solutions of polymers LXI are determined by the nature of the phenol residue: irrespective of the nature of the Ar-unit, the derivatives of β-naphthol are characterized by an orange fluorescence with λ_{max} 585–595 nm, whereas the chlorophenol derivatives do not fluoresce. The occurrence of the fluorescence band in polymers of the first type is associated with the charge transfer states between the electron-donating residue of β-naphthol and the electron deficient isoindole ring. In polymers of the second type, the chlorine atom causes fluorescence quenching as a result of the heavy atom effect.

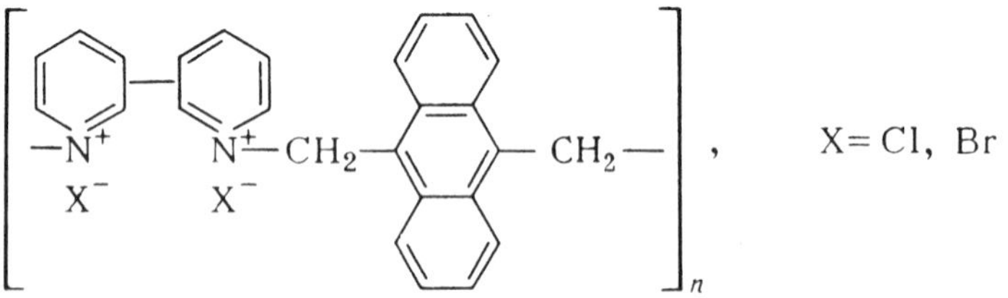

LXII

The absorption and fluorescence properties of the polymer salts of dipyridilium (polyviologens) LXII are covered in [228]. Polymers LXII were synthesized by the Menschutkin reaction of 2,2'- or 4,4'-dipyridyl with 9,10-dichloro-[DCh-MA] and 9,10-dibromomethylanthracene (DBMA) in a polar solvent at 30–70°C for 6–8 h. The polymers obtained are yellow or dark yellow powders, soluble in water, alcohols, and other polar solvents. The values of [η] for polymers LXII are listed in Table 4.4.

Table 4.4. Composition and some properties of polymers LXII

Initial monomers		Solvent	T /°C	ε	$[\eta]^a$ /10^2 cm^3 g^{-1}
anthracene	dipyridyl				
DBMA		Dimethylsulphoxide	70	45.00	0.047
	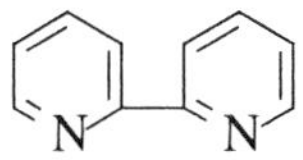	Dimethylformamide	70		0.063
		Methanol	70		0.052
			50		0.040
			30		0.033
DBMA	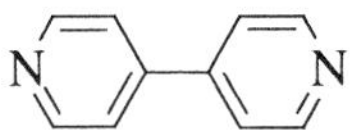	Dimethylformamide	70	37.60	0.067
DChMA	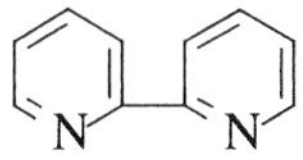	Methanol	70	32.65	0.044
DChMA	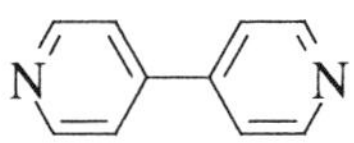	Dimethylformamide	70	37.60	0.055

^{a} Measured in 0.1 m aqueous solution of KBr at 25°C.

As seen from Table 4.4, the viscosity of polymers LXII are higher for bromides than for chlorides of similar structure. Polymers with the highest RMM were prepared in a dimethylformamide medium. Using methanol as solvent, it was shown that with a decreasing reaction temperature, all other factors being the same, the RMM and yield of polymers LXII are reduced.

The absorption and emission spectra of solutions of polymers LXII, the model 9,10-dimethyleneanthracenepyridinium halides LXIII and the initial 9,10-dihalomethylanthracenes are illustrated in Figs 4.25 and 4.26.

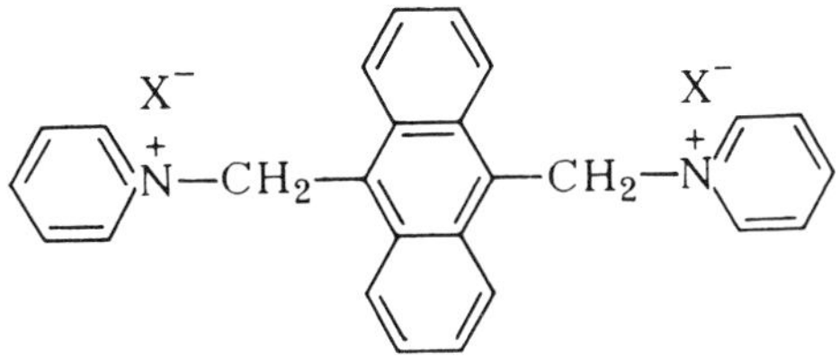

LXIII

The absorption spectra of all six compounds are seen to exhibit the (allowed) structure typical of anthracene derivatives; they are rather similar to each other in the shape and arrangement of the bands. Conversely, the fluorescence properties of the monomers, models LXIII and polymers LXII, differ noticeably. For example, while 9,10-dichloromethylanthracene and the corresponding model compound have a comparatively weak fluorescence when they are in the form of crystals and solutions, the polyviologens fluoresce intensely both in the solid state and in solution. Moreover, 9,10-dibromomethylanthracene and the corresponding model LXIII do not fluoresce at

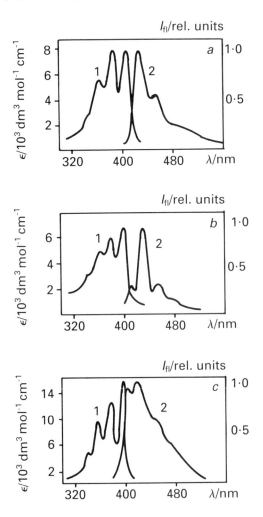

Fig. 4.25. Absorption (1) and fluorescence (2) spectra of 9,10-dichloromethylanthracene (a), model compound LXIII (X = Cl) (b) and polyviologen LXII (X = Cl) (c) in dimethylformamide solution.

all at room temperature; the emission spectra of these compounds have been measured at 77 K. Polyviologen LXII (X = Br) exhibits an intense fluorescence even at room temperature.

The authors of [228] explain the above differences in the following way: in dichloromethylanthracene, and particularly in dibromomethylanthracene, owing to the strong spin–orbital interaction, the probability of the nonradiative transition from S_1 to T_1, responsible for fluorescence quenching, increases. In polymers, the halide ions are located near the nitrogen atoms of the dipyridyl groups, which imposes a reduction of the vibrational degrees of freedom and probability of nonradiative transitions, thus enhancing the fluorescence quantum yield. The noticeable differences in the emission spectra of polymers LXII and the corresponding models LXIII are apparently associated

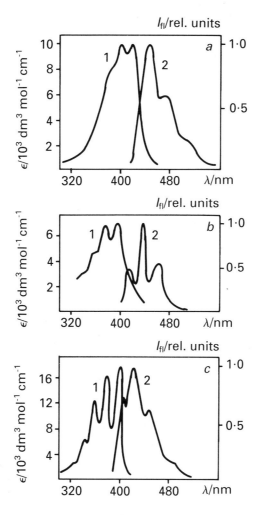

Fig. 4.26. Absorption (1) and fluorescence (2) spectra of 9,10-dibromomethylanthracene at 77 K (a), model compound LXIII (X = Br) at 77 K (b), and polyviologen LXII (X = Br) at 293 K (c) in dimethyl-formamide solution.

with the fact that the dipyridylium salts, containing no anthracene moieties, unlike the pyridinium salts, have a fluorescence spectrum with a variety of intense bands in the 400–450 nm region.

The above examples show that of the heterocyclic aromatic polymers, those having received some spectroscopic analysis are the polyimides. The properties of other poly-heteroarylenes have either been studied very little (polybenzimidazoles, polybenzoxa-zoles) or not at all (polyoxadiazoles, polyquinoxalines, polyquinazolones, etc.). At the same time, investigation of the fluorescence characteristics of these polymers is of inter-est both from the theoretical and practical points of view since it is known [5] p. 80, that many of their low molecular analogues exhibit intense fluorescence.

4.4 PRODUCTS OF POLYMER-ANALOGOUS CONVERSIONS

The preparation of fluorescent polymers by polymer-analogous conversions was described for the first time in [232]. Polymers containing an amine group (such as poly(vinyl amine), poly(p-aminostyrene), polyethyleneimine) were reacted with 1-dimethylaminonaphthalene sulfonyl chloride. The reaction was conducted by mixing aqueous or tetrahydrofuran solutions of the polymer with an acetone solution of the sulfonyl chloride. Depending on the proportion of the reagents, fluorescent polymers were formed containing different numbers of naphthalene groups in the chain.

In later studies 9-anthrylmethyl isocyanate was used to attach anthracene groups to amine-containing polymers [233–235]. In particular, [235] is concerned with the fluorescence properties of products resulting from the reaction of different amounts of 9-anthrylmethyl isocyanate with the water soluble copolymer of vinyl pyrrolidone and vinyl amine. It was shown [235] that the appearance of the long wavelength structureless fluorescence band of these polymers obtained does not depend on their concentration in solution, but is determined solely by the anthracene group content of the polymer chain. According to [235], the structureless band appears to be due to the formation of intramolecular excimers at the expense of the hydrophobic interaction of anthracene rings. The formation of excimers is promoted by the hydrogen bonds available between the side chains bearing the anthracene group, as well as by the mobility of the side chains, the bearers of anthracene groups, by the compactness of the macromolecular coil, and the attachment of the anthracene rings to the nearby amine groups of the polymer chain.

To site luminescent units in the polymer side chain, they were reacted with carboxyl [126, 236–239], chloroanhydride [241], and hydroxyl groups [233, 242–246], benzene rings [126, 247], chloro- and bromomethyl groups [247–255], nitrile [256], and anhydride groups [257].

The interaction of polymeric acids such as poly(acrylic), poly(methacrylic) or their copolymers with such luminophors as 9-anthryldiazomethane, 9-methyl-10-anthryldiazomethane or diazomethyl-α-naphthylketone was examined in [126, 236, 237]. The reaction was conducted in a homogeneous medium (methanol) and nonblending solvents (water-hexane) at 20–55°C to give, for example, LXIV.

LXIV

The concentration of fluorescent group dopants can be varied widely by changing the polymer : diazomethane ratio. The fluorescent properties of polymers prepared in this way are similar to those of the corresponding model compounds (Fig. 4.27). As seen in Fig. 4.27, featuring the emission spectra of solutions of poly(methacrylic acid) containing one 9-anthrylacyloxymethane group per 1400 polymer units, and of the model 9-anthrylacetoxymethane, both compounds are characterized by a monomer fluorescence free from any traces of excimer emission.

Weakly fluorescing polymers prepared by the interaction of poly(ethylene-4-thiobenzoyl(thio)-2-acetic acid) or the copolymers of the initial thioester with styrene, methyl

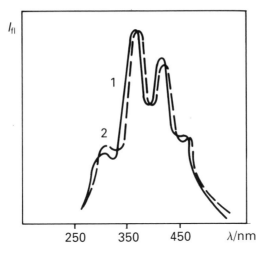

Fig. 4.27. Fluorescence spectra of benzene solution of 9-anthrylacetoxymethane (1) and aqueous solution of poly(methylacrylic acid) (2) containing one 9-anthrylacetoxymethane group per 1400 polymeric units.

methacrylate, or methacrylic acid, with diamines such as acridine yellow, 1-amino-anthraquinone, or cresyl blue are described in [238]. The reaction was conducted in a dimethylformamide-benzene mixture (1:1, v/v) and a molar ratio of polymer to amine of 2:1 at room temperature for 24 h. Some properties of copolymers LXV are given in Table 4.5.

The fluorescent properties of copolymers LXV have not been investigated thoroughly. It has been noted only that the emission intensity of the initial dyes such as acridine yellow reduces appreciably after bonding with a polymer. The authors of [238] offer no explanation for this effect.

Table 4.5. Absorption properties of copolymers LXV [202]

Basic monomer	Thioester unit content/% (mol)	Dye	Solvent	Absorption	
				λ_{max}/nm	log ε_{max}
Styrene	0.8	Acridine yellow	Chloroform	430	1.08
				377	1.10
Methyl methacrylate	10		Dimethyl-formamide	425	1.92
				380	2.07
Styrene	0.8	1-Aminoan-thraquinone	Chloroform	387	0.90
		Cresyl blue	Methylene chloride	625	1.40
				525	1.50
				450	1.55
Methyl methacrylate	10		Chloroform	625	2.66
				525	2.81
				450	2.91
Methacrylic acid	2.0		Dimethyl-formamide	625	1.60
				510	1.90
				450	2.08

In our opinion, the quenching of the luminophor fluorescence in copolymers LXV can be explained by the formation of hydrogen bonds between the diaminoacridine moiety and the thioglycolic acid present in the system. According to the authors of [240], the hydrogen bond connecting two π-systems promotes fluorescence quenching in the pairs where both the luminophor and quencher function as hydrogen donors.

As in [126, 236–238], where carboxyl groups were used in polymers to form strong covalent bonds with a fluorescent molecule, in systems LXVI featured in [239], an ionic linkage is reported to be formed between poly(acrylic acid) and poly(methacrylic acid) and the amine luminophor.

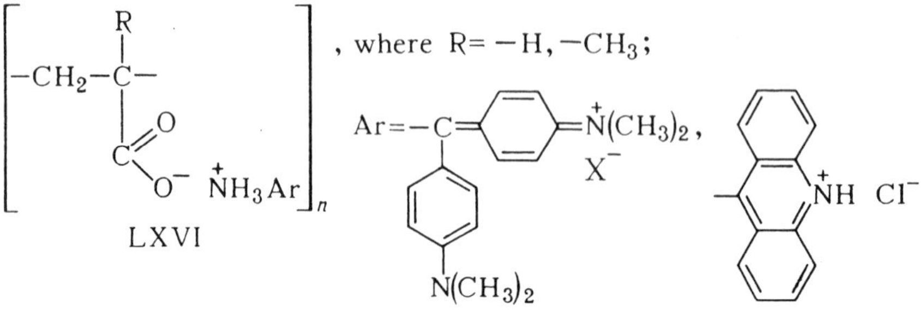

, where R= $-$H, $-$CH$_3$;

Ar= ... $\stackrel{+}{N}$(CH$_3$)$_2$, X$^-$

LXVI

In the preparation of polymeric salts, auramine, which shows no dimerization and has a weak fluorescence, and 9-aminoacridine hydrochloride, capable of forming dimers and intensely fluorescing, are used as cationic dyes. It has been shown that the process of bonding of the dye is determined by three factors: (i) the nature and conformational state of the polyions (the degree of ionization α), (ii) the molar ratio between polymer and dye, and (iii) the ionic strength. The emission intensity of auramine increases 100 times in the presence of poly(methacrylic acid). For this compound, as the polymer : dye ratio increases, the fluorescence quantum yield and relative content of bound dye also increase.

An increased degree of ionization of the polymeric acid brings about a reduction, in the case of auramine, and, conversely, an increase, in the case of 9-aminoacridine hydrochloride, of the efficiency of binding between the dye and the polymer. It is of interest that at low α, the addition of poly(methacrylic acid) to 9-aminoacridine hydrochloride leads to a reduction in the fluorescence intensity. The authors of [239] attribute the differences observed in the behaviour of the luminophors to the different rigidities of their molecular frameworks and the presence in them of different side groups. We add that, as in the example discussed above, hydrogen bonds seem to make some contribution to the fluorescence quenching of the acridine derivative.

Fluorescein isothiocyanate, readily engaging in polymer-analogous conversion reactions with hydroxyl-containing polymers, has been used to interact with amylase of RMM *ca* 1700 [242]. In the polymer containing the units LXVII, measurements were made of the fluorescence lifetimes for different macromolecular concentrations in the solution and varying values of the medium pH.

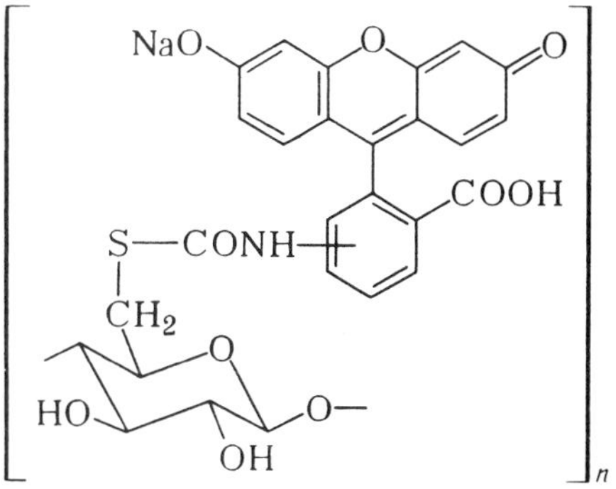

LXVII

Refs [243, 245] describe the absorption and fluorescence properties of films of poly(vinyl alcohol) containing, as side groups, rhodamine B, diphenyloxazole, diphenylanthracene, and pyrene.

Polymers LXVIIIa–d were synthesized by treating poly(vinyl alcohol) with the corresponding sulfonyl chloride or chloroaldehyde in *N*-methylpyrrolidone, firstly at room temperature and then at 60–70°C. Depending on the ratio of the initial reagents, the polymers obtained had from 1 to 4 luminophoric moieties per 200 repeat units.

Fig. 4.28 shows the absorption and fluorescence spectra of polymers LXVIIIa–LXVIIIc cast from *N*-methylpyrrolidone solution into thin films. A comparison of these spectra with those of initial luminophor monomers [5] p. 285, confirms the similarity of their fluorescent characteristics and hence the monomer nature of the emission, free from any detectable traces of excimer states.

The fluorescence spectra of pyrene-containing polymers LXVIIId are more complex (Fig. 4.29). It is seen that, depending on the solution concentration, the spectra feature either the monomer fluorescence band of pyrene groups with λ_{max} 376, 395, 415 nm, or, along with these, an excimer band with λ_{max} 495 nm. The authors of [245] discovered a correlation between I_M/I_E and the concentration of an aqueous solution of polymer LXVIIId. The replacement of water by the more viscous ethylene glycol causes a sharp reduction in the excimer emission intensity, the intensity of monomer fluorescence bands being unchanged. It is evident that as the viscosity of the solvent increases, the rate constant of intramolecular excimer formation decreases, in accordance with the diffusion controlled mechanism of the process.

$$\left[\left(\begin{matrix}-CH_2-CH-\\ |\\ OH\end{matrix}\right)_m\left(\begin{matrix}-CH_2-CH-\\ |\\ OR\end{matrix}\right)_p\right]_n$$

LXVIII

where R=

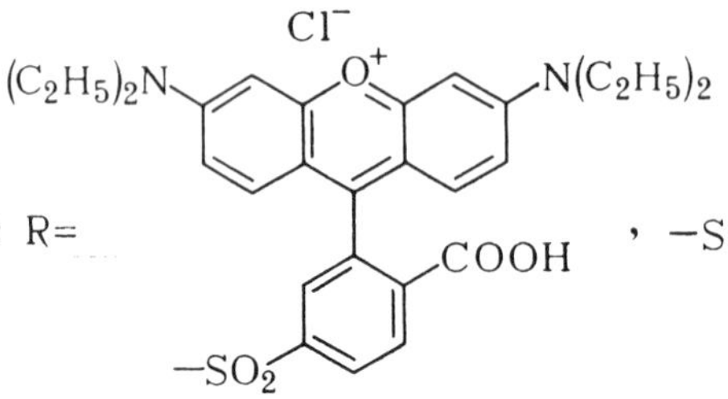

LXVIIIa 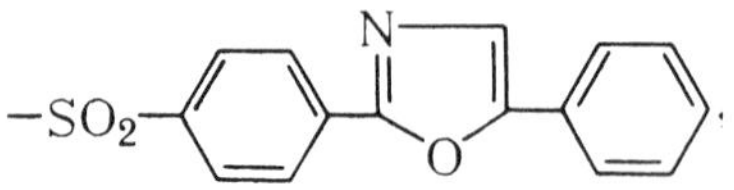   LXVIIIb

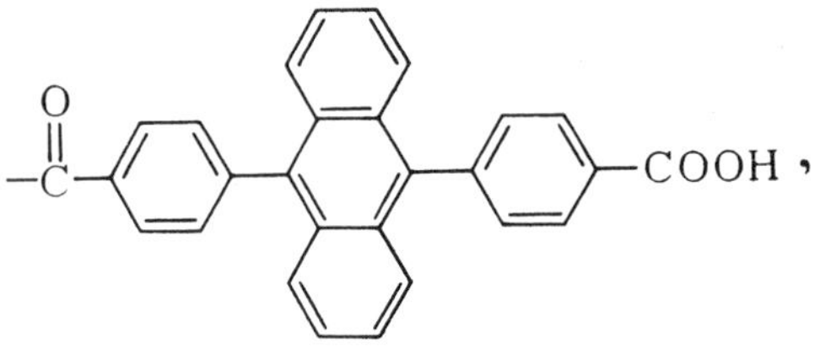

LXVIIIc 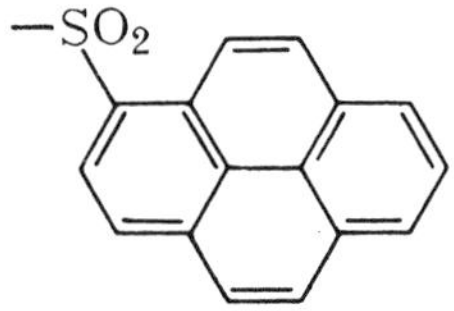   LXVIIId

The above deduction is supported by [244] dealing with the treatment of partially hydrolysed poly(vinyl acetate) with 4-(3-pyrenyl)butyric acid in tetrahydrofuran solvent in the presence of trifluoroacetic anhydride at 50°C. In the polymer obtained, the average distance between the pyrenyl groups was some 200 repeat units.

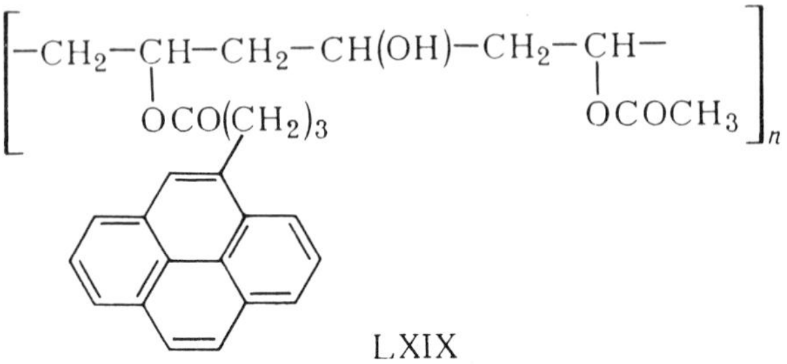

$$\left[\begin{matrix}-CH_2-CH-CH_2-CH(OH)-CH_2-CH-\\ |\qquad\qquad\qquad\qquad\qquad |\\ OCO(CH_2)_3\qquad\qquad\quad OCOCH_3\end{matrix}\right]_n$$

LXIX

From the fluorescence spectra of polymer LXIX prepared in a solution of methanol

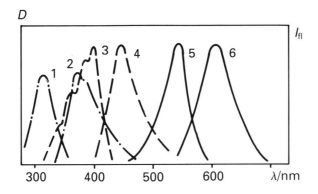

Fig. 4.28. Absorption (1, 3, 5) and fluorescence (2, 4, 6) spectra of polymers LXVIIIb (1, 2), LXVIIIc (3, 4) and LXVIIIa (5, 6) as films.

and in ethylene glycol, as illustrated in Fig. 4.30, it follows that in the former solvent the monomer and excimer emissions have maxima at 396 and 480 nm respectively. In the more viscous ethylene glycol, the excimer fluorescence is absent. For mixtures of solvents of similar nature, the authors of [244] have established the dependence of I_E/I_M on the viscosity of the medium. In all cases, as the viscosity of the solvent increases, the rate of intramolecular formation of excimers decreases.

The introduction of luminophoric anthracene moieties into the side chain of poly(vinyl alcohol) is described in [245, 246]; [245] shows the possibility of applying the fluorescence techniques for monitoring the condensation of OH-groups of the polymer with 9-anthraldehyde (for details, see Chapter 6), and [246] suggests the preparation of anthracene-containing copolymers of vinyl alcohol with vinyl acetate by treating a solution of the initial copolymer in dimethylacetamide with 9-chloromethylanthracene at 70 to 130°C. This

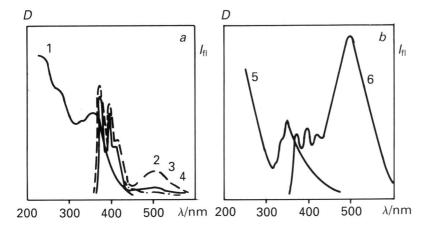

Fig. 4.29. Absorption (1, 5) and fluorescence (2, 3, 4, 6) spectra of aqueous solutions of different concentration (a) and of film (b) of polymer LXVIIId: 1, 4, 0.017 mol dm^{-3}; 2, 0.28 mol dm^{-3}; 3, 0.14 mol dm^{-3}; 5, 6, film.

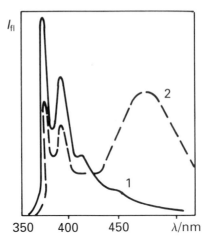

Fig. 4.30. Fluorescence spectra of polymer LXIXc in ethylene glycol (1) and methanol (2) solutions.

method enables the attachment of luminophoric groups only to the vinyl alcohol residues in the copolymer. This selectivity of attachment of anthracene groups makes it possible to apply the polarized luminescence method in a separate study of relaxation processes in the vinyl alcohol and vinyl acetate sites of the copolymers with account of the microsequential distribution of separate units. The same authors [246] have developed methods for assessing the microsequences of copolymers of similar structure on the basis of luminescent labels whose luminescence parameters are sensitive to change in the length of the sequences of vinyl acetate units. To indicate the nonuniform distribution of polymer units, the suggestion was made of using either poly(ethylene glycol) macromolecules labelled with an anthracene group or the molecules of the low molecular luminescent dye, acridine orange.

Polymer-analogous conversions via chloro- and bromomethyl groups permit doping the side chain of polymers with naphthalene [248] and anthracene [249] rings, flavin [250, 251], and xanthene [252–254] rings and tertiary aromatic amine residues [255]. Thus, [248] is concerned with the synthesis of the naphthalene-containing copolymer of styrene and vinylbenzyl chloride by the interaction of the initial copolymer with the product of reaction of 5-dimethylamino-1-naphthalene sulfonyl chloride and n-butylamine in N,N-dimethylformamide solvent in the presence of sodium hydride, and subsequent treatment of the reaction solution with trimethylamine.

LXX

Depending on the ratio $n:m$, aqueous solutions of copolymer LXX are characterized by a fluorescence spectrum with a band having its maximum at 512 ($n:m = 2:1$) to 536 ($n:m = 59:1$) nm. The fluorescence technique was applied to study the specific polyelectrolytic behaviour of copolymer LXX in aqueous solution [248].

The interaction between chloromethylated polystyrene with 10-hydroxymethyl-bis-9-anthrylmethyl ester was used to prepare a polymer containing 3 anthracene units per 10 macromolecules [249]. Absorption and fluorescence spectroscopy were applied to study the photodimerization of anthracene moieties occurring at various degrees of conversion in the solution or film of these polymers.

Flavin-containing copolymers of styrene were synthesized by the reaction of 10-ethylisoalloxazine and triethylamine with a partially chloromethylated styrene in N,N-dimethylformamide [290].

The fluorescence spectra of polymers LXXI exhibited a broad structureless band with λ_{max} 507–514 nm (depending on the values of n and m), whose form and position was similar to the fluorescence spectral band of the initial isoalloxazine. It was discovered [251] that the fluorescence quantum yield and the average lifetime of the excited state decrease with an increasing ammonium ion content of the copolymer.

$$\left[-CH-CH_2-\right]_n-\left[-CH-CH_2-\right]_m-\left[-CH-CH_2-\right]_{1-n-m},$$

$$n = 0.004 - 0.077$$
$$m = 0.19 - 0.95$$

LXXI

Interesting polymers LXXII were prepared by the interaction of poly(chloromethylstyrene) [252] or the copolymer of a mixture of m- and p-chloromethylstyrene and styrene (253) with methyleosine (sodium salt) and Rose Bengal dye respectively, as well as polystyrene containing bromomethyl end groups, with Rose Bengal (disodium salt) [254]. For the two former polymers, the reaction was conducted in N,N-dimethylformamide solvent at 24–30°C in the presence of tetrabutylammonium bromide as catalyst (LXXIIa) or at 80°C without any catalyst (LXXIIb). In [252], polymer LXXIIa was used for subsequent reaction with excess potassium crotonate to replace the chloromethyl groups by crotonoyl groups (see next page).

The absorption spectra of the polymer LXXIIb formed show the bands with λ_{max} 345 and 550 nm, whose intensity varies on illumination with light with $\lambda < 470$ nm, which is associated with the opening of the $C=C$ bonds in the crotonoyl residue. A comparison of the spectra of polymers LXXIIb, exposed to light, with the spectra of polymers of similar structure containing no methyleosine groups and illuminated under the same conditions, shows the latter to slow down the rate of photochemical conversion when exposed to light with λ_{max} from 390 to 470 nm.

A specific feature of polymer LXXIIc is the dependence of its absorption spectrum on the concentration of Rose Bengal moieties (Fig. 4.31) [253]. As seen from Fig. 4.31, the long wavelength absorption band of polymer LXXIIc undergoes a bathochromic shift in comparison with the band of the initial dye. The quantum yield of singlet oxygen generated on photo-oxidation of the dopant was determined for solid mixtures of polymer LXXIIc and 2,3-diphenyl-*p*-dioxene and dichloromethane solutions of these compounds. It has been shown [253] that the above quantum yield depends on the efficiency of intra- and inter-molecular transfer of electronic energy in the Rose Bengal moieties of copolymer LXXIIc.

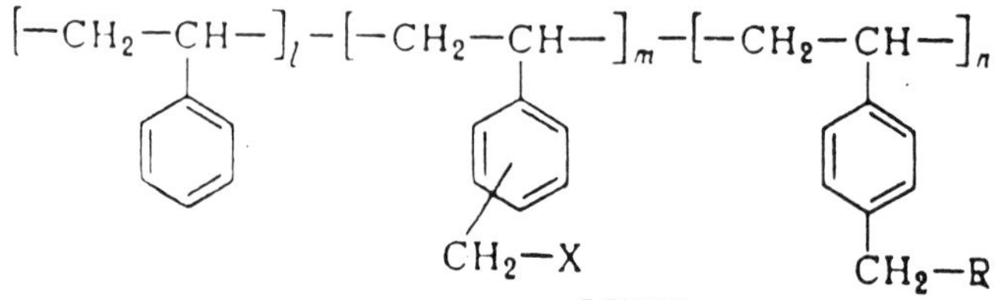

LXXII

where $l = 0$, X= Cl, R =

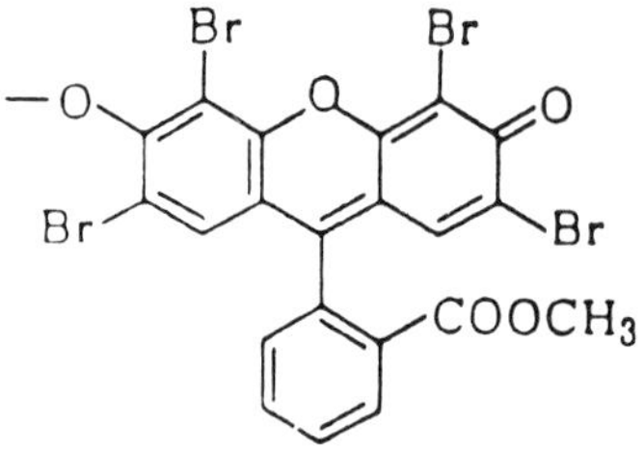

LXXIIa

$X = -OCOCH{=}CH-CH_3$, R— same   LXXIIb

$l:m=2:1$; R= Cl—   LXXIIc

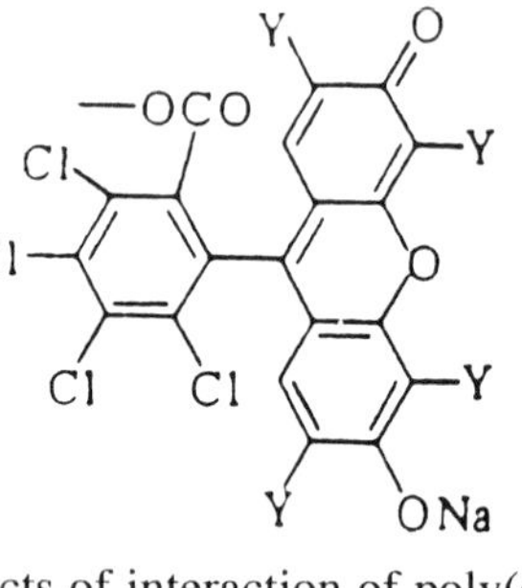

Ref. [255] describes the products of interaction of poly(epichlorohydrin) with tertiary aromatic amines such as 2,2'-dipyridyl, 4-methylquinoline, and 8-hydroxyquinoline as a result of which polymer LXXIII is formed.

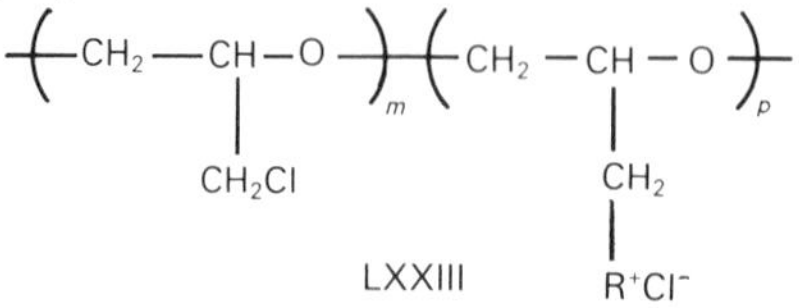

LXXIII

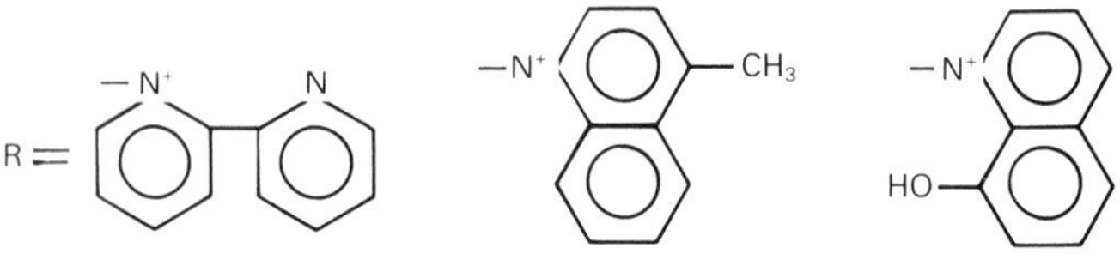

R=

LXXIIIa   LXXIIIb   LXXIIIc

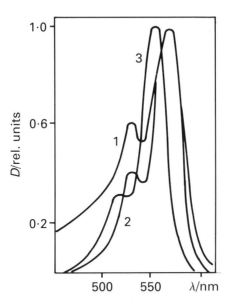

Fig. 4.31. Absorption spectra of copolymers LXXIIc in methanol solutions with $(\ell + m):n = 1.07:0.61$ (1) and 1.07:0.152 (2), of Rose Bengal dye (3).

The synthesis of the latter polymers was performed at a molar ratio of poly(epichloro-hydrin) and amine from 1:0.25 to 1:1 in a N-methylpyrrolidone medium at 100–120°C for 2–24 h. The degree of quaternization was determined by the results of a titrimetric chloride ion analysis by the Moor method (Table 4.6). At short reaction times, a consid-erable portion of the tertiary amines fails to enter into reaction and the degree of quater-nization does not exceed 9–14%. As shown by the example of the reaction mixture of poly(epichlorohydrin) and 8-hydroxyquinoline, irrespective of the molar ratio of the reactants, the degree of quaternization remains low and increases only when the process is significantly extended.

As follows from a comparative study of the spectral and luminescence properties of solutions and films of polymers LXXIII and corresponding model compounds such as octyl-2,2'-dipyridylium chloride, octyl-4-methylquinolinium chloride and octyl-8-hydroxyquinolinium chloride, the long wavelength absorption in the 280–320 nm region and fluorescence in the 315–390 nm region are attributed to electronic transitions in the cations of the corresponding quaternary ammonium salts [255]. A distinctive feature of the fluorescence spectra of polymers LXXIIIc is the presence of an intense band with λ_{max} 505–510 nm. The origin of this band is connected with photodissociation of a pro-ton from hydroxyquinoline units in their lowest singlet excited state, as a result of which they become converted into anions exhibiting long wavelength emission.

To prepare fluorescent polymers by polymer-analogous conversions, anthracene derivatives and other condensed aromatic rings took part in Friedel–Crafts reactions [126, 247]. As anthracene compounds, derivatives of type LXXIV were used.

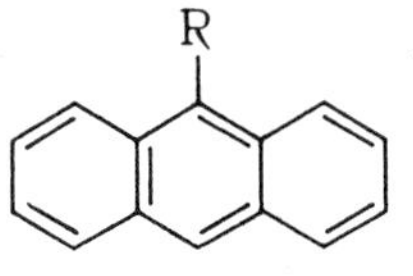

where R= $-CH_2Cl$, $-CH_2-OH$, $-\overset{CH_3}{\underset{|}{CH}}-Cl$, $-CH=CH_2$.

LXXIV

The reactions of these derivatives with phenyl-group containing polymers such as polystyrene, poly(α-methylstyrene), poly(vinyl benzyl), or poly(propenyl benzyl) ester were conducted at room temperature or at $-78°C$ in the presence of $SnCl_4$ or $BF_3 \cdot Et_2O$ as catalyst, for 0.5–360 h [247]. The polymers obtained contained from 0.2–83 anthracene groups per 1000 polymer units and displayed fluorescent properties similar to those of the model benzylanthracene (Fig. 4.32).

Table 4.6. Synthesis and spectral properties of chromophore-containing poly-(epichlorohydrin)

Polymer	Proportion[a] /mol	Time /h	Degree of quaternization /%	$\lambda_{max}^{abs[b]}$ /nm	$\lambda_{max}^{fl[b]}$ /nm
LXXIIIa	1:0.8	2	9.2	282	315,325
LXXIIIa	1:0.8	20	21.1	282	315,325
LXXIIIb	1:0.8	3	13.9	284,312	382
LXXIIIb	1:0.8	22	23.7	284,312	382
LXXIIIc	1:0.25	4	10.7		
LXXIIIc	1:0.5	4	11.2		
LXXIIIc	1:0.8	4	9.8	257,320	390,505
LXXIIIc	1:0.8	24	20.3		
LXXIIIc	1:1	24	21.8		

[a] Proportion poly(epichlorohydrin): amine.
[b] Long-wavelength absorption and fluorescence bands of N-methylpyrrolidone solutions.

It was noted in [247] that, simultaneously with the normal alkylation proceeding under the conditions of Friedel–Crafts reactions, the following polycondensation processes can be observed.

During homopolycondensation of 9-chloromethylanthracene in the absence of a phenyl-containing polymer, an 85% yield of polymer LXXV ($M_n \approx 1500$) is obtained. The emission spectrum of this polymer is shown in Fig. 4.32. Homopolycondensation reduces the yield of alkylation and may cause attachment of groups LXXV to the phenyl-containing polymer, which has a bearing on the fluorescence spectra.

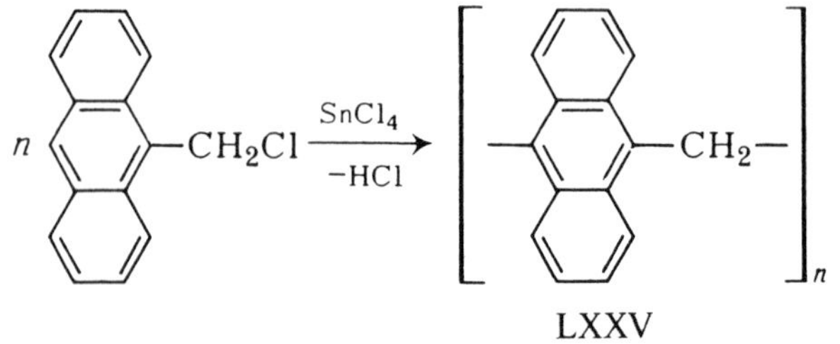

LXXV

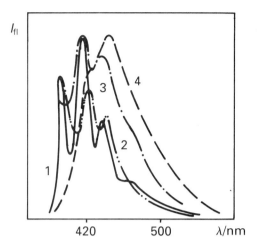

Fig. 4.32. Fluorescence spectra of toluene solutions of 9-benzylanthracene (1), polystyrene with 0.14 mol % anthracene groups (2), polystyrene with 0.25 mol % anthracene groups (3) and 9-chloromethyl-anthracene homopolymer (4).

To introduce anthracene and naphthalene groups into the side chain of the copolymer of styrene and acrylonitrile, [256] suggests using the reaction of comparatively unreactive nitrile groups with 9-methylanthryl lithium and α-methylnaphthyl lithium. The mechanism of this reaction and the spectral properties of the resulting polymers have been investigated.

Polymers with superior adhesion properties, containing 2.7 anthracene molecules per macromolecule, were prepared by reacting the copolymer of styrene and maleic anhydride with 9-hydroxymethylanthracene [257]. The rotational correlation times of toluene solutions of the resulting systems were measured by fluorescence depolarization.

Interesting fluorescent polymers whose synthesis can be regarded with some degree of approximation as a polymer-analogous conversion are covered in [258]. These compounds are chelate complexes of europium with macromolecules containing as ligands β-diketone units, namely poly(p-benzylstyrene) LXXVI and poly(aryl-β-diketone) LXXVII.

$$\left[-CH-CH_2-\right]_m - \left[-CH-CH_2-\right]_n$$

LXXVI $n/m=5$

LXXVII

Polymer LXXVI was synthesized by reacting partially acetylated styrene with methyl benzoate in the presence of sodium amide in tetrahydrofuran for 90 h at 50°C. The complexes with Eu^{3+} were prepared by mixing 1–2% solutions of the polymers in tetrahydrofuran with a europium trichloride solution in a tetrahydrofuran–methanol mixture.

Polymers LXXVI and LXXVII and their chelate complexes are heat resistant and start decomposing at 350°C After excitation, Eu^{3+} (ligated to the polymer complex) exhibits radiative deactivation, at a frequency typical of inorganic crystalline systems. The fluorescence spectra of the complexes show two bands: one with λ_{max} 613 nm caused by the $^5D_0 \rightarrow {}^7F_2$ transition, and the other (weak) band λ_{max} 580 nm associated with the forbidden transition $^5D_0 \rightarrow {}^7F_0$. The fluorescence intensity of complexes based on polymer LXXVI increases linearly with increasing Eu^{3+} content up to 3%, and then remains constant. The emission intensity of these complexes is 20–30 times higher than that of complexes with polymer LXXVII (Eu^{3+} content 1%). This is explained [258] by the fact that on introducing the β-diketone group into the polymer chain, the complexation between the Eu ions and the β-diketone may be impeded by steric hindrance and reduction of the degree of rotational freedom.

A series of poly(methyl methacrylates) with pendant crown ether groups attached via the ester group, i.e. $-[CH_2\text{-}CMe(CO_2CH_2\text{-}crown)\text{-}]_n$ have been prepared with crown = 18-crown-6 and 15-crown-5. These ligate Eu^{2+} ions in the crown cavity and display strong visible luminescence from the Eu^{2+} centres on UV excitation [258A].

Fluorescent polymer complexes, similar to those described above, have been prepared [259] by treating poly(6-vinyl-2,2'-bipyridine) with $Ru(bpy)_3Cl_2 \cdot 2H_2O$ in a boiling mixture of 1-propanol and dioxane.

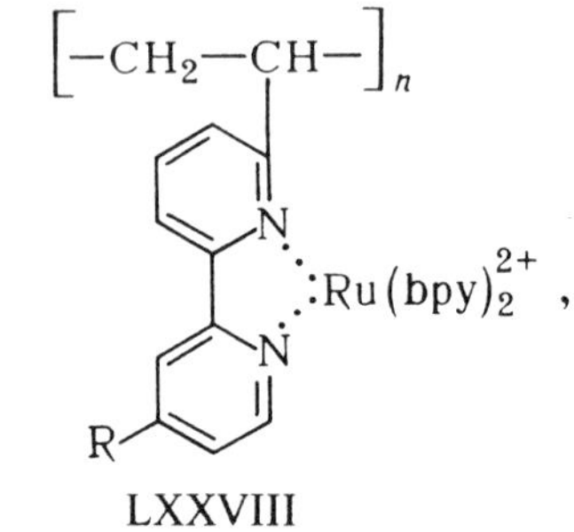

where $R = H, CH_3$

LXXVIII

It has been established that the luminescence spectra of aqueous solutions of polymer LXXVIII have a band with λ_{max} 617–618 nm, whereas the initial ruthenium complex luminesces with a maximum at 610 nm. The luminescence quantum yields and luminescence lifetimes of polymers LXXVIII have also been measured (quantum yield up to 0.062). The temperature dependences of the luminescence lifetimes of ethanol solutions of the initial complex and of the polymers obtained have been studied.

Copolymers LXXIX described in [260] are interesting as regards the processes of electronic energy transfer from the organic luminophore to the rare-earth ion.

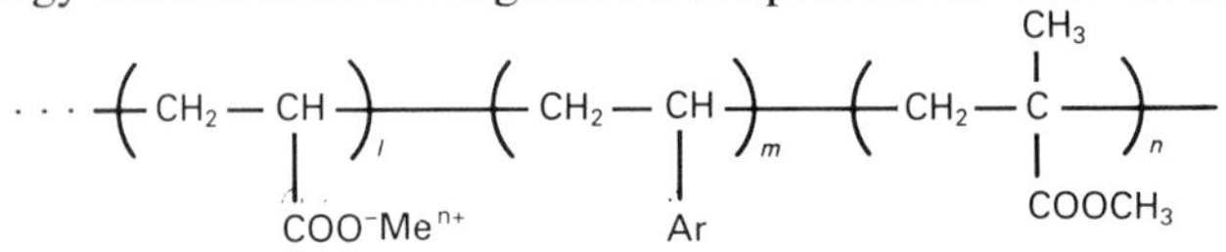

LXXIX

Me = Sm, Eu ; Ar =

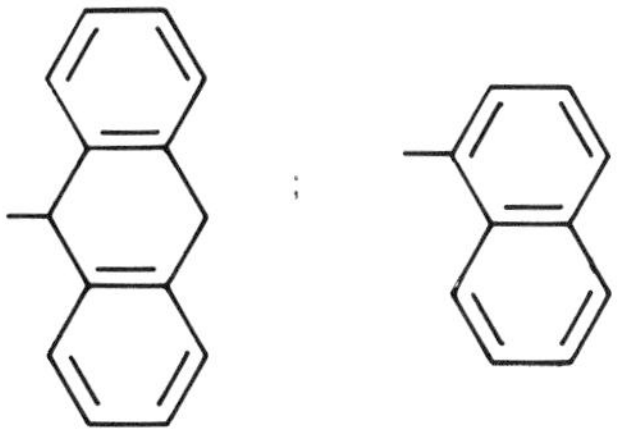

These were synthesized by the reactions between naphthalene- or anthracene-containing copolymers of methyl methacrylate and acrylic acid with the acetates of rare earth metals such as europium and samarium.

The attachment of polypyridyl complexes of Ru(II) and Os(II) to polystyrene has been achieved by reacting poly(styrene-co-4-chloromethylstyrene) with, for example, $[Ru(bipy)_2(bipyCH_2OH)]^{2+}$ ions to give a pendant luminophor bonded via an ether linkage [260A]. It has also proved possible to attach both Ru(II) and Os(II) centres onto a single chain and to induce electron transfer and energy transfer along the chain [260A].

5

Natural polymers

Many natural polymers (natural rubber, cellulose, etc.) have some capacity to luminesce under UV radiation [260]. In most cases, however, such fluorescence does not emanate from the macromolecules of these polymeric systems but is attributed to impurities. Thus it is expedient to examine only one group of natural polymers namely the proteins, which unambiguously exhibit their own emission.

5.1. PROTEINS WITH NATURAL FLUORESCENCE

The near-UV fluorescence of proteins was discovered in 1956 by Shore & Pardee [261] who showed that the most efficient fluorescence was excited in the absorption region of the aromatic amino acid residues such as tryptophan, tyrosine, and phenylalanine.

   The fluorescence excitation spectra of proteins and aromatic amino acids have the same maxima as the respective absorption spectra. The measurements have revealed rather a high level of fluorescence quantum efficiency in phenylalanine (λ_{max} 282 nm), tryptophan (λ_{max} 348–353 nm), and tyrosine (λ_{max} 303 nm), being 0.04, 0.20 and 0.21 respectively.

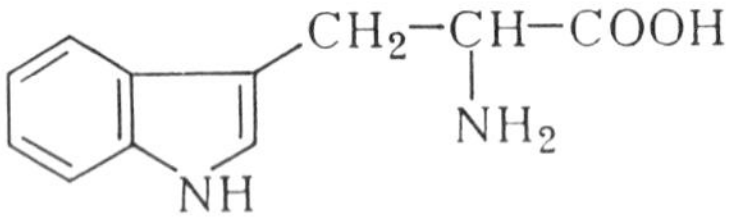

I

tryptophan

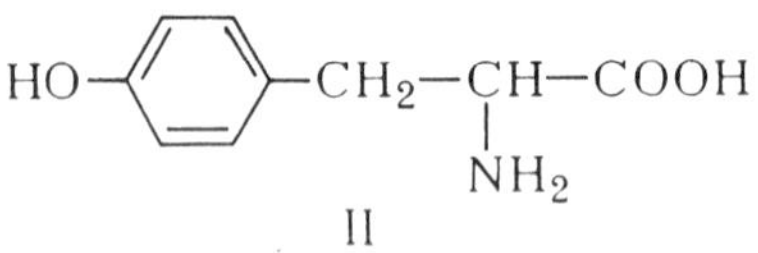

II

tyrosine

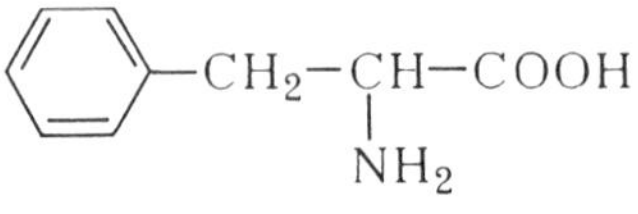

III

phenylalanine

Study of the luminescent properties of proteins offers valuable information about the structure of biopolymers, and the photochemical processes and structural and physico-chemical conversions occurring in them. The measurements of different fluorescence parameters permit determination of the exact localization of aromatic groups in protein globules and the protein content of biological preparations.

All these points have been covered in sufficient detail in two monographs by Burstein [262, 263]. In our treatment we are solely concerned with the absorption and fluorescent features of some proteins.

The first quantitative measurements of the absorption and fluorescence spectra of proteins were taken in 1957 and 1958 [278]. The data on the absorption spectra of the separate chromophoric groups permit assessment of their relative contribution to the total absorption of proteins (Figs 5.1, 5.2).

It should be noted that the natural absorption of proteins as neutral solutions is in the region up to 310–315 nm. Some authors have examined the amino acid content of a hypothetical 'average globular protein' and the contribution of the separate amino acid residues and peptide groups to the total protein absorption [262, 264]. Thus, monograph [264] presents the following values of molar extinction coefficients for different wavelengths in protein absorption spectra: 205 nm, $\varepsilon = 3.84 \times 10^6$; 220 nm, $\varepsilon = 9.5 \times 10^5$; 250 nm, $\varepsilon = 3.39 \times 10^4$; 230 nm, $\varepsilon = 3.79 \times 10^4$.

In real proteins the contributions of the separate chromophores to the total absorption may differ markedly from the average statistical values, and at specified wavelengths will deviate significantly from the above values. We point out that in their chromophore composition, human fibrinogen and gamma globulin are closest to the 'average' protein.

As shown in earlier studies [278], the fluorescence spectra of proteins containing tyrosine (and not containing tryptophan) are very similar to the spectrum of tyrosine. In the

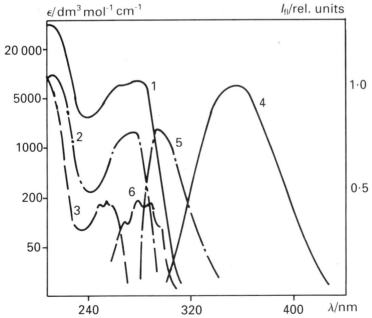

Fig. 5.1. Absorption (1, 2, 3) and fluorescence (4, 5, 6) spectra of aqueous solutions of tryptophan (1, 4), tyrosine (2, 5) and phenylalanine (3, 6).

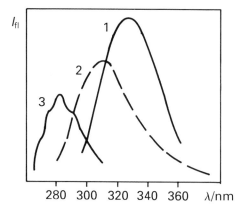

Fig. 5.2. Fluorescence spectra: (1) aqueous ribonuclease, pH = 6 (tryptophan fluorescence); (2) aqueous-ethanolic albumin (tyrosine fluorescence); (3) aqueous parvalbumin (phenylalanine fluorescence).

spectra of proteins containing tryptophan, only one long wavelength fluorescence maximum typical of tryptophan was observed. For aqueous solutions of proteins, the fluorescence quantum efficiency varied from 0.017 (ribonuclease) to 0.15 (bovine serum albumin) [265]. The author of this study failed to find the fluorescence of any other protein molecules, other than tyrosine or tryptophan, in protein luminescence. It was not until early in the 1970s that the comparatively rare proteins such as parvalbumins containing neither tyrosine nor tryptophan, and displaying phenylalanine emission in their fluorescence spectra, were obtained from the muscles of fish [266]. The fluorescence quantum yield of proteins was found to be unexpectedly high: for carp parvalbumin it was 0.265 which is 7 times as much as that of free phenylalanine. In the opinion of the author of [262], such a sharp increase is associated with the exclusion of strong quenching by water which is observed in solutions of low molecular compounds. It is also confirmed by the following fact: on replacement of only one of 10 phenylalanine residues in carp parvalbumin by tyrosine, the fluorescence of phenylalanine cannot be measured, owing to the effective energy transfer.

Apart from the nature of the amino acid residues, the positions of the maxima in the fluorescence spectra of proteins depend on many factors: the pH of the medium, the specific environment of the luminophoric moiety in the macromolecule, the wavelength of the exciting light, the spatial configuration of the protein macromolecule, etc. Table 5.1 illustrates the positions of the most intense maxima in the fluorescence spectra of separate representatives of each of the three types of protein.

Their low fluorescence quantum efficiency, the excitation wavelength dependence of the spectrum, and the location in the UV region (demanding the use of high power quartz monochromators and UV sensitive photomultipliers), considerably complicate, and restrict the possibilities of the study of proteins by their natural fluorescence. In a number of cases, more valuable information about the structure of protein molecules may be obtained by measurements of low temperature luminescence spectra. Low temperatures (77 K) markedly decrease the influence of dynamic factors such as the universal orientational interactions, and the dynamic quenching on collision with neighbouring groups which, under normal conditions, perturbs the properties of the chromophore during its lifetime in the excited state.

Conversely, a number of factors which are negligible at room temperature profoundly affect the luminescent properties on freezing. In particular, the influence of complexes formed by hydrogen bonding in the ground state is generally intensified, the effect of the physical anisotropy of the medium corresponding to the most advantageous configuration of groups in the local environment is conspicuous, etc.

Table 5.1. Position of fluorescence maxima in emission spectra of proteins of different structure

Protein	Luminophoric group	λ_{max}/nm
Chymotrypsinogen A (pH = 5.9)	Tryptophan	332
Ribonuclease C_2 (pH = 6.3)		327
α-Chymotrypsin (pH = 5.4)		336
Papain (pH = 4.5)		340
Papain (pH = 8.0)		347
Papain (pH = 9.5)		350
Myosine (rabbit)		338
Zein	Tyrosine	310
L-Asparaginase (pH = 4.0)		323
Carp parvalbumin (pH = 6.4) (component 2)		304
Parvalbumin	Phenylalanine	284
Carp parvalbumin (components 3 and 5a)		282

All these factors lead to a poor correlation between the low temperature luminescence parameters and those measured at room temperature. Thus, the fluorescence spectra of tyrosine residues undergo a short wavelength shift by 6–8 nm and have λ_{max} 296–298 nm at 77 K [267]. These findings point to a considerable effect of the universal orientational perturbations on the Stokes shift of the fluorescence of tyrosine-containing proteins [262]. For proteins of this type, the existence of two kinds of band in the fluorescence spectra of frozen solutions is typical, depending on whether or not they are in contact with the disulfide residues of amino acids [268]. For instance, a protein such as insulin, in which tyrosine residues are in contact with disulfide groups, has a fluorescence spectrum with λ_{max} 410–420 nm (compare with the spectrum of free tyrosine with λ_{max} 387 nm). Proteins of the second type have a fluorescence spectrum which does not differ significantly from that of tyrosine.

In the low temperature luminescence spectra of proteins containing tryptophan residues, along with tyrosine residues, the dominant feature is the emission of tyrosine whose fluorescence spectrum shows λ_{max} 413 nm. The authors of [268] explain this by the effecient energy transfer from the tryptophan residues to tyrosine in the singlet state, and, in the opposite direction, from tyrosinates to tryptophan residues in the triplet state.

The low temperature fluorescence spectra of proteins containing only tryptophan moieties undergo a hypsochromic shift unlike the room temperature spectra. The fluorescence bands of the majority of proteins show their maxima at 320–332 nm, being characterized by the absence of a fine structure which reflects the interaction between the chromophore and the polar groups in a frozen protein. The fluorescence spectra of tryptophan-containing proteins exhibit two types of band: a short wavelength band with λ_{max} 407–418 nm and a long wavelength band with λ_{max} 434–447 nm. In contrast to model compounds, such proteins reveal no correlation between the positions of the phosphorescence and absorption spectra, which, apparently, can be attributed to the action of dispersion forces. This feature indicates that the nature of the phosphorescence spectrum provides information about the qualitative properties of the protein environment other than given by the fluorescence. According to [267], the red shift of the phosphorescence spectrum of proteins, as compared with that of tryptophan, can be attributed to anisotropic interactions between chromophore moieties and neighbouring polar groups. The direction of these interactions, for example in hydrogen bonds in which the nitrogen atom of the indole ring acts as a proton acceptor, is close to the direction of vector of the transition moment of the phosphorescence, that is, normal to the ring plane.

Ref. [269] reviewed the fluorescence characteristics of a wide range of biopolymers comparatively recently.

It is known [262] that the natural fluorescence and phosphorescence of proteins do not necessarily provide an unambiguous picture of the structure of a protein molecule. There are a number of situations in which the most convenient method for studying a protein is its pretreatment with luminophor-containing reactive groups to form a covalent bond with the macromolecule.

5.2 PROTEIN LABELLING WITH FLUORESCENT DYES

The presence of functional groups (—SH, OH, —NH$_2$, etc.) in protein molecules permits bonding of the dyes to a protein substrate. To increase the sensitivity of protein dyeing methods, Critch & Johns [270] suggested the use of such a fluorescent dye with an active group such as 1,2-benzanthryl isocyanate.

Later [279], binding of proteins with luminescent dyes came into wide use in immunological research, the measurement of protein RMMs, studies aimed at obtaining data on the asymmetry of protein macromolecules, etc.

Dyes bound to proteins are expected to meet a number of requirements: to have high fluorescence quantum yield, sufficiently reactive functional groups (generally, —SO$_2$Cl, —NCO, —NCS, etc.), and minimum absorption in the maximum region of the excitation spectrum of the protein (270–290 nm). The reactive groups of dyes are fixed to the protein molecules mainly via the amine group, or, to be exact, via the ε-amine groups of lysine residues. In addition, the isocyanate and thioisocyanate groups can be reacted with thiol residues:

$$\text{Protein}\begin{array}{l}\diagup \text{NH}_2 \\ \diagdown \text{SH}\end{array} + \text{O=C=N—R} \longrightarrow \text{Protein}\begin{array}{l}\diagup \text{NH—CO—NH—R} \\ \diagdown \underset{\overset{\|}{O}}{\text{S—C—NH—R}}\end{array}$$

IV

Table 5.2. Active fluorescent dyes for protein staining

Dye	Formula	Conjugate emission colour	Source
1	2	3	4
1-Dimethylaminonaphtha-lene-5-sulfonyl chloride	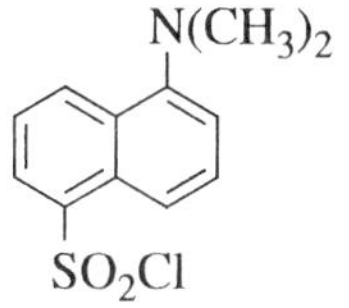 (structure: naphthalene with $N(CH_3)_2$ and SO_2Cl)	Yellow-green	[273]
9-Cyanatoanthracene	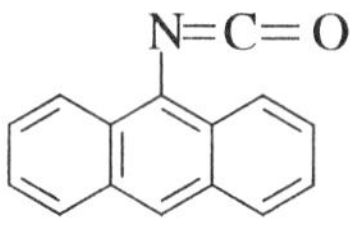 (structure: anthracene with $N{=}C{=}O$)	Blue	[271]
Sulfonyl chloride of rhodamine B	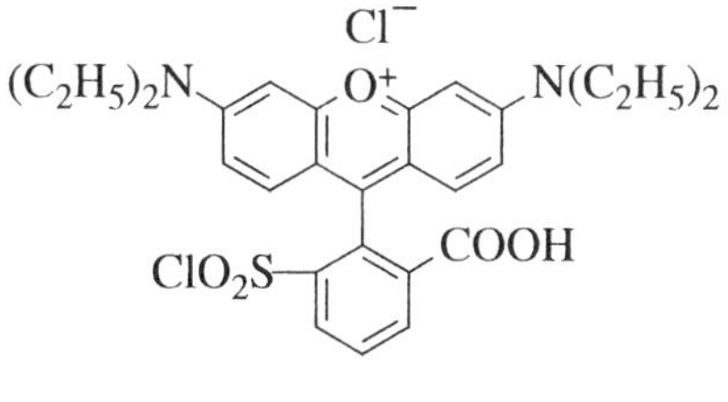 (structure: rhodamine B, $(C_2H_5)_2N$, Cl^-, O^+, $N(C_2H_5)_2$, ClO_2S, $COOH$)	Red	[271]
Fluorescein isocyanate	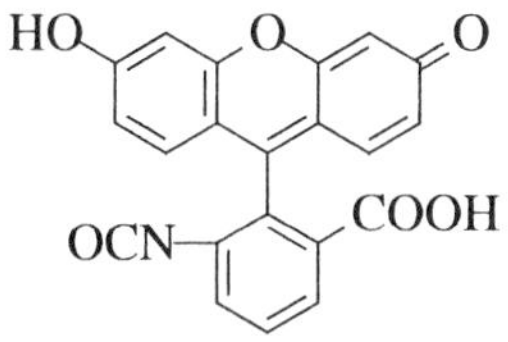 (structure: fluorescein, HO, O, O, OCN, $COOH$)	Green	[272]
3-Hydroxypyrene-5,8,10-trisulfonyl chloride	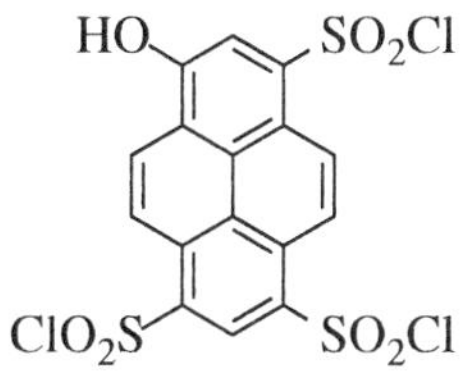 (structure: pyrene with HO, SO_2Cl, ClO_2S, SO_2Cl)	Green	[273]

Table 5.2. *Continued*

Dye	Formula	Conjugate emission colour	Source
1	2	3	4
Isothiocyanate of rhodamine B		Orange-red	[273]
Fluorescein isothiocyanate		Green	[272]
Dichlorotrianizinyl-aminofluorescein		Green	[274]
Fluorescamin		Bluish-green	[275]

with diethylamine ethylacrylate as the fluorescence quenching agent. In these experiments the authors worked on the assumption that the fluorescence quenching of an excited molecule is determined by the reciprocal diffusion rate of the luminophor–quencher pair; assessment of the rate constants of interaction between the luminophor and quencher molecules gives quantitative information about diffusion processes at different degrees of conversion.

A luminescent organometallic probe, *fac*-ClRe(CO)$_3$(Ph$_2$phen), has been used to monitor the progress of curing of epoxy resins [288A].

Assessment of the rate constants of fluorescence quenching by the Stern–Volmer equation and their comparison with the bimolecular rate constants of diffusion controlled reactions led to the following deduction [287]. In the polymer–monomer system studied, at concentrations of polymer up to 60%, its presence does not noticeably hamper the diffusion of low molecular particles either to each other or to macromolecules. Then, over a quite narrow concentration range, hindrance to diffusion emerges, leading to autoretardation of the radical polymerization.

A study of pyrene fluorescence quenching by triethylamine during methyl methacrylate polymerization in the presence of 25% methanol [288] established that in the solid polymer formed, which was transparent, there are microregions where the luminophor–quencher pair can interact as in pure methanol.

In other studies [285, 286], polymerization was examined via the relation between the growth of fluorescence intensity and reaction time. For instance, in [285] in the case of polymerization of methyl methacrylate, certain fluorescent probes, [(*N,N*-dialkylamino) benzylidene malononitrile] derivatives, were added to the monomer:

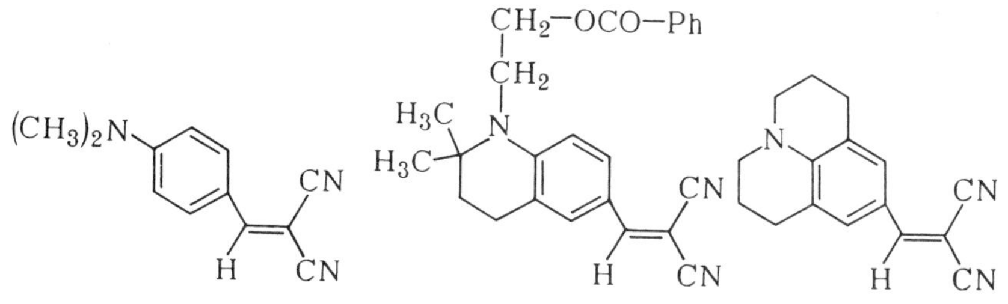

$$\lambda^{\,fl}_{max} = 472\ nm \qquad \lambda^{\,fl}_{max} = 490\ nm \qquad \lambda^{\,fl}_{max} = 493\ nm$$

Polymerization was carried out at 50–90°C in the bulk monomer with addition of 10^{-5} mol dm^{-3} of luminophor in the presence of azobisisobutylonitrile as initiator. It was shown that the fluorescence intensity of the samples steadily increased (approx. 1.5 times) with up to 60% conversion (Fig. 6.1).

Further increase of the degree of conversion caused a sharp (20 to 40-fold) increment in the fluorescence, which reached its limiting value at maximum conversion. This discontinuity of the fluorescence intensity occurs when the polymerizing system reaches the glassy state. In this region, the mobility of the macromolecules is severely limited, and internal molecular relaxation in the system is controlled by the microscopic free volume of the

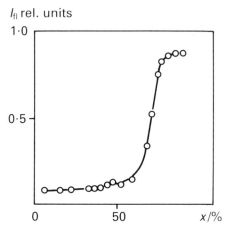

I_{fl} rel. units

Fig. 6.1. Probe II fluorescence intensity vs degree of conversion of methyl methacrylate at 70°C.

polymer, leading to a decreased contribution of radiationless processes and an increased emission intensity. The relationship between the limiting fluorescence efficiency and the limiting conversion, which reflects the influence of free volume on the molecular relaxation of the polymer, was determined.

Application of these methods of fluorescence study of polymerization processes often implies usage of polarization measurements. It is known [91] p. 501, that the degree of polarization is determined by the relation

$$P = \left(I_{\parallel} - I_{\perp}\right)/\left(I_{\parallel} + I_{\perp}\right) \tag{6.1}$$

where $I_{\parallel}$ and $I_{\perp}$ are the emission intensities with the electromagnetic vectors parallel and perpendicular to the excitation light vector respectively.

As shown in [91] p. 511, for a completely random distribution of fluorescent molecules, the degree of polarization differs from zero, depending on the relative orientation of the absorbing and emitting oscillators, thus the limiting value of P changes from $+\frac{1}{2}$ (parallel absorbing and emitting oscillators) to $-\frac{1}{3}$ (perpendicular oscillators).

Three characteristics are used to describe the polarization of emission spectra: the polarization spectra (the degree of polarization P, vs the excitation wavelength), polarization diagrams (the degree of polarization P, vs the direction of observation and the orientation of the electric vector of the excitation light), and limiting values of the degree of polarization determined in the case of linear oscillators by the formula

$$P_{lim} = \left(3\cos^2\alpha - 1\right)/\left(\cos^2\alpha + 3\right) \tag{6.2}$$

where α is the angle between the absorbing and emitting oscillators.

In [291] the radical copolymerization of styrene with divinylbenzene was studied in the presence of 9-vinylanthracene as a fluorescent label. Copolymerization was carried out in benzene solution at 60°C in the presence of benzoyl peroxide. The process was monitored by fluorescence polarization. On gel formation, in systems containing different amounts of divinylbenzene, the degree of polarization increases owing to chain branching. With increasing conversion, the length of the elastic sections of the network decreases, and the highest density of segments is observed at the highest degrees of conversion. It was established that the most uniform networks have a given degree of polarization for a given degree of swelling.

The observation of an increasing fluorescence intensity with λ_{max} = 345 nm during the course of reaction was used in [292] for monitoring the rate of radical copolymerization of diethylene glycol bisallylcarbonate with 0.1–3.0% (mass) 2-hydroxynaphthalene methacryl ester. According to the fluorescence data obtained, the copolymerization rate decreases by 14–16% with increase in the concentration of the naphthalene comonomer by an order of magnitude, which agrees with the data obtained by dilatometry.

Fluorescence techniques are applied to determining the conversions of polymers, by monitoring their chemical reactions [198, 245, 293–301], photochemical conversion [302–308], and thermo-oxidative ageing [309–314]. For instance, in [293] fluorescence polarization was used to study the interaction of poly(ethylene glycol) (PEG) with poly(acrylic acid) (PAC) and poly(methacrylic acid) (PMAC) containing, in their side chains, one luminophoric anthracene label per 1000 monomer units. This permitted direct monitoring of the course of reaction in solution without formation of solid products. It was also shown that formation of polyacid complexes with PEG is accompanied by a considerable decrease in the intramolecular mobility of the macromolecular components.

Complex formation in the PAC–PEG system in which PEG was characterized by the presence of end pyrene groups was studied by the fluorescence technique (by examining the intensity ratio of excimer and monomer fluorescence) in [294].

The possibility of using the UV and fluorescence spectroscopic methods was studied in the evaluation of the extent of the polycyclodehydration reaction of acridine- and stilbene-containing polyamide acids into polyimides [222]. It was found that to calculate the degree of polycyclodehydration it was necessary to determine the optical densities of films at 262 and 285 nm. The data obtained agreed well with the results of IR analysis. The fluorescence method of determining the degree of polycyclodehydration entails measurements of fluorescence intensity in films at λ_{max} 530 and 435 nm. This method is highly accurate when the degree of polycyclodehydration reaches 50%. At higher degrees of polycyclodehydration, the fluorescence technique is difficult to apply because of the nonlinear effects of suppression of fluorescence caused by the intersystem crossing in the pyromellitimide moieties of the polymer formed.

The enhancement of fluorescence during the polycyclodehydration of weakly luminescent polyphenolic amides into highly fluorescent polybenzoxazoles [220] can apparently be used to monitor this reaction.

Interesting results of a kinetic study of the preparation of crosslinked polystyrene from 9,10-bis(chloromethyl)anthracene and linear polystyrene by UV spectroscopy and polarized luminescence are cited in [295]. The reaction is conducted in an inert solvent medium of dichlorobenzene in the presence of $SnCl_4$ catalyst, and it leads to formation of singly (IV) and doubly meshed (V) anthracene groups.

The total content of structures IV and V was determined by spectrophotometry, and the contribution of each structure, differing in mobility, by measuring the luminescence polarization and lifetime of the excited states of the respective moieties. The following relation was used to compute the concentrations of structures IV and V:

$$\frac{1}{\tau} = \frac{\alpha_{IV}}{\tau_{IV}} + \frac{1 - \alpha_{IV}}{\tau_V} \tag{6.3}$$

where α_{IV} and $\alpha_V = (1 - \alpha_{IV})$ are the fractions of structures IV and V, τ_{IV} and τ_V are the lifetimes reflecting the mobility of chain sections without (τ_{IV}), or with (τ_V), bridge bonds. According to earlier data by the same authors, $\tau_{IV} = 5.4$ ns and $\tau_V = 16$ ns.

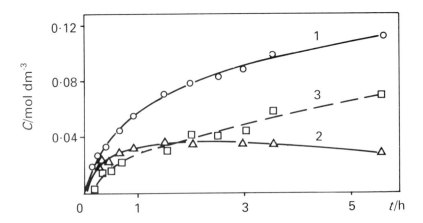

The results of the kinetic studies are shown in Fig. 6.2. It is clear that the content of pendant anthracene groups increases rapidly at first (for 1.0–1.5 h), and then slightly decreases. This can be explained on the basis that, upon reaching a certain level of structures IV in the crosslinking polymer, a reaction begins in which the pendant chloromethyl groups are destroyed and the bridged structures formed. The curve describing this

Fig. 6.2. Content of anthracene fragments in polymer vs reaction time of polystyrene (PS) with 9,10-bis(chloromethyl)anthracene (BCMA): 1, total content of anthracene groups (BCMA fragments) in polymer; 2, content of single-meshed BCMA fragment, structure IV; 3, total content of cross-link bridge structure V. Reaction conditions: [PS] = 5 gpp; [BCMA] = 0.16 mol per 100 PS units; solvent – o-dichlorobenzene; [SnCl₄]; [BCMA] = 3:1; 60°C.

process has the S-shape typical of sequential reactions. The consumption of IV-groups is compensated, at first completely and then partly, by the first step of the reaction of styrene with 9,10-bis(chloromethyl)anthracene.

The fluorescence and spectrophotometric monitoring of the reaction of poly(vinyl alcohol) with 9-anthraldehyde is described in [245].

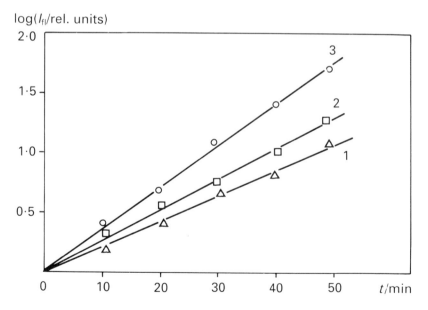

It was found that 9-anthraldehyde does not fluoresce in solution in polar solvents, while polymer VI has a distinct fluorescence with a strongly marked anthracene-type vibronic structure. Thus, observation of fluorescence enhancement provides a suitable method for the kinetic study of polymer-analogous conversions. In particular, a linear relationship was established between the logarithm of the fluorescence intensity of reaction solutions in N,N-dimethylacetamide and the reaction time over the range from 100–140°C (Fig. 6.3).

Fig. 6.3. Logarithm of fluorescence intensity of poly(vinyl alcohol) (PVA) and 9-anthraldehyde (ANTA) reaction solutions in dimethylacetamide (100-fold prediluted) vs reaction time at 100°(1), 120° (2), and 140°C (3). Reaction conditions: [PVA] = 0.6 base-mol dm^{-3}, [ANTA] = 0.03 mol dm^{-3}; solvent, dimethylacetamide + 5 mass% + p-toluenesulfonic acid.

The spectrophotometric study of this reaction (by measuring the decrease in the optical density of reaction solutions at $\lambda = 420$ nm) showed that in the presence of a considerable excess of the polymer component, it closely follows second order kinetics. The values of the rate constants computed from the initial sections of the kinetic curves at 100, 120, and 140°C are 0.97×10^{-3}, 1.59×10^{-3} and 2.56×10^{-3} dm^3 mol^{-1} s^{-1}) respectively.

Fluorescent methods open additional possibilities for monitoring the hardening kinetics of epoxy resins [198, 297–301]. The luminophoric additive introduced in the epoxy oligomer before hardening can be chemically inert and not included in the composition of the polymer formed [297, 298]. For monitoring the hardening rate of the resin, the suggestion was made to use, in this case, the incremental intensity of the excimer fluorescence of dye additives, as determined by the microviscosity, which is dependent on the mobility of the polymer chains. The application of luminescent labels entering a crosslinked polymer during hardening suggests the presence in their structure of glycidyl groups [198] or sufficiently reactive amine groups [299–301].

The results of kinetic studies were taken into account in the hardening of mixtures of oligomer ED-20 (i.e. an oligomer of the diglycidyl ether of bisphenol-A) with 1.0–10.0% (mass) of 9-hydroxyanthracene glycidyl ester using stoichiometric amounts of diethylenetriamine [198]. It was shown that the (second order) rate constants determined fluorimetrically, increase significantly with the content of anthracene monoglycidyl derivative in the initial mixture. This effect is related to fewer diffusion hindrances occurring during the decrease in crosslink density of the network polymer formed. The kinetic results were found to agree with those obtained calorimetrically.

Interesting results were obtained in the process of hardening the diglycidyl ether of bisphenol A or butanediol by means of 4,4'-diaminodiphenyl sulfone in the presence of small amounts of 4,4'-diaminoazobenzene which gives rise to the fluorescence of oxirane polymers with λ_{max} 560 nm [299, 300]. On reaching the gel point, the fluorescence quantum yield of the polymer rises sharply. The formation of meshed structures can be followed by the rate of rise in the quantum yield. The observation of an epoxy composition including the diglycidyl ether of butanediol and 4,4'-diaminodiphenylmethane during hardening in the presence of 0.1% (mass) of 4,4'-diaminostilbene used as a label, shows a red shift of the fluorescence band [301]. The amount of the shift and the increment in the fluorescence intensity with λ_{max} 418 nm with increasing time of crosslinking permit control of the formation of three-dimensional structures especially at the final stages of reaction.

Luminescent monitoring of the photochemical conversions of polymers was employed, for instance, during the photosensitized crosslinking of polyethylene [302], the photo-oxidation of poly(N-vinylcarbazole) [303, 304], polyalkenes [305], aliphatic polyamides [306, 307], and melaminoformaldehyde resins [308].

Ref. [302] shows the possibility of employing fluorescence for monitoring the photochemical crosslinking reaction of xanthone sensitized polyethylene. It was found that irradiation of the polymer with UV increases the luminescence spectral intensity with maxima at 415–510 nm, testifying to the photoreduction of xanthone to xanthydrol. The process of crosslinking of polyethylene is determined mainly by the reactions of macroradicals and products forming in the photochemical reaction.

The particular features of the photo-oxidation of poly(N-vinylcarbazole) were studied in [303, 304]. It was established that the combined action of oxygen and UV irradiation (290–320 nm) on solutions of this polymer in dioxane leads to quenching of both the monomer fluorescence with λ_{max} 370 nm, and the excimer fluorescence with λ_{max} 415 nm. The authors of [303] have shown that excimer fluorescence intensity decreases owing to the 'removal' of monomer species in the photochemical process, as well as to singlet exciton migration, followed by trapping by photo-oxidized polymer units.

In the process of photoageing, a new fluorescence peak, λ_{max} 480–490 nm, appears in the polymer solution, being related apparently to the products of photo-oxidation.

Phosphorescence monitoring of photochemical oxidation of polyethylene and polypropylene is discussed in [305]. Under UV irradiation (Xenotest-150) of polymer films, their fluorescence spectra blue shifted (with λ_{max} 465–445 nm for polypropylene), while the phosphorescence excitation spectra red shifted (with λ_{max} 270–290 nm) with a decreasing phosphorescence lifetime. It has been pointed out [305] that the reason for the occurrence of phosphorescence in industrial polyalkene samples is the presence of carbonyl impurity groups which initiate photodegradation.

The authors of [306, 307] studied the photodegradation of nylon-6,6 films both untreated and preoxidized in air for 1 h at 130°C. In the latter case [306], irradiation caused an increase in the mean lifetime and a decrease in the phosphorescence intensity (excited at 280–290 nm) as well as a decreased absorption in the 250–400 nm range. It was shown [305] that the yellow colouring of the polymer on prolonged irradiation is not related to the formation of phosphorescent residues. If the polymer is not prethermo-oxidized, then on photo-oxidation the phosphorescence lifetime decreases, and its spectrum shifts to shorter wavelengths (λ_{max} 475–420 nm) [307]. The authors attribute this change in the character of the phosphorescence to photodegradative processes of the type

$$\cdots\text{—CONH—CH}_2\text{—CO—CH}_2\text{—}\cdots \xrightarrow{h\nu} \cdots\text{—CONH—CH}_2\text{—}\overset{\bullet}{\text{C}}\text{O} + \overset{\bullet}{\text{C}}\text{H}_2\text{—}\cdots$$

$$\text{VII}$$

Luminescence techniques are quite an appropriate tool for monitoring thermo-oxidative ageing of polyalkenes [310] and aliphatic polyamides [284, 307, 309, 312–314].

The authors of [310] studied the phosphorescence spectra of polypropylene powders oxidized in air at 120°C. Under irradiation at λ_{max} 300 nm, the phosphorescence intensity increased at λ_{max} 465–520 nm and the phosphorescence lifetime decreased with growing levels of oxidation. The authors of [310] explain this by the formation, during oxidation, of carbonyl and peroxide groups whose accumulation promotes development of thermodegradative processes.

In [284, 309], as well as in [307, 311–313], the effect was studied of thermo-oxidation on the luminescent properties of aliphatic polyamides such as polycaproamide, nylon-6,6, and nylons -11 and -12. According to [307], the phosphorescence spectrum of an industrial sample of nylon-6,6 has two maxima at 420 and 465 nm with lifetimes of 2.06 and 0.29 s respectively. The position of the phosphorescence maxima, determined by the presence of carbonyl groups in the polymer, depends on the excitation wavelength. Resulting from thermo-oxidation, the number of ketone groups decreases and the concentration of aldehyde groups increases, owing to the following reactions:

$$\cdots-\overset{\overset{\displaystyle O}{\|}}{C}-NH-CH_2-\overset{\overset{\displaystyle O}{\|}}{C}-CH_2-\cdots$$

$$+$$

$$\cdots-CH_2-N\overset{\displaystyle H}{\underset{\displaystyle H}{<}}$$

$$\xrightarrow[-H_2O]{}$$

$$\cdots-\overset{\overset{\displaystyle O}{\|}}{C}-NH-CH_2-\overset{\overset{\displaystyle N-CH_2-\cdots}{}}{C}-CH_2-\cdots$$

VIII

$$\cdots-R-CO-NH-CH_2-R-\cdots$$

$$\xrightarrow[-R'-C\overset{O}{\underset{H}{\diagdown}}]{}\quad \cdots-R-CO-NH-C\overset{O}{\underset{H}{\diagdown}}$$

IX

$$\xrightarrow[-R''-C\overset{O}{\underset{H}{\diagdown}}]{}\quad \cdots-R-CO-NH_2$$

X

As a result, the phosphorescence intensity with λ_{max} 420 nm decreases during thermo-oxidation, and the rate of this process increases with the temperature of the thermal treatment [307].

The nature of the phosphorescence of aliphatic polyamides produced on thermo-oxidation has not been completely defined [284, 311–314]. It was assumed, in particular in [284], that the fluorescence of polycaproamide with λ_{max} 452 nm refers to the formation of defective ketoimide structures in the polymer on heating in air. The fluorescence of nylon-6,6 with λ_{max} 415 nm was ascribed to the same type of structure [314].

It was shown in [313] that the fluorescence with λ_{max} 326 nm (for λ_{excit} 290 nm), or λ_{max} 390 and 420 nm (for λ_{excit} 340 nm) as well as the phosphorescence, with λ_{max} 407, 430, and 450 nm, of thermo-oxidized nylon-6,6 films are related to the aldol condensation of polymer moieties to cyclopentanone derivatives:

$$\cdots-R-NH-CO-(CH_2)_4-COOH \longrightarrow$$

$$\longrightarrow \cdots-R-NH-CO-HC\underset{H_2C-CH_2}{\overset{CO}{\diagup\diagdown}}CH_2 + H_2$$

XI

The fluorescence technique is so appropriate a tool for monitoring the thermo-oxidative ageing processes of polycaproamide that it has been suggested as a means of polymer control in industry [309]. As shown in [309], from the fluorescence intensity (λ_{max} 452 nm) of this polymer, one can assess the degree of thermo-oxidative degradation, that is, the reduction in the quality of the polymeric product (Fig. 6.4).

The universal character of fluorescence monitoring is indicated by the observation that the polymer emission spectrum does not change after addition of the thermostabilizer H-1 [309]; its addition to polycaproamide prohibits the use of UV spectroscopy which is normally used for quality assessment of the polymer.

Another application of the fluorescence technique is in the determination of RMM. Although the fluorescence technique has not yet been widely employed for this purpose, in some cases it is no worse than other techniques. The use of polymer luminescence for RMM determination was first described in [315]. It was noted [315] that a great variety of vinyl polymers of the polystyrene and polyacrylonitrile series, of poly(methacrylic

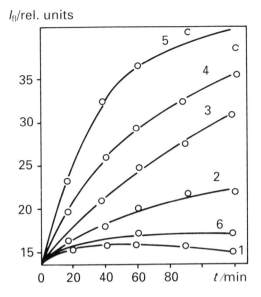

Fig. 6.4. Fluorescence intensity of polycaproamide (PCA) vs duration of thermo-oxidation in air at 100°
(1), 120° (2), 130° (3), 140° (4), and 150°C (5, 6): 1–5, non-thermostabilized PCA; 6, PCA with
addition of 0.2 mass % H-1.

acid) derivatives, and highly conjugated polymers of the polyphenylacetylene type
exhibit 'universal fluorescence'. The spectra of these polymers studied by the author
were similar and were characterized by the presence of two types of bands: a short wavelength group with a maximum in the 420–450 nm region, and a long wavelength group
with a maximum in the 540–600 nm region.

As an example, Fig. 6.5 shows the fluorescence spectra of poly(*p*-carbethoxyphenyl
methacrylamide) films of different RMMs.

These publications make the assumption that the observed fluorescence of the polymers refers not to monomeric species but to certain defect structures, such as end groups
[315]. The author of [315], not ascertaining the nature of the luminescence, made some
observations that enabled him to relate the fluorescence characteristics of polymers to
their RMMs. For such characteristics he chose (i) the short wavelength fluorescence peak
intensity, (ii) the position of the fluorescence maximum, and (iii) the intensity ratio of
short and long wavelength fluorescence peaks.

Fig. 6.6 shows the relation between these characteristics and the RMM of poly(*p*-car-
bethoxyphenyl methacrylamide).

Ref. [315] analyses the possibility of applying these relationships to RMM determination of specific polymers. Thus, for polyvinyls, methods (i) and (iii) are more efficient,
while for polyconjugated polymers, featuring a strong relation between fluorescence and
RMM, method (ii) is better. The author believes that the technique he has suggested has
a number of advantages over the other methods of RMM measurements, particularly, the
possibility of studying insoluble polymers. We think, however, that the undefined nature
of the observed fluorescence as well as a number of reservations and discrepancies in
[315] considerably reduce the value of the results obtained.

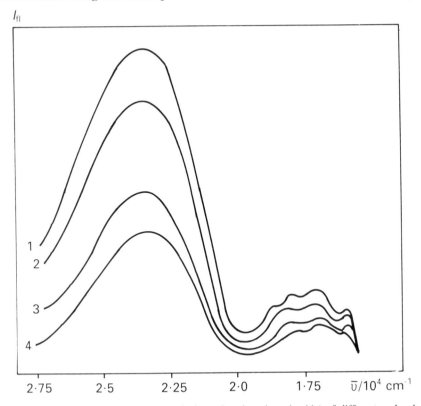

Fig. 6.5. Fluorescence spectra of poly(*p*-carbethoxyphenyl methacrylamide) of different molecular
masses: 1, 2.2×10^5; 2, 3.55×10^5; 3, 5.9×10^5; 4, 8.7×10^5.

A more promising application of fluorescence spectroscopy to RMM measurements
of polymers seems to be the method described by Baumbach in [316]. This technique
uses the tendency of vinyl polymers such as polystyrene, polyvinylpyrrolidinone, etc. to
quench the fluorescence of solutions of some xanthene dyes, such as rhodamine B.

The methology is as follows: to a dilute solution of rhodamine B in water or chloro-
form ($C = 5 \times 10^{-2}\text{--}4 \times 10^{-1}$ mg cm^{-3}) with fluorescence intensity F_0, an acidified
solution of the test polymer in the same solvent ($C = 1 \times 10^{-3}\text{--}3 \times 10^{-3}$ g cm^{-3}) is added.
Upon standing for 24 h, the rhodamine B fluorescence intensity decreases and equals 1.

The value of the number average molecular mass M_n of the polymer is computed by
eqn (6.4) [316]

$$F_0/F - 1 = \phi(M_n/M_0)^{1/2} C + \text{const},\qquad(6.4)$$

where C is the solution concentration of the polymer in g dm^{-3}, M_0 is the RMM of the
repeat unit and ϕ is the fluorescence constant, depending mainly on the type of solvent.
For chloroform $\phi = 3.32 \pm 0.04$ while for water $\phi = 1.15 \pm 0.02$.

In employing eqn 6.4, one should bear in mind that all measurements must be performed with one type of equipment. It was found [317] that in the measurement of the
RMM of polystyrene in chloroform, $\phi = 1.15 \pm 0.02$.

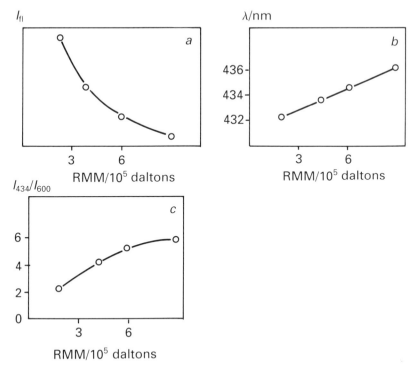

Fig. 6.6. Calibration ratios for definition of molecular masses of poly(p-carbethoxyphenyl methacryl-amide): a, by changes of fluorescence intensity with λ_{max} 434 nm; b, by shift of the fluorescence maximu-min the 432–436 nm region; c, by the ratio of fluorescence intensities with maxima at 434 and 600 nm.

Also worth noting are the studies of the intensity relationship between the excimer (or exciplex) and monomer emissions of poly(1-vinylnaphthalene) and poly(1-naphthyl methacrylate) [131], polyesters with side ω-(9-carbazoyl)butylene groups [181] and polystyrene [318]. It was shown [318], in particular, that ratios $I_E/I_M = f(M)$ for solvents such as cyclohexane, ethyl acetate, and tetrahydrofuran have a similar form in the RMM range from 500 000 to 612 000. It was found that $I_E/I_M = f(M)$ increases with RMM, reaching a plateau at $M \approx 10\ 000$, irrespective of the nature of the solvent. According to [318], the ratio $I_E/I_M = f(M)$ is independent of the thermodynamic properties of the solvent, and quantitative assessment of the RMM of polystyrene requires account to be taken of the motion of the phenyl group on the side chain of the macromolecule. As shown in [131], the intensity ratio of the excimer (λ_{max} 385 nm) and monomer (λ_{max} 335 nm) fluorescence, $I_E/I_M = f(M)$, for poly(1-naphthyl methacrylate) steadily increases from 0.61 to 0.73 with the polymer RMM increasing from 40 000 to 360 000.

The opposite picture, that is, enhancement of the excimer fluorescence intensity with decreasing RMM, has been noted in [319] for poly(ethylene glycol) containing pyrenyl groups. The authors attribute this to a higher probability of excimer formation due to a strong hydrophobic interaction between the molecules of pyrene located at the ends of the polymer chain.

A strong dependence of the formation of fluorescent excimers in solutions of proto-nated poly(2-vinylpyridine) on the RMM of the latter is noted in [320, 321].

Let us consider the abilities of fluorescence spectroscopy for studying the structural features and properties of polymers. It is known that the lifetime, τ, of a fluorescent molecule in the excited state is within 10^{-9}–10^{-7} s. Therefore any relaxation process occurring in the vicinity of this molecule can be studied within this period. Most fluorescence techniques for polymer studies suggest determination of the ratio of the molecular fluorescence lifetime and the relaxation time of the polymer system. The latter lies in the range from 10^{-9} s to seconds and minutes, for example the short range motions of polyalkene segments have values of τ of from 10^{-9} to 10^{-7} s, that is, referring to the region of micro-Brownian motion.

The oriented phase of rigid chain macromolecules and polymers in the condensed state have much longer relaxation times. Thus, in studying the creepage and crystallization of polymers, the segmental motions are rather limited and relaxation times are from 10^{-4} s to several seconds.

The application of luminescence methods, particularly luminescence polarization, can provide significant information concerning relaxation and other processes in polymers. These methods of studying polymers can be applied from two aspects—kinetic and structural. Let us consider them in turn.

6.2 FLUORESCENCE SPECTROSCOPY FOR STUDYING SPECIAL FEATURES OF POLYMER STRUCTURE AND PROPERTIES

6.2.1 Kinetic aspects of the application of luminescence techniques

Of the three main quantitative characteristics of luminescence, namely the quantum yield, the excited state lifetime, τ_ϕ, and the polarization, P, the last is the most sensitive to the macromolecular structure and to its intramolecular mobility and interaction [120]. When oscillators (luminophoric labels in polymers) are excited by light, their distribution is characterized by a certain anisotropy. If reorientation of the luminophor, related to the polymer chain or segmental motion, is possible, then with the molecule remaining in the excited state for time τ_ϕ, the oscillator's position becomes less anisotropic. The decreasing luminescence polarization, that is, luminescence depolarization, can be used as a measure of the participation of the chain related luminophor in the micro-Brownian motion of the polymer chain.

In practice, it is convenient to employ the values of the reduced depolarization which, in the case of rotational Brownian of an emitting oscillator (under luminescence excitation by natural light), is determined from the formula [125] p. 1433:

$$Y = \frac{1/P + 1/3}{1/P_0 + 1/3} = \frac{2/3}{\dfrac{1}{\tau_\phi} \displaystyle\int_0^\infty \left\langle \cos^2 \theta(t) \right\rangle e^{-t/\tau_\phi} \, dt - \dfrac{1}{3}} \tag{6.5}$$

where $\left\langle \cos^2 \theta(t) \right\rangle$ is the average squared cosine of the oscillator reorientation angle at time t, τ_ϕ is the average oscillator lifetime in the excited state (assuming exponential decay), and P is the polarization.

The value of τ_ϕ determines the average time interval within which reorientation of the luminescing oscillator occurs as recorded by the luminescence polarization measurement.

The value of τ_ϕ provides a time scale against which the times of micro-Brownian motions of the oscillators are compared. The value of P_0 in eqn (6.5) is the limiting luminescence polarization observed on the rigid fixation of the luminescing oscillators at $T/\eta \rightarrow 0$, that is, the freezing of Brownian motion. The value of P_0 is determined by the chemical structure of the luminophor and the relative positions of the absorbing and emitting oscillators.

It can be shown that in the case of the rotational Brownian motion of spherical particles of volume V in a liquid with viscosity η, eqn (6.5) transforms into the Levshin–Perrin formula [125] p. 1434:

$$Y = 1 + \frac{\tau_\phi kT}{\eta V} \tag{6.6}$$

where T is the absolute temperature and k is Boltzmann's constant.

In the more general case, without a linear relation between V and the reciprocal relaxation time $1/\tau$ (where τ is the rotational diffusion time), the following relation between the luminescence depolarization and the spectrum of relaxation times τ_j, and individual relaxation contributions f_i, is true:

$$Y = 1 \bigg/ \sum_i \frac{f_i}{1 + 3\left(\tau_\phi/\tau_j\right)} \tag{6.7}$$

The term τ_ϕ/τ_j in eqn (6.7) determines the impact of this or some other relaxation process on the luminescence polarization. The greater the difference between them, the lesser is the impact of the τ_jth relaxation process on the luminescence polarization.

It is known [19] p. 79, that for relaxation processes manifested in polarized luminescence, the times τ_j change proportionally with the solvent viscosity η. Therefore, for the experimental study of various relaxation processes, it is advisable to change the solvent viscosity on using different dopants, and to study the relationships

$$1/P = f(T/\eta), \quad \text{or} \quad Y = Y(T/\eta)$$

Typical relationships $Y(T/\eta)$ have the form of curves convex to the positive ordinate. The initial slope of the $1/P = f(T/\eta)$ curve (as $T/\eta \rightarrow 0$), the limiting asymptote slope to the $1/P = f(T/\eta)$ curve (as $T/\eta \rightarrow \infty$), and the ratio of the $1/P_0'$ section cut by the asymptote on the ordinate and the value of $1/P_0$, are related to the principal, averaged characteristics of the relaxation spectrum τ_j, and the average suspended relaxation time

$$\tau_w = \sum f_j \tau_j^2 \bigg/ \sum f_j \tau_j = \left\langle \tau^2 \right\rangle \bigg/ \left\langle \tau \right\rangle$$

by certain ratios cited in [208].

In the experimentally accessible region, we can already observe values reaching the asymptote to the $Y(T/\eta)$ curve, demonstrating rather a limited range of relaxation times. We have the possibility of a formal analysis of the experimental data based on the assumption of the existence of only two most probable times (or regions of relaxation) τ_1 and τ_2 with weightings f_1 and $f_2 = 1 - f_1$. In this simplest version, the system is characterized by three parameters: τ_1, τ_2, and f_1 which are unambiguously determined from the initial and limiting slopes $Y(T/\eta)$ and Y_0'.

In most cases, to define the relaxation time τ_w characterizing the intramolecular mobility of the polymer, it is sufficient to measure the luminescence polarization of labelled polymer solutions with various dopants, and by changing their viscosity.

To eliminate the influence of temperature on the rotational mobility of the fluorescent label and polymer chains, measurements are performed at constant temperature.

A number of original and review papers [19, 19A, 56B, 125, 322, 323] cite data on research using polarized luminescence of the impact of such factors as the chemical structure of the polymer chain, the RMM of the polymer, the structure of the fluorescent label, and the mode of its bonding to the macromolecule, and on the intramolecular mobility of the polymers.

Studies of aqueous solutions of poly(acrylic acid) and poly(methacrylic acid) labelled with 9-anthrylacyloxymethane or acylaminonaphthalene sulfonic acid groups, demonstrated that the nature of the luminescent labels located on the side chains of polymers only slightly alters the relaxation time τ_w.

Comparison of τ_w values for solutions, containing luminescent labels in the main and side chains, of poly(methacrylic acid) and its methyl ester in water and in methyl acetate solution shows their essential difference [323]. It means that these labels permit the study of various relaxation processes. While label XIII enables one to study the motion of the main chain sections, label XII refers to the relative reciprocal motion of the main chain section with adjacent side groups and the side methylanthracene group.

XII   XIII

The dependence of τ_w on the position of the label in the polymer chain indicates that the contribution to τ_w of rotational motions in the side group is crucial. The influence of the chemical structure of the polymer chain on the intramolecular mobility was studied with such polymers as poly(acrylic acid) and poly(methacrylic acid) and their alkyl esters, polystyrene, poly(α-methylstyrene), polyvinyl- and polypropenylbenzyl esters, and polydimethoxyethylene [19]. Interesting data on the relation between the composition of random copolymers and intramolecular mobility were obtained with the copolymer of methyl methacrylate with methyl acrylate and of styrene with α-methylstyrene; while for polymers of the first type, with the methyl acrylate content increasing from 0 to 50 mol %, a monotonic increase in $1/\tau_w$ from 22×10^7 to 37×10^7 s^{-1} is observed, for styrene copolymers $1/\tau_w$ values reach a maximum of 32×10^7 s^{-1} at 30 mol % α-methylstyrene content.

The fluorescence anisotropy of macromolecules containing a low molecular luminescent probe or label attached to the polymer chain was observed in a study of the molecular mobility of acrylic polymers [324, 325] and poly(γ-benzyl-L-glutamate) [326].

Such factors as the formation of intramolecular hydrogen bonds and intramolecular structurization in polymers significantly affect intramolecular mobility [19, 327]. It was shown that in nonaqueous solutions of poly(acrylic acid) and poly(methacrylic acid), the intramolecular formation of hydrogen bonds significantly reduces intramolecular mobility. A sharp decrease in the relaxation times and a corresponding change in the intramolecular mobility were noted on structurization in aqueous-alcohol solutions of poly(methacrylic acid) [327] and aqueous solutions of polydimethoxy-ethylene [19].

The polarized luminescence technique is quite a suitable tool for monitoring intramolecular structurization occurring under the action of metal ions in carboxyl-containing polymers [328]. This technique provides valuable information on trends in the formation of crosslinked and branched polymers, on the molecular mobility of net sites of polymers, and of linear sections between sites.

An undoubted advantage of the polarized luminescence technique is the possibility of studying completely crosslinked and insoluble polymers. For such measurements, a suspension of a stable, uniformly luminescing polymer species (up to 0.5 µm) is prepared in a suitable solvent. Polarized luminescence enables determination of the molecular parameters of the linear sections of the net. When the luminescent label is directly at the net site, it is possible to define the mobility of the site itself.

A feature of the polarized luminescence technique is the possibility of obtaining information from a label introduced into a predefined section of the polymer chain (into the main chain, side group, or chain edge). It enables us to study the mobility of the individual sections of the polymer chain and their role in the structurization (Table 6.1).

Table 6.1. Relaxation times (ns) for polymers with different types of fluorescent label (25°C, $M = 10^5$) [19]

Polymer	Solvent	XII	XIII	XIV[a]
Poly(methyl methacrylate)	Methyl acetate	3.9	8.3	3.2
Polystyrene	Toluene	5.2	8.9	4.1
Poly(methacrylic acid)	Water	77	120	–
	Methanol	7.4	19	–

[a] XIV – refers to label

It has been found, for example [329], that during the formation of mesophase nuclei in comb-like polymers with mesogenic cholesterol groups, there occurs the introduction of side chains into structured sections, and only then does the intramolecular retardation of the main polymer chain increase.

The same technique was used to study the structural transitions in polymer solutions exemplified by the cholesterol ester of poly(methacryloyl-ω-hydroxyundecanoic acid) [330]. It was shown that for comb-like liquid crystal polymers with mesogenic cholesterol groups in the side chains, the structural transition proceeds in two stages: ball→ isotropic liquid globule, and isotropic-liquid globule→liquid crystalline globule.

The simplicity of the polarized luminescence technique promotes its application in the study of macromolecular structures, intramolecular structurization, and other problems related to the study of intramolecular mobility. The polarization measuring unit comprising a birefringent prism and two photoelectron multiplexors is described in detail in [19, 331].

6.2.2 Structural aspects of the application of the luminescence technique

By introducing luminescent labels and probes into solid polymers, we can study structural peculiarities of the polymeric phase. In particular, in [11] the transition temperatures in polymers were determined, and in [332] a new method for their recording was suggested. To define these transitions in styrene and methyl methacrylate copolymers with vinylcarbazole, naphthyl methacrylate and some other comonomers, the temperature dependences of the phosphorescence of the above polymers were studied in the temperature range 77–300 K.

Fluorescence spectroscopy permits detailed study of the conformational and configurational features of macromolecular structures as determined by the nature of hydrogen bond [333] and other factors [334, 335]. The polarized luminescence technique was suggested, for example, for studying configurational changes of polypeptide chains containing the fluorescent tryptophan residue (in the form of sheep wool) in the polymer on stretching under the action of steam [334]. The data obtained are in good agreement with X-ray diffraction results, showing that the transition of the wool keratin polypeptide chain from its α- to β-configuration is reversible.

Studies of solid polymers with the luminescence technique allows the monitoring of structural changes occurring in deformation processes [336–339]. As shown in [336], the fluorescence spectrum of luminescent dopants; for example anthracene doped in polyethylene comprising amorphous and crystalline phases, is the summation of two spectra. These spectra correspond to the introduction of anthracene into the crystalline and amorphous regions, and are therefore sensitive to structural transformations of the polymer matrix.

As follows from Fig. 6.7, irreversible stretching of polymer samples leads to a sharply increasing fluorescence intensity of the crystalline (I_{cr}) sites of anthracene in the 24 000–24 300 cm^{-1} region [337]. It applies to polymers with an initially low level of crystallinity. Conversely, for initially highly crystalline polyethylene, irreversible stretching brings about a decrease in the relative intensity of I_{cr}.

Thus, according to [336, 337], destruction and restoration of the structural formations of polyethylene are observed in the process of stretching, and the degree of crystallinity (if the relative intensity I_{cr} is taken as its average characteristic) may increase or decrease, depending on its initial value.

On the basis of X-ray diffraction and other data, some researchers believe that molecules doped into a polymer enter only the amorphous and not the crystalline regions [338], which casts doubt on the conclusions of [336, 337].

Ref. [339] considers the effect of deformation on the structure of low pressure polyethylene and its blends with poly(methylene oxide) containing anthracene as a molecular probe. As test samples, (i) limiting oriented 12-fold extended samples, or (ii) samples continuously loaded for 15–20 days were chosen. For the samples of type (i) an increased packing density of the macromolecules due to their high orientation was initially observed. On the subsequent action of continuous loading on the limiting oriented

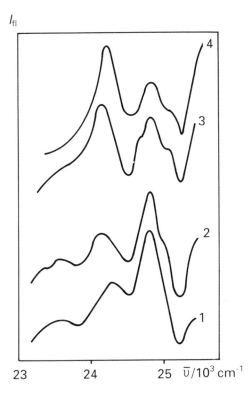

Fig. 6.7. Fluorescence spectra of solid solutions of anthracene in polyethylene at 77 K: 1, initial sample; 2–4,—samples stretched 2- (2), 4- (3), and 5-fold (4).

samples, a slight hypsochromic shift of the peaks occurred in the fluorescence spectra of the probe molecules, indicating layering in the system. Samples of type (ii) showed a noticeably greater blue shift of the fluorescence peaks.

A significant heterogeneity appearing in the polymer structure in the process of stretching polyethylene at constant load was revealed by the fluorescence technique [339]. The existence of regions with density differing by 3% from the average was detected, demonstrating that the restructuring of polyethylene occurs on the level of mesomorphic structures of one- or two-dimensional ordering.

In [340, 341] the polarized fluorescence technique was used to monitor the orientation of low molecular luminophors in poly(ethylene terephthalate) fibres and cellulose films; it showed a difference in the behaviour of the amorphous and crystalline regions of the polymer. Ref. [340] demonstrated the abilities of this technique for studying disorientation processes of amorphous regions on the thermal treatment of fibres. The relationship between the intensities of the polarized components ($I_{\rm I}$ and $I_{\rm II}$) vs the angle between the axis of stretching of the films and fibres and the direction of the linearly polarized electric vector of the excitation light permitted [341] definition of the functional group characterizing the degree of orientation of the polymer molecules.

Polarized fluorescence was also used to study molecular orientation in poly(vinyl alcohol) [342, 343] and polystyrene [344]. Thus, [342] suggested a method to determine the ordering of polymer chains based on dichroism and polarization studies of the fluorescence of luminophor molecules, doped into the polymer studied, whose orientation is known to be unambiguously determined by their macromolecular orientation. According to this method, the degree of molecular orientation is characterized by two orientational parameters, $A_1 = \overline{\cos^2 \theta}$ and $A_2 = \overline{\cos^4 \theta}$, that are respectively the second and fourth moments of the function of angular molecular distribution. A relationship was established between the shape of the dopant molecules and the orientation parameters determined by the measured dichroism and polarization. This technique has high sensitivity and precision, and ensures more detailed information on the distribution function compared with other methods.

As stated in [343], a uniaxial deformation of low crystalline poly(vinyl alcohol) at 100°C in water vapour has shown that deformed films possess fewer chain segments oriented perpendicular to the stretching direction than follows from consideration of athenic deformation. The data obtained were interpreted by the author as a manifestation of the preferential decomposition of crystallites oriented perpendicular to the direction of stretching.

In a study of orientation processes in polystyrene with an anthracene dopant, the authors of [344] used a model of diffusional motion of molecular chains under viscoelastic conditions along the 'tube', formed by neighbouring chains, whose shape depends on strains in the polymer. Under load and on heating to 116–135°C, polystyrene samples underwent deformation typical of the glassy state at low tension, and highly elastic deformation at high tension with a noticeable contribution of flow deformation at high temperatures. The author of [344] studied the relationship between the degree of fluorescence polarization and deformation at different temperatures and RMMs of polystyrene. The degree of orientation increases with RMM, since in the range of temperatures and deformation rates studied, the longest macromolecules undergo highly elastic extension, and the short ones transfer to flow conditions.

The influence of orientation of the polymer chain in polyethylene on the fluorescence anisotropy of dopants was studied in [345].

In some instances fluorescence spectroscopy has been applied to studies of orientation processes in macromolecules containing luminophoric moieties in their chains [346, 348]. In these examples, measurements of fluorescence depolarization were used to study segmental orientation in nonstressed polyisoprene networks with fluorescent anthracene groups in the chain, XV.

$$\cdots -CH_2-CH{=}C-CH_2-CH_2{-}\underset{\displaystyle CH_3}{\overset{\displaystyle CH_3}{|}}\ \ CH_2-CH_2-\underset{}{\overset{\displaystyle CH_3}{|}}C{=}CH-CH_2-\cdots$$

XV

To prepare this polymer 9,10-bis(bromomethyl)anthracene was added to a multifunctional 'living' chain of polyisoprene, produced via anionic polymerization, having a RMM of 300 000, with the resulting polymer having a RMM of 600 000 with the anthracene label in the centre. The authors of [346] studied the relationship between the stress in the polymer and the elongation at different temperatures. They found two divergences from the classical theory of elasticity: at small and medium elongation, super-elongation of the dry polymer film increases close to the glass transition temperature, and significant decreases during swelling; at greater elongation they observed orientation saturation, especially in the swollen state.

The former divergence is satisfactorily explained by a modified treatment of weak nematic interaction between segments, but the authors of [346] have not found an explanation for the latter.

The segmental motion of polypropylene during the transition of the polymer to the glassy state was monitored in [347] by the change in the emission intensity of the intramolecular excimers of the low molecular luminophor—meso-2,4-di-(N-carbazolyl)pentane doped into the polymer. This approach was later extended to pressure effects on cis-isoprene rubber and poly(propylene oxide) [347A].

Shear-induced molecular orientation in a melt of anthracene-tagged polybutadiene has been monitored by fluorescence anisotropy measurements [347B].

Relaxation processes are extremely important in the physical study of polymers. The deformation and mechanical properties of polymers are closely related to them. Luminescence spectroscopy, alongside other physicochemical methods, can reveal the special features of relaxation phenomena in solid polymers [19B, 348, 349].

The relaxation properties of crosslinked polyisoprene containing three fluorescence probes—trans-diphenylbutadiene, trans-diphenylhexatriene and trans-diphenyloctate-traene—were studied in [348]. In particular, the dependence of the degree of fluorescence polarization on the cross-link density was considered. It was found that the rotational mobility of the probes commences at 10°C, that is, 50°C above the glass transition temperature. With increasing crosslink density, the onset of label mobility shifts to higher temperatures. Application of the model of orientational diffusion along the macromolecular main chain made possible determination of the relaxation time ρ, reflecting the minimal segmental motion (that is, of five monomer units). The slopes of graphs of $\log \rho$ vs temperature and the temperature relation $\log a_T$—given by the Williams–Landell–Ferri equation—are approximately equal in the range of 20–80°C:

$$\log a_T = -C_1 \frac{T - T_g}{C_2 + \left(T - T_g\right)}$$

where C_1 and C_2 are empirical constants and T_g is the glass transition temperature.

In the region of low mobility, corresponding to the temperature range from 10 to 20°C, the slope of the graph of $\log \rho$ slope vs temperature is smaller, indicating lower activation energies, probably owing to the emergence of the second transition.

Monnerie and colleagues and other groups have extended study of molecular mobility by fluorescence quenching and polarization methods to other rubbers [348A], copolymers [348B] and blends [348C, D], enabling the determination of miscibility of blends. [348D] gives a valuable number of papers in this topic area. Theoretical treatment of the dynamics of conformational transitions leading to intramolecular excimer formation in aromatic polyesters with methylene or oxymethylene spacers [348E] and to dimer models of polystyrene [348F] are due to Bahar and Mattice.

Ref. [349] illustrates the abilities of the depolarized phosphorescence technique in studies of relaxation processes in solid polymers like poly(methyl methacrylate) with a low content of naphthalene as luminophor or of naphthalene units in the chain. In particular, the absence of luminescence depolarization due to the rotation of the luminescent probes below the onset of the β-relaxation temperature was noted. Research into the temperature dependence of the triplet state lifetime allowed determination of the relaxation transition temperatures in the copolymers studied.

Luminescence spectroscopy can also be applied to studies of some diffusion processes [350–354]. The most suitable technique for this type of research is the monitoring of fluorescence quenching.

The authors of [350] studied diffusion processes in aqueous-alcoholic solutions of 4-vinylpyridine and anthrylmethyl methacrylate (PVP-F) copolymers containing one anthracene moiety per 500 polymer units. Diffusion constants were determined by processing data on the fluorescence quenching of these copolymers with quaternized pyridine derivatives, using the Stern–Volmer equation in the form

$$\tau_0/\tau = 1 + K_q \, [Q]$$

where τ_0 and τ are the excited state mean lifetimes in the absence and presence of quencher, respectively, K_q is the Stern–Volmer constant related to the rate constant of bimolecular encounters between luminophor and quencher $K_q = \tau_0 k_D$, and [Q] is the quencher concentration in solution.

As quencher they used either poly(vinylpyridine) quaternized with dimethyl sulfate (PVP-Q) or the copolymer containing both the fluorescent label and quencher (PVP-FQ).

The constant k_D for the PVP-FQ polymer was determined from the Stern–Volmer equation (Fig. 6.8a). The value of this constant was 0.9×10^9 dm^3 mol^{-1} s^{-1}. This means that the 'internal' viscosity, determined by the impedance to the motions of chain units due to the barriers of internal rotation [351], cannot limit the rate of chemical reactions between remote units of one chain.

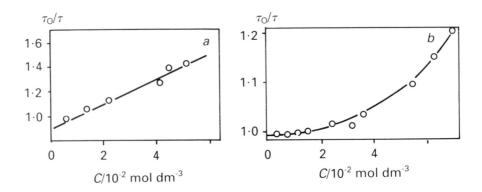

Fig. 6.8. Dependence of ratio τ_0/τ on concentration of quencher: (a) for polymer PVP-FQ on local concentration of quencher units ($C_{pol} = 2 \times 10^{-3}$ mol dm^{-3}); (b) for polymer PVP-F ($C_{pol} = 1 \times 10^{-3}$ mol dm^{-3}) on concentration of quencher PVP-Q (degree of quaternization 50%).

As shown in Fig. 6.8b, no fluorescence quenching of chromophores occurs up to a certain concentration of quenching units. On further addition of PVP-Q macromolecules, however, quenching commences and the ratio τ/τ_0 starts to increase with quencher concentration. It follows that the susceptibility of macromolecular balls to reciprocal permeation changes stepwise with increasing polymer concentration in solution: at low concentrations it is insignificant, but increases considerably on reaching a certain limiting concentration. The values of k_D computed by the ascending slope of the curve in Fig. 6.8b approach those characteristic of intramolecular encounters.

The fluorescence quenching technique can provide useful information on the structure and relaxation properties of hydrogels such as polyacrylamide and poly(vinyl alcohol) with 30–100% polymer content [353]. Acridinium cation and bromide anion were used as the luminophor quenching agent pair. Rate constants for the fluorescence quenching of acridinium cation by bromide anion in aqueous solution are close to diffusion control. The quenching pattern follows the Stern–Volmer equation. This result indicates the absence of static quenching of fluorescence in the systems studied because of the formation of stable ion-pair complexes by acridinium cations and bromide anions. The quenching rate constant diminishes monotonically with increasing polymer concentration in the gel up to 60%, and sharply decreases with further increase of the polymer content up to 70%; the authors of [353] explain this in terms of the temperature approaching the glass transition region. For both types of gel, they noted an exponential decrease of the fluorescence intensity of an excited acridinium molecule, indicating a similar luminophor microenvironment in the sample as a consequence of the segmental motions of the polymer chain and diffusion of the luminophor molecules.

The difference in the relaxation behaviour of macromolecules in polyacrylamide and poly(vinyl alcohol) gels originates, in the first case, from the relatively long times of small scale motions; the main diffusion mechanism, according to [353], is the motion of ions in the water in the relatively motionless polymer 'frame'. It can be assumed that, in the second case, the voluminous acridinium cations are expelled from the hydrogen bonded associated units of poly(vinyl alcohol), hence their microenvironment is represented by the relatively polymer-depleted structureless regions of a polymer gel where reciprocal diffusion of the luminophor and quencher occurs. These phenomena accord with the fact that at equal polymer concentration, the fluorescence quenching rate constant in the poly(vinyl alcohol) gel medium is higher than that in the polyacrylamide gel.

The fluorescence quenching technique is also a suitable tool for studying the kinetics of exchange of polyelectrolytes in nonstoichiometric water soluble complexes [355], as well as for determining the reaction mechanism of macromolecular substitution in polyelectrolyte solutions [356] and the interaction of luminophor-containing polystyrene sulfonate with methyl viologen and copper ions [357]. Self-quenching has been detected in ionomers of poly(styrene-co-acrylic acid) containing Eu^{3+} ions as a luminophor, and a general review of interactions between lanthanide ions and macromolecules is due to Okamoto and Kido [357A].

7

Applications of fluorescent polymers

Fluorescent polymers and polymer formulations are used extensively in various fields of science and technology.

The development of nuclear power engineering, the exploration of outer space, the application of radioactive isotopes in chemistry, biology, medicine, and the search for mineral resources require totally reliable methods of ionizing radiation detection. One of the methods which holds great promise is the scintillation method. It is based on the transformation of ionizing radiation energy into light energy in the detector, with the subsequent recording of the light energy by a photomultiplier tube. The scintillator acts as the detector of ionizing radiation [46, 358]. Along with monocrystal and liquid scintillators, plastic scintillators which are polymer scintillation formulations or compounds with a great variety of assets, are much used.

Fairly recently, the intricately shaped light guides were superseded in scintillation engineering by converting polymer compounds (spectrum mixers) for the collection of light from the large areas and volumes of ionizing radiation detectors.

In recent years there has been a growing interest in the use of solar energy since the reserves of conventional organic fuel will last for 100 years at most, and the use of nuclear power, as confirmed by recent events, is fraught with a number of technical limitations and restrictions associated with environmental protection. At present, photo-electric generators are, unfortunately, uncompetitive with regard to the cost of 1 kWh of electric power, and the efficiency of photoelectric converters is low. To improve the efficiency, solar energy concentrators of various designs, including the luminescent polymer solar concentrators (LSC), are widely used [362]. The advantage of LSCs is that they concentrate both direct sun and diffuse light, thereby functioning irrespective of atmospheric conditions. In this case, the need for follow-up systems is obviated and the required level of concentration is provided for by the stationary collector.

In recent years fluorescent polymers have acquired much importance because of their application in radioelectronics as the active elements for quantum generators. Similar to the liquid lasers based on organic luminophors, polymer luminescent compounds permit, in principle, a gradual change from the coherent radiation frequency to practically any frequency of the visible and near infrared spectrum region [13, 363, 364].

It should also be noted that fluorescent polymer compounds find extensive application in the paint and varnish industries, as well as in the manufacture of plastics and textiles.

In any of the above fields, polymeric luminescent compounds are expected to meet specific requirements with respect to their spectral and luminescent properties.

7.1 IONIZING RADIATION RECORDING MATERIALS

7.1.1 Plastic scintillators

Plastic scintillators represent a complex system composed of a polymer (polymer base) and an organic luminescent compound (luminescent additive). The main characteristics of plastic scintillators are their scintillation efficiency and light yield T_s [12].

The scintillation efficiency of a polymeric compound depends on the properties of both the luminescent additives and the polymer base. The use of polymers in scintillation compounds is determined by their ability to convert electronically excited energy into light energy, which happens because of the presence of the π-conjugated system in the macromolecule, and also by the transmission by polymers of the light of their natural luminescence and radiation of the luminescent additives.

The main requirements imposed on luminescent additives are a high quantum yield and maximum overlap of the absorption and luminescence spectra of the polymer base. To increase the transmission of the whole system, another luminescent additive, the spectrum mixer, is introduced in the polymeric scintillation compound as a result of which the efficiency of light flash recording by the photomultiplier tube also increases.

The first plastic scintillators (PS) were produced on the basis of polystyrene and p-terphenyl in 1950 [365]. The purpose of further investigations was to increase the scintillation efficiency of PS by suitable selection of luminescent additives (LA) and the polymer base. A large number of organic luminophors such as polyphenyls, aromatic condensed ring compounds, phenyl substituted derivatives of ethylene and dienes, and heterocyclic compounds, namely derivatives of 1,3-oxazole, 1,3,4-oxadiazole, and Δ^2-pyrazoline, have been studied as LAs [366], as has 2-(2'-hydroxyphenyl)benzothiazole [366A].

Various aromatic vinyl polymers have been investigated as polymer bases.

As a result of investigations of the scintillation efficiency and radiation resistance of PSs, the suggestion was made to classify polymeric scintillation compounds on the basis of the structure of the polymer repeat unit [12]. Plastic scintillators are divided into four groups:

(i) scintillators based on polystyrene and its derivatives;
(ii) scintillators based on vinyl monomers with polyphenyl, condensed aromatic, and heterocyclic rings;
(iii) compositions with an inactive polymer base (π-conjugation is absent in the repeat unit) and a 'secondary solvent' (a low molecular compound with aromatic rings);
(iv) scintillators based on copolymers of various compositions (Table 7.1).

Polystyrene and its ring-methylated derivatives are most widely used as the PS base. Refs [367, 368] establish a relationship between an increasing number of methyl

Table 7.1. Classification of plastic scintillators (light yield measured relative to polystyrene scintillators whose T_s is taken as 100%)

Group I		Group II	
Substituted derivatives of polystyrene	$T_s{}^a$/%	Monomer-based polymers	$T_s{}^b$/%
-2-CH$_3$	117	Vinyldiphenyl	120
-3-CH$_3$	105	3,3-Dimethyl-4-vinyldiphenyl	180
-4-CH$_3$	125	α-Vinylnaphthalene	130
Mixture of -2-CH$_3$	120	β-Vinylnaphthalene	120
and -4-CH$_3$		β-Vinylnaphthalene	144^c
-2,4-CH$_3$	160	N-Vinylcarbazole	100
-2,5-CH$_3$	150	2-Vinylfluorene	60
-3,4-CH$_3$	140	2-Vinylbenzofuran	60
-2,4,5-CH$_3$	170		
-α-CH$_3$	86		
-α-CH$_3$, -4-C$_3$H$_7$	106		
Mixture of isomers			
-α-CH$_3$, -2-CH$_3$	111		
-α-CH$_3$, -3-CH$_3$			
-α-CH$_3$, -4-CH$_3$			

Group III			Group IV Copolymers		
Secondary solvent	LA/%	T_s/%	monomer 1/%	monomer 2/%	T_s^a/%
Poly(methyl methacrylate)			MMA 100	—	10
Naphthalene 15	POPOP 0.4	40	MMA 60	Styrene 40	80
Phenanthrene 20	POPOP 0.4	40	MMA 10	Vinylcarbazole 90	100
Naphthalene			MMA 30	Vinylcarbazole 70	120
8	0.08	30	Vinyldiphenyl		
10	PPO 1.5 POPOP 0.08	40	10	90	100
			20	80	120
			30	70	140
15	PPO 1.5 POPOP 0.08	60	50	50	120
			Vinylterphenyl	Styrene	
			10	90	130[d]
			20	80	138
Poly(deuteromethyl methacrylate)			α-Vinylnaphthalene	Styrene	
Deuteronaphthalene 10	POPOP 0.08	40	30	70	136[a]
			β-Vinylnaphthalene	Styrene	
			30	70	156[a]

[a] p-Terphenyl (TP) 2%, 1,4-di(5-phenyloxazolyl-2) benzene (POPOP) 0.1%.
[b] 2,5-Diphenyloxazole-1,3 (PPO) 1%, POPOP 0.1%.
[c] 2,5-[(4-Methoxyphenyl)(4-dimethylaminophenyl)]-oxadiazole-1,3,4 1.5%.
[d] POPOP 0.1%.

groups, their location in the benzene ring, and the efficiency of the polymer base. A number of studies describe highly efficient scintillators based on polymethyl-, dimethyl-, and trimethyl styrenes [369, 370]. An increase in the size of the alkyl groups leads to an insignificant reduction in the scintillation efficiency. Other electron-donating substituents such as methoxy, increase the light yield, but less than the methyl group [367]. The presence of halogens (Cl, Br, I) in the benzene ring causes a sharp decrease in the scintillation efficiency. Thus, the CH_3 group is the most effective substituent in the benzene ring. On introducing the methyl group into the polymer chain, at the α-position in particular, the scintillation efficiency of PSs does not improve. Thus, it has been shown in [367] that the light yield of PSs based on poly(α-methylstyrene) is practically zero. Further investigations of α-methylstyrene and its homologues [371] indicate that with the introduction of the methyl group in the α-position, the scintillation efficiency decreases insignificantly. As in the case of polystyrene, the introduction of methyl groups in the benzene ring of poly(α-methylstyrene) leads to an increase in the light yield of the scintillator (see Table 7.1).

The scintillation efficiency of polymeric systems is also improved by increasing the number of phenyl groups in the repeat unit. Thus, the light yield of polyvinyldiphenyl- and polyvinylnaphthalene based scintillators exceeds that of polystyrene samples by some 20 and 40% respectively (US Patent 2824841) [372, 373]. A high scintillation efficiency is characteristic of PSs formed by a polyvinyltetrahydronaphthalene base and a long-wavelength LA [374].

Study of vinyldiphenyl derivatives as scintillator bases confirms the generality displayed by styrene derivatives (see Table 7.1).

However, in spite of the fact that an increase in π-conjugation results in a higher light yield of PSs, the polymer bases included in group II are of no practical value since, during bulk polymerization, they form rather fragile polymers with low RMM and transmission (the initial monomers are mainly solid substances which are difficult to clean). All this necessitates the reprecipitation of the finished polymer which restricts the manufacturing possibilities to film scintillators. The most promising is the application of the copolymers of given monomers either with styrene or methyl derivatives of styrene in the benzene ring or with methyl methacrylate (group IV). Of the inactive polymer bases, poly(methyl methacrylate) is most popular owing to its good physical and mechanical properties and high transmission. However, because of the absence in the repeat unit of PMMA of π-conjugation, PMMA based scintillators have low scintillation efficiency, not exceeding 10% compared with that of polystyrene. The light yield of inactive bases is increased by introducing a secondary solvent. Satisfactory results are obtained with such secondary solvents as naphthalene (US Patent No. 3068178, USSR Inventor's Certificate 203229) and phenanthrene [375]. The scintillation efficiency of such systems is determined by the amount of activator, the nature of the primary LA, and the presence of a spectrum mixer; it can reach approximately 60% of the scintillation efficiency of polystyrene scintillators. However, being influenced by the sample transmission, the light yield of large scintillators based on PMMA is no less than that of polystyrene based scintillators.

Refs [376, 377] point out the possibility of producing macromolecular scintillators on the basis of the MMA copolymer and vinyl-*tert*-butyl-2-phenyl-5-(4-biphenylyl)-1,3,4-oxadiazole [376], and poly(isobutyl methacrylate) [377].

To solve some nuclear physics problems, in fast neutron spectrometry in particular, deuterium-containing PSs based on deuteromethyl methacrylate and deuteronaphthalene [378], and an activation neutron detector based on gadolinium-containing PMMA [379], have been developed.

To impart elasticity to scintillation systems, use is made of inactive monomers such as methyl and butyl methacrylates [12]. Also known are epoxy resin scintillators, in which xylene is used as the secondary solvent, and scintillators based on organosilicon compounds and rubber [380–382].

The natural luminescence transmission of scintillators also increases when copolymers of aromatic vinyl monomers, styrene and methylstyrene, with methyl methacrylate, are used as the base. The light yield of such scintillation systems depends on the copolymer composition. MMA-methylstyrene plastic scintillators show rather high scintillation efficiency (FRG Patent 1645215).

Ref. [12] describes scintillators based on MMA and dimethyl styrene, vinyldiphenyl, and N-vinylcarbazole. Of interest are scintillators made on the basis of styrene and methyl derivatives of styrene, vinyl monomers with polyphenyl groups, and condensed aromatic rings [113, 325], styrene, and 4'-vinyl-3-hydroxyflavone [383]. The scintillators based on vinyldiphenyl, α-, β-vinylnaphthalene–styrene copolymers have a higher light yield than those having the homopolymers of vinyldiphenyl and vinylnaphthalene [373].

To improve the heat resistance of scintillators, the suggestion has been made to use crosslinked polymers as the base [298]. Thus the softening temperature of poly(2,4-dimethyl styrene) crosslinked with diisopropenyl benzene is 10–15°C higher than that of linear poly(2,4-dimethyl styrene), the light yield being unchanged.

The light yield of scintillators is increased by copolymerization of styrene or 2,4-dimethyl styrene with aromatic vinyl monomers containing chromophore groups such as terphenyl, diphenyloxazole, and diphenyloxadiazole [115, 116] (USSR Inventor's Certificate 172040).

One of the advantages of polymeric scintillation materials over monocrystal and, to a certain degree, liquid scintillators, is the possibility of varying their composition not only by using polymers and LAs of different types but also by adding to the polymer different organic and inorganic compounds to impart new properties to the ionizing radiation detectors.

PSs have a low effective atomic number ($Z_{eff} = 5.8$) determined by their elemental composition (C ~ 93%, H ~ 7%). For some purposes such as the detection of low-level γ- and X-ray radiation and monitoring γ- and X-ray radiation, compounds with heavy atoms ('weighting agents' or 'compensators') need to be incorporated into the scintillating polymer (polymer + LA). The presence in PSs of elements whose nuclei are capable of entrapping neutrons also makes possible the construction of slow neutron detectors.

The main feature of the scintillator, owing to which it is used as a detector in the control of X-ray and γ-radiation, is the minimum dependence of the dosimeter sensitivity on the X-ray or γ-radiation energy (the linear energy response). This feature determines the dependence of the relative detector efficiency, that is, the ratio between the scintillator light flux and the gas chamber readings (I/D) on the radiation energy E.

To obtain correct dosimetric information, it is necessary that the absorption of ionizing radiation by the scintillator should correspond to the absorption in the medium under study (air, biological tissue) which happens when the detector and the test medium have equal effective atomic numbers. In dosimetry such detectors are called air and tissue equivalent. None of the organic and inorganic scintillators currently available is suitable for monitoring X-ray and γ-radiation over a wide energy range because the effective atomic numbers of such scintillators are either below or above that of air ($Z_{eff} = 7.64$) or tissue (muscular $Z_{eff} = 7.4$; adipose $Z_{eff} = 6.0$; osseous $Z_{eff} = 13.8$). The correction of the 'linear response' of such detectors at energies below 100 keV is fraught with difficulties.

Element containing compounds, both soluble in the polymer (homogeneous systems) and those forming a disperse phase (heterogeneous systems) can be incorporated into polymeric scintillation compositions, as a result of which the Z_{eff} of the entire composition will be changed.

To obtain minimum dependence of the dosimeter sensitivity I/D on the X-ray and γ-radiation energy E, ZnS(Ag) powder, which is a compound with scintillation properties, is added as a compensator to the heterogeneous PSs. Since this substance is suspended in the polymeric system, the transmission of the detector decreases, depending on the amount and degree of dispersity of the ZnS(Ag).

Fig. 7.1 illustrates the linear response for a polystyrene based scintillation composition. The low energy region shows a significant departure from linearity which is explained by the fact that the effective number of the PS ($Z_{eff} = 5.8$) is below that of air ($Z_{eff} = 7.6$). Only in the Compton interaction region, starting from about 100 keV, is the $I/D–E$ relation linear. With introduction into the PS of a compound with heavy atoms, the curve straightens, and the efficiency in the low energy region improves. Heterogeneous detectors are suitable for dosimetry starting from 30 keV [386] (see Fig. 7.1).

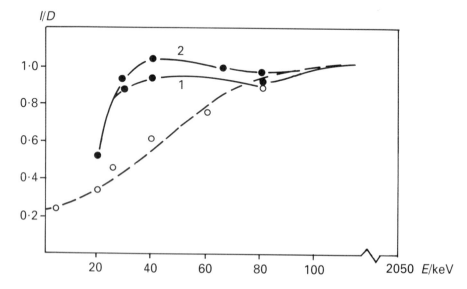

Fig. 7.1. 'Linear energy response' of scintillators based on polystyrene and polystyrene copolymers. 1, polystyrene with 0.5% ZnS (Ag) (mass), 2% TP (mass) and 0.1% POPOP (mass), 2, styrene copolymer with 10% vinylnaphthalene and 20% *m*-chlorostyrene with 1.5% PPO (mass) and 0.06% POPOP (mass) (circles, experimental data, dashes, calculated data).

The possibility of using compensators which are soluble in the polymer base and form a homogeneous system but have no luminescent properties is related to scintillation quenching by heavy atoms. The organic compounds of heavy metals (Sn, Pb, Hg, As, Se), which are soluble in the polymer, are unsuitable for dosimetry because they cause luminescence quenching and, consequently, a reduction in the light yield of the PS. Also in the low-energy γ-radiation region, metal-containing PSs have a comparatively high 'linear energy response' owing to the absorption jump caused by the photoeffect in the 10–100 keV region of heavy metals [387, 388]. It is the strong photoeffect peculiar to this region that determines the application of organic compounds of heavy elements as weighting agents to improve the efficiency of low energy γ- and X-ray radiation absorption by polymer scintillators [389, 390].

Halogen substituted organic compounds offer promise as compensators for homogeneous air equivalent plastic scintillators (USSR Inventor's Certificate 173409). The *p*-dibromobenzene scintillator has a comparatively high light yield, while at the same time the linear response of this scintillator is within ±8% in the energy range from 30 to 1250 keV.

Of particular interest are the scintillators based on the styrene–*p*-chlorostyrene copolymer which has found application in the radiation control of the low energy region. The main drawback of such scintillators is their low light yield, which is explained by the quenching effect of chlorine whose concentration is rather high (up to 20%).

The light yield of chlorine-containing plastic scintillators goes up when more effective monomers such as methylstyrene and 2,4-dimethyl styrene are used for copolymerization with chlorostyrene. However, the highest T_s is displayed by scintillators which have naphthalene as the secondary solvent or the copolymers of styrene and vinylnaphthalene or vinyldiphenyl with the additives PPO and POPOP as the polymer base. The light yield of these detectors reaches 80–95% of that of a conventional polystyrene scintillator, which is approximately 30% higher than the T_s of the air equivalent scintillator without naphthalene ($T_s = 55\%$) (USSR Inventor's Certificate 380666) [391].

The advantages of air equivalent detectors based on the ternary copolymer of halogen derivatives of styrene, vinyldiphenyl (vinylnaphthalene), and styrene (methyl-substituted compounds of styrene) are their high heat resistance and the stability over time of the light yield.

Ref. [392] describes PSs equivalent to tissues in effective atomic number. Muscular and adipose tissue equivalent PSs have a rather high light yield and satisfactory linear energy response both with ZnS(Ag) and with *p*-dibromobenzene.

It is known [393, 394] that element-containing compounds incorporated in PSs have a quenching effect mainly on the polymer base rather than the luminescent additives. For this reason, in [393] the heavy atom was introduced directly into the luminescent additive PPO and PPD. As a result, these compounds increase the effective atomic number of the PS and simultaneously function as the primary LA. The introduction of chlorine atoms in the LA does not affect the strength of its fluorescence which fully corresponds to the high light yield of the PS with given luminescent additives.

Table 7.2 compares the light yields of PSs containing chlorine derivatives of PPO and PPD as the primary LAs and of PSs with the halogen derivatives of styrene; in all the cases the secondary LA is POPOP. The concentration of element-containing compounds in PSs of both types corresponds to the concentration of the compensator which is responsible for the air equivalence of the detector.

Table 7.2. Light yield of halogen-containing plastic scintillators

Halogen-containing compounds	Compound concentration/%	Halogen concentration/%	T_s/%
Standard[a]	–	–	100
Cl—PPO—Cl	20.0	5.0	115
Cl—PPD—Cl	20.0	5.0	100
p-Chlorostyrene	20.0	5.0	50

[a] Based on polystyrene with 2% p-terphenyl and 0.1% POPOP.

It was noted above that the introduction of metal-containing derivatives of Sn, Pb, Hg, Bi in polystyrene with luminescent additives changes the capacity of PSs to absorb low energy γ- and X-ray radiation owing to the photoeffect produced on heavy atoms. It permits the selection of appropriate compositions for monitoring low energy radiation. It is shown in [389, 390] that the coefficient of soft γ-radiation absorption by plastic scintillators becomes higher with increasing atomic number of the element included in the metal-containing additive or with increase in the additive concentration. As a result, the light yield of the PS is simultaneously reduced. Phenyl derivatives of Sn, Pb, Hg, Bi have been studied as weighting agents in [389, 390, 395]. To improve their solubility in the polymer, in a number of cases alkyl groups were introduced in the phenyl group, or the phenyl group was replaced by naphthyl [396].

The light yield of metal-containing PSs was increased by incorporating naphthyl and diphenyl groups in the polymer chain through the copolymerization of vinylnaphthalene and vinyldiphenyl with styrene or 2,4-dimethyl styrene, which allows variation of the concentration of the secondary solvent over a wide range.

7.1.1.1 Principal properties of plastic scintillators

One of the factors determining the light yield of a PS is the primary interaction of ionizing radiation with the scintillator material [397, 398], the efficiency of which depends on the type of radiation.

On passing through the substance in its process of interaction with atoms, each particle of charged corpuscular radiation loses energy in a more or less monotonic way. The decisive characteristic, in this case, is the free path of the particle which is dependent on the mass and charge of the particle and the density and elemental composition of the substance.

If the radiation in question is electromagnetic in type, only a certain part of it interacts with the substance, as a result of which γ-quanta are likely to disappear completely or their energy may be sharply reduced. The intensity of visible electromagnetic and γ-radiation decreases strictly according to the exponential law $I = I_0\,e^{-\mu\ell}$ (where I_0 and I are the intensities of the incident radiation and transmitted radiation, μ is the linear absorption coefficient peculiar to the given substance and radiation energy, and ℓ is the length of the sample).

In the passage of γ- and X-ray radiation through the substance, three processes of interaction can be observed, depending on the energy of the quanta and the nature of the absorbent. These are the photoelectric effect, Compton interaction, and pair produc-tion. The Compton effect dominates in PSs when the radiation energy is within the range from 0.1 to 100 MeV. Photoelectric absorption is observed below 100 keV, and it is insignificant, owing to the absence in the scintillator of heavy atoms possessing a large number of core shell electrons. Pair production occurs with high level ionizing radiation (above 1.02 MeV) and the process increases with increase in the quantum energy [397].

Occurrence of all the above processes in PSs leads to the formation of secondary elec-trons, which induce the ionization and excitation of molecules in scintillators [46].

A special feature of γ-radiation is its deep penetration which depends directly on the energy of the γ-quanta. This is why to monitor γ-radiation, especially that of high energy, PSs of great thickness are needed. For this reason rather severe requirements are imposed on them with regard to natural luminescence transmission.

To record high γ-radiation fluxes, use is made of small detectors with a high light yield. Since low level γ-radiation recording is effected solely via the photoeffect, small metallorganic PSs are used as detectors of γ-radiation of up to 100 keV.

Electrons interact with orbital electrons and cause the ionization and excitation of atoms. The free path of an electron is directly proportional to its energy and is inversely proportional to the density of the irradiated sample. Hence, the depth of penetration of electrons into the PS substance is practically equal to their free path in water owing to the practically equal density of water and the polymer. The maximum ionization created by electrons in the medium is distributed unevenly: at first it grows, then it decays expo-nentially. As in the case of γ-radiation, further processes occurring in plastic scintillators are caused by the secondary electrons formed near the track of the primary particle [397].

On interaction with PSs, the protons and α-particles induce the excitation and ioniza-tion of macromolecules and the molecules of luminescent additives, which also leads to the formation of secondary electrons. Owing to their comparatively large mass and rela-tively low velocities, the density of ionization created by α-particles and protons is high and the value of the free path is small.

On the excitation of a PS by heavy particles, one may observe a rather intense, slow component of the emission (about 300 to 400 ns) accounting for approximately 10–15% of the total radioluminescence intensity. The nature of the slow component is determined by the presence of radicals and molecules in the triplet state. Since the intensity of the slow component increases notably with increasing ionization density, these components of the emission allow recording of heavy particles on the strong scintillation background created by electrons.

On interaction with the polymer base of PSs, fast neutrons, whose kinetic energy reaches 0.5 MeV and over, cause the formation of recoil protons with an energy suffi-cient for their recording. Recoil protons have a very short range but they contribute to the intense excitation and ionization of molecules. The higher the hydrogen content of the polymer, the more effective is the scintillator in the recording of fast neutrons.

Neutrons with a kinetic energy below tens of eV are slow and incapable of inducing the ionization and excitation of ions by themselves. The recording of slow neutrons is based on their capture by the nuclei of some elements. The nuclei formed as a result of neutron capture are likely to be unstable and undergo nuclear transformations accompanied by the emission of a γ-quantum, electron, α-particle, or proton. This is the secondary radiation which is responsible for the ionization and excitation in the polymer base. The occurrence of such processes depends on the elemental composition of the sample. Practically no slow neutrons are captured in aromatic vinyl polymers. Thus, the PSs of normal composition are unsuitable for recording slow neutrons. To initiate the capture of slow neutrons, the PSs are doped with special compounds containing boron, lithium, halogens, and nitrogen [399, 400].

When recording heavy particles (protons and α-particles) the light yield of plastic scintillators is much lower than it is during β- or γ-excitation. To characterize the light yield of scintillators in terms of the type of excitation, the ratios α/β and p/β are introduced, showing the value of the light yield resulting from excitation by α-particles, fast neutrons, and protons.

For all organic scintillators, the ratio α/β is virtually independent of the individual properties of the scintillator and is within 0.07–0.15 [358]. For PSs, the α/β ratio is about 0.1. PSs show the following spread in the p/β ratio: 0.14; 0.24; 0.26. This is likely to be connected with the excitation by protons of various energy levels. Ref. [401] presents values of the p/β ratio for all types of Russian-made PSs. For scintillators based on polystyrene, poly(methyl- and poly(2,4-dimethyl styrene), p/β equals 0.2 at $E_p = 1.3$ MeV and $E_e = 0.624$ MeV. With decreasing proton energy, the p/β ratio also decreases to 0.12 ($E_p = 0.58$ MeV). Very different is the p/β ratio for the PMMA scintillators ($p/\beta = 0.041$, $E_p = 1.3$ MeV).

The low radioluminescence yield during excitation with heavy particles can be explained by the formation along the track of free radicals which make a significant contribution to scintillation quenching [402]. In [403] scintillation quenching is attributed to the local increase in the temperature of the charged heavy particle track.

The application of plastic scintillators of large sizes and lengths for monitoring γ-radiation and elementary high energy particles makes special demands on their transmission and luminescence time [404]. Not only do the small light losses have a bearing on the light yield in PSs, but they also facilitate the discrimination of the background and separate detection of particles emitting energy of various types in the scintillator, which ensures a better energy resolution in large total absorption counters [405].

The parameter determining the application of scintillators in fast coincidence circuits is the ratio between the scintillation efficiency and the time of scintillation emission, s/τ. The higher the ratio, the more effective is the scintillator in fast coincidence circuits [406].

The transmission of luminescence by polymeric systems is determined both by the optical properties of the polymer base and the spectral characteristics of the luminescent additives.

The polymer base should transmit in the emission region not only radiation from the secondary but also the primary luminescent additive, because a portion of energy (up to 50%) is transmitted from the primary to the secondary LA by radiation. The transmission of the polymer base is determined by the structure of the repeat unit, the quality of the initial material, and the conditions of polymerization [404].

The highest transmission of PSs is provided by luminescent additives used as spectrum mixers in which the minimum overlap between the absorption and luminescence spectra occurs, otherwise the reabsorption and re-emission of light by the LA is observed. As a result, the luminescence spectrum of the spectrum mixer shifts to the longer wavelengths and its technical quantum yield decreases. To eliminate this, the concentration of the spectrum mixer is usually reduced compared with that approximating to unit efficiency of nonradiative energy transfer from the primary LA to the spectrum mixer.

Together with the above factors, the transmission of a polymeric scintillation system depends on the shape and geometrical size of the sample, the quality of the surface, and the presence and nature of the coating effecting light collection [407].

The transmission of the scintillator can be characterized by the length corresponding to 50% transmission, the length corresponding to light loss by a factor e or the absorption index κ, which is calculated from Beer's law:

$$I = I_0 e^{-\kappa \ell} \tag{7.1}$$

The absorption of radiation in the scintillator depends on its wavelength. Short wavelength radiation is absorbed by the first layers of the scintillator, which is why the absorption index of the short samples is much higher than that of the longer scintillators where account is taken largely of the absorption of the long wavelength component.

The highest transmission in the range of 250–400 nm is displayed by PMMA, owing to the absence of conjugation in its repeat unit. The presence of the benzene ring in polystyrene and its derivatives reduces transmission at $\lambda > 250$ nm. However, because PMMA is an inactive polymer base, to make it act as a scintillator, it has to be doped with a secondary solvent such as naphthalene, which absorbs at longer wavelengths, as a result of which the transmission of the polymer compound worsens.

Use of a secondary LA, the luminescence spectrum maximum of which is shifted significantly to longer wavelengths, increases the effective transmission of PSs. Higher transmission of PSs is also achieved by improving monomer purification methods [404].

To increase the light yield and maintain the high transmission of PMMA-based scintillation systems, a series of LAs was selected to suit the specified polymer base. One of the requirements imposed on the LAs is their high solubility in the monomer at the polymerization temperature (18–20°C). This requirement is met by some oxazole derivatives such as PPO, 2-(4-biphenylyl)-5-phenyloxazole (BPO) and biphenyl-α-naphthyloxazole (BαNO).

In [378] (USSR Inventor's Certificate 203229) the dependence of transmission on the polymerization temperature is reported for PMMA-based scintillators. The advantage of low temperature polymerization (18–20°C) is seen from Table 7.3.

Table 7.3. Relationship between effective absorption index of PMMA scintillators and polymerization conditions [377]

Concentration/%		Polymerization conditions/°C	κ/cm^{-1}
Naphthalene	POPOP		
8	0.06	80	0.0111
8	0.06	18 to 20	0.0055
15	0.08	80	0.0100
15	0.08	18 to 20	0.0060

For a fixed or similar chemical composition of polymeric systems, the transmission of PSs is also dependent on the quality of the surface finish, especially when the polymer layers are very thin (<10 mm) [405].

The surface quality of detectors stems from grinding and polishing. The T_s of detectors also depends on the light collection coefficient which, along with the natural luminescence transmission of the system, is determined by the reflection, refraction, and absorption of light at the scintillator–medium interface. The relationship between T_s and surface quality is ambiguous, being dependent on the geometry of the sample and the h/r ratio.

According to the type of surface finish, detectors have a reflectivity of different types: diffuse reflection on matt surfaces and total internal reflection on polished surfaces.

For cylindrical detectors with h/r ratio <10, T_s is higher if the surface is matt (diffuse reflection); the same holds true for polished surfaces when $h/r > 10$ (total internal reflection). Whatever the height of the parallelepiped detectors may be, the highest T_s is displayed by polished samples [407].

Light collection in PSs and consequently, the light yield, are increased by applying a reflective coating. In large detectors, total internal reflection is best attained on a well polished surface without a reflector. For this reason, layered detectors are used without a coating.

PSs are the fastest detectors; their luminescence of several ns is explained by the mechanism of the scintillation process [408].

The shape of the light pulse generated by the PS is presented in Fig. 7.2. The flash time (pulse rise time τ_r) is determined by the rate of the excitation energy transfer process [409]. The excitation of luminescence is associated with the deactivation of higher level solvent molecules excited by radiation. This process precedes energy transfer. The

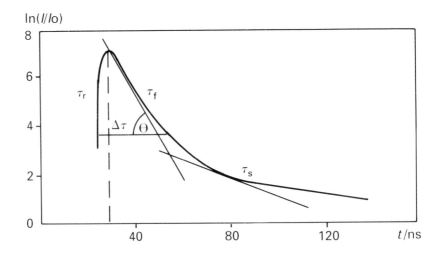

Fig. 7.2. Scintillation pulse shape. τ_r, pulse rise time; τ_f, fast component decay time; $\Delta\tau$, full width at half maximum.

duration of luminescence buildup is determined by the time during which the pulse amplitude increases in its value from 10 to 90% [409].

The transfer of the LA molecule from its lower excited singlet state to its ground state determines the decay time τ_f of the fast scintillation component, which is shown as the slope of the linear part of the given component $\tan \theta$.

In [410] the value τ_f was determined by the time of reduction of the pulse by a factor of e. τ_f is also taken to be the time during which the pulse amplitude decreases from 90 to 10% [411].

The decay time τ_s of the slow component depends on the triplet–triplet annihilation process. The decay constant of the slow component is large compared with τ_f and reaches *ca* 200 ns, its contribution to the light yield being small.

To characterize the kinetic properties and time resolution of PSs, the concept of the *full pulse width at half maximum* is introduced ($\Delta \tau$) (see Fig. 7.2). The full width at half maximum depends on the scincillation rise and decay times.

The study of scintillation polymeric systems by the single photon counting method [412], with the subsequent approximation of the experimental values by the least squares technique, permits determination of the shape of the PS pulse and factorization of the medium component (τ_m = 20 ns) from the fast component (τ_f = 2–5 ns). The relationship between the fast and medium components depends on the LA concentration in the polymer. The fast component is absent at low concentration of the LA and dominates when its concentration is optimal. The medium component is observed at practically any concentration of the LA, even zero, and its time constant is 18–20 ns irrespective of the LA concentration. The decay time of the medium component corresponds to that of the residual monomer. The luminescence time of the fast component decreases with increasing concentration of the LA. The fast component dominates at the optimal concentration; its light yield exceeds that of the medium component by approximately an order of magnitude, thus determining the light yield of the scintillator. The scintillation pulse shape is identical for PSs based on polystyrene with various LAs; their difference lies in the light yield and the time constant of the fast component of luminescence.

In some cases, when using PSs in fast response electronic equipment, efforts should be made to reduce the emission time of the scintillator; at the same time it is desirable to maintain the light yield at a sufficiently high level. One way to reduce the time of PS emission is to increase the concentration of the primary LA, which, however, may result in self-quenching. Increase in the concentration of luminescent additive is closely related to its solubility in the polymer base; the solubility is usually improved by introducing methyl groups into the molecules of the luminescent additive [413].

The luminescence time can be reduced not only by concentration quenching but also by adding quenching agents such as benzophenone or piperidine [414] and LAs containing heavy atoms [413, 415]. Table 7.4 gives the luminescence time and light yield of some scintillators from various countries.

As is evident from Table 7.4, the shortest decay time τ_f is displayed by Pilot U, while the least full width at half maximum $\Delta \tau$ is characteristic of NE-111 and PS-330. With reduction in the luminescence time, the light yield of the scintillator drops correspondingly.

Table 7.4. Time characteristics (ns) of proprietary scintillators

Scintillator	τ_r	τ_f	$\Delta\tau$	$T_s/\%$
NE-111	0.13–0.4	1.66	1.38	50
Pilot U	0.5	1.36	1.8	58
Pilot B	0.7	1.69	2.4	60
Naton-136	0.7	1.70	2.3	45
NE-102A	0.9	2.2	3.2	58
PS-131	0.6–0.9	2.1–2.5	3.4	40
PS-331	–	–	1.9	68
PS-330	–	–	1.35	50
PS-305	–	–	1.57	56
PPS-B9	0.4	1.7	1.5	30–40

Ref. [416] investigates systems based on poly(2,4-dimethyl styrene) with one LA which is a highly conjugated system including 1-phenyl-2-(2,5-diphenyloxadiazolyl)ethylene and 1-phenyl-2-(2-phenyl-5-biphenyloxadiazolyl)ethylene. Owing to the presence of the oxadiazole ring, their absorption band is similar to that observed in PPO and PBD, while the increased conjugation promotes the shift of the luminescence maximum to longer wavelengths, which means one can use these additives without a spectrum mixer. The $\Delta\tau$ of such scintillators is 1.57 ns, their T_s is at the level shown by scintillator NE-102A.

An important performance characteristic of scintillators is their resistance to ionizing radiation. Of all types of scintillator, PSs show the best radiation resistance. Thus, for polystyrene scintillators, the absorbed dose, at which the light yield reduces by half, exceeds by an order of magnitude the dose causing similar damage to the anthracene monocrystal [417, 418]. Much research [419–422] is devoted to the resistance of PSs to hard ionizing radiation. In these papers most attention is given to the resistance of polystyrene scintillators to continuous γ-radiation, and few authors are concerned with the resistance of scintillators to α- and β-particles.

Ref. [423] studies the radiation resistance of PSs based on methyl derivatives of polystyrene. The luminescence and EPR spectra of irradiated polymers reveal that the introduction of methyl groups into the benzene ring results in worse damage to the polymer base. The degradation constants have been determined for a number of these polymers. Thus, for polystyrene, polymethylstyrene, and poly(2,4-dimethyl styrene), the degradation constants are 0.04, 0.14, and 0.2 respectively.

The radiation resistance of PMMA based scintillators is determined by the amount of secondary solvent (naphthalene) present in them. When naphthalene accounts for 15 wt% of PMMA, the radiation resistance of the material is similar to that of the polystyrene scintillator [424]. An important feature of PSs is their capacity to regain their original light yield after irradiation (the 'post-effect'). The highest rate of regaining light yield is displayed by the poly(2,4-dimethyl styrene) scintillators. The chemical nature of the LA has a significant bearing on the level of restoration.

The structure of LA molecules influences the level of damage, and, consequently, the reduction in light yield of polymer-based scintilllators of each particular type. The best resistance is observed in the PSs which use as LAs compounds of the polyphenyl (ter-phenyl) or tetraphenylbutadiene type. Oxazole LAs (PPO) promote a faster degradation of the polymer base. As a rule, the presence of the secondary LA (the spectrum mixer) increases the radiation resistance of plastic scintillators [425].

The high radiation resistance of polystyrene and LAs of the polyphenyl type containing a large number of aromatic rings is explained by their unique 'sponge' effect. The molecular orbitals associated with the phenyl group provide a series of energy levels which effect the fast dissipation of electronic energy without any residual effects. Increasing conjugation in the molecules of aromatic compounds increases their radiation resistance.

Ref. [426] suggests the improvement of the radiation resistance of polystyrene scintillators by adding naphthalene. However, the solubility of naphthalene in polystyrene is limited (about 4%). Highly radiation resistant polymer scintillators are based on styrene with vinyl derivatives of naphthalene and diphenyl [427].

Pulsed γ-radiation acts on PSs in a different way [420]. Irradiation with 2×10^4 rad causes significant damage to the system within a short time (1 µs), as a result of which severe deterioration in the natural luminescence transmission of the system is observed.

7.1.1.2 Production of plastic scintillators

The main method for producing plastic scintillators is bulk polymerization of the LA solution in the monomer. PSs can be also obtained by dissolving the LA in the polymer melt, by pressure melting the polymer–LA mixture at a high temperature, and by pressure moulding and extrusion [404].

The most generally employed is radical bulk polymerization, which forms the basis for the commercial production of PSs. The advantages of this method are the most uniform distribution of the LA in the polymer base and the possibility of manufacturing scintillators of large sizes and different shapes.

Promising methods for producing small (about 100 cm^3) scintillation detectors in the form of cylinders, hollow cylinders, glasses, etc., as well as scintillation films, fibres, capillary tubes, and thin plates, are transfer moulding and extrusion. These methods can be used for the manufacture of scintillators on the basis of thermoplastic polymers obtained solely by ionic polymerization.

The high temperature of bulk polymerization (thermal polymerization proceeds at 200°C) [404] ensures a high production rate, the absence of by-products (products of initiator decay) in the sample, and insignificant amounts of residual monomer (ca 1%). At the same time, the RMM of the polymer is relatively small (approximately 100 000). However, the high temperature mode cannot be used to manufacture large blocks because of the exothermic nature of the process and the difficulties of heat dissipation.

During thermal polymerization at low temperatures (50–60°C), the RMM of the polymer increases to 400 000–600 000. At the same time, the production rate falls perceptibly and the content of residual monomer increases (up to 3%) [404, 428]. Increase of the reaction rate at low temperatures requires the use of a polymerization initiator.

Polymerization at an average temperature (140–170°C) in the presence of an initiator [428] is widely practised. This method is used to obtain large polymer scintillation detectors.

Quite promising is two-stage polymerization [429]. The prepolymer is formed (30–40% conversion) during the first stage at 80–90°C. The additional polymerization of the prepolymer is performed during the second stage at 160°C.

During initiated polymerization the light yield is significantly affected by the initiator. Benzoyl peroxide is used heavily in the manufacture of plastic scintillators; however, the data on the effect of benzoyl peroxide on the light yield of the scintillator are rather contradictory. In a number of studies it has been established that benzoyl peroxide has a quenching effect, even at low concentrations. However, it is shown in [428] that low concentrations of this initiator have practically no effect on the light yield of the scintillator. Even at high concentrations of azobisisobutyronitrile, the light yield of PSs does not fall; however, its presence reduces the RMM.

It has been noted in [430] that during polymerization in air, the light yield of scintillators falls by 20–30%. A high light yield is attained in nitrogen or *in vacuo*. It follows that during the manufacture of PSs, it is important to exclude all oxidation processes which lead to the formation of compounds causing luminescence quenching or absorption.

As noted above, polymer scintillators may be produced by processing masses of plastic [431].

Extrusion is used to obtain scintillation fibres and capillaries. The necessary conditions for producing high transmission scintillation fibres are the correct selection of the PS composition, high surface quality and sufficient elasticity of the material, no contact between the melt and air, and uniform extrusion.

To obtain film plastic scintillators, the extrusion and varnish methods are used. This method consists in the application of the LA solvent and the polymer onto a glass or PMMA support in an organic solvent (toluene or methylisobutyl ketone) with subsequent evaporation. To make the film elastic, the solution is doped with a plasticizer, and, to improve the light collection, the film surface is matted [432]. To improve the transmission and thermal resistance of film scintillators, a procedure for grafting styrene to poly(ethylene terephthalate) has been developed. The thermal resistance of films was also increased by using isotactic polystyrene.

As shown in [431, 433], plastic processing is the most efficient method for obtaining thin plates, scintillation films, fibres, capillaries, etc. The advantages of this method are the high production rate, the ease of manufacture, the possibility of producing long items of uniform thickness with high surface quality. The extrusion method induces a preferred orientation of the macromolecules, which increases the scintillation efficiency of the system.

7.1.1.3 Fields of application of plastic scintillators

Depending on their composition, dimensions, and structure, plastic scintillators possess a distinct grouping of properties which determines their fields of application [12]. To detect elementary high energy particles, which are part of cosmic radiation or are obtained artificially with accelerators, use is made of plastic scintillation detectors of large area, while large volume detectors are used to detect high energy γ-radiation and fast neutrons.

The investigation of cosmic radiation (and of extended aerial showers in particular), including determination of their composition, anisotropy, and energy spectrum, the search for the neutrino, the detection of muons, the generation of the neutrino, and the seach for long lived heavy particles of plutonium and quarks, are all performed both at ground level space stations and underground at a depth of up to 8000 m of water equivalent. In a number of experimental installations, the total area of the detectors composed of polymer scintillation plates reaches many tens of km^2.

PSs of large area are used in spectrometers for recording charged particles in flight. Plastic scintillators in the form of strips, rods, and fibres find application in hodoscopic devices designed to determine the trajectory of the particles obtained from accelerators.

Large plastic scintillation detectors based on PMMA have proved to be very efficient in the protection of low background detectors from cosmic irradiation. In their low background properties, scintillation plates based on PMMA are superior to their polystyrene analogues because their natural background as regards natural radioactivity is approximately a third of that of scintillators based on polystyrene.

The main requirement of PSs of large area, ensuring uniform light collection, is their high self-radiation transmission. Owing to this, background discrimination is easier in large full absorption counters. The highest transmission is displayed by PMMA based scintillators with a 10% naphthalene content (mass). In a study of the dependence of the maximum uniformity of light collection by polystyrene and PMMA based scintillators on the nature of the reflectors and light guides, and the arrangement of different types of photoelectric multipliers, it has been established that the amplitude nonuniformity factor is ±18% for PMMA scintillators and ±30% for polystyrene scintillators. In scintillators of the type NE-102A, the nonuniformity factor is within ±12 to ±32%.

Measurements of low energy cosmic radiation and weak neutron fluxes, as well as the study of X-ray radiation of cosmic origin, are conducted directly in space on Earth satellites, at orbital space stations, and in the atmosphere by means of rockets, balloons, and aircraft. For this purpose, various combined detectors are used.

Along with their self-radiation transmission, a short emission time is one of the most significant properties of plastic scintillators. PSs are among the fastest detectors. Poly(2,4-dimethyl styrene) (polyvinylxylene) holds much promise as a polymer base for scintillators with a short emission time. Detectors based on poly(2,4-dimethyl styrene) have been used in the manufacture of a positron lifetime spectrometer.

To determine low activities, for example the γ-activity of a living organism, PSs of large volumes (rings, hollow cylinders) are used.

To detect medium energy γ-radiation, in particular in geological prospecting, plastic scintillation cylinders of diameter up to 300 mm and length 500 mm are used. Detectors of small volume with a large light yield and short emission time are required for intense γ-radiation fluxes. For this purpose, polyvinylxylene scintillators are best suited.

To determine the absorbed dose of γ- and X-ray radiation, in medicine and biology, use is made of detectors equivalent to air and tissue as regards atomic number and density. PSs equivalent to tissue in effective atomic number have found application in clinical dosimetry [392]. In a number of biological investigations, especially into the action of neutrons on living organisms, there is a need for PSs equivalent to tissue not only in atomic number but also in elemental composition.

The penetrating capacity of electrons is much poorer than that of γ-radiation, and it depends on their energy and the nature of the irradiated substance. This is why to detect electrons of β-active isotopes, use is generally made of PSs whose thickness does not exceed 1–5 mm.

The currently available β-detectors have a wide variety of formats. To determine low β-activities (surface contamination), use is made of detectors in the form of thin scintillation plates (in the absence of a γ-background) and film scintillators (in the presence of γ-radiation). In the latter case, the scintillation film is applied to a support made of an inert transparent polymer.

To determine the specific activity of aqueous media, for example potable, sea and ground water, the best detector is one with an extended surface based on scintillation plates, with considerable contact with the object under study. Flow detectors are essentially a coiled capillary made from scintillation plastic.

Counters have been developed for measuring the radioactivity of gases, such as air, carbon dioxide (with $^{14}CO_2$ impurity), and some aggressive gases. Such detectors are usually manufactured in the form of hollow cylinders, spheres, or jointed disks. In some cases, PSs are combined with counters of a different type to determine low activity in solid samples.

Of special note is the application of PSs for internal measurements in liquids. When measuring the β-activity of dilute solutions, ion exchange scintillation granules are used on the PS surface to increase the isotope concentration.

Plastic scintillators are used as detectors of secondary and elastically reflected electrons in the electron microscope, and as detectors of β-radiation in radioisotopes, thickness gauges, and densimeters.

The low atomic number of PSs permits their use, along with organic monocrystals such as anthracene, as detectors in β-spectrometry. For this purpose, scintillators are based on poly(2,4-dimethyl styrene) which has the lowest Z (5.7) and a high light yield.

To detect high energy neutrons, long detectors are employed in time-of-flight spectrometers in the absence of a γ-background. Use is also made of directed scintillation detectors in the form of parallel fibres, rods, or plates.

One topical problem of scintillation engineering is the selective detection of different types of radiation excited in mixed fields (for example, during space exploration, from accelerators, or during radioactive isotope contamination control). In this case, one type of radiation is separated from its companions; γ-background neutron detectors are examples of these.

The simultaneous (combined) detection of several types of radiation is a more complex task. Detectors of two types are used: a homogeneous radiation detector providing for the discrimination of radiation by pulse shape, and combined detectors composed of scintillators of different types and properties.

A detector for the separate recording of α- and β-radiation is based on a layer of polycrystalline ZnS(Ag) applied to a scintillation polymer plate. The highly efficient scintillation compositions based on poly(2,4-dimethyl styrene) in combination with organic monocrystals (naphthalene, stilbene) are used for the detection of γ- and β-radiation. Pulse separation is based on the difference in luminescence times of the two scintillators. The separate detection of neutrons and γ-radiation is achieved by the combined use of liquid and plastic scintillators.

The literature describes the design and principle of operation of combined inorganic monocrystal detectors of the types NaI(Tl), CsI(T1), and PS in which inorganic monocrystals act as intrusion scintillators operating on the anticoincidence principle. In this case, PSs are made in the form of hollow cylinders or well-type cylinders of varied configuration. Such detectors find application in the investigations of cosmic radiation. PSs are also combined with other types of detector such as semiconductors and ionization and spark chambers.

7.1.2 Polymer compositions for conversion

The method of conversion is designed for collecting light from sources of any area and volume. During conversion, along with the reduction in the cross-section of the light beam, its spectrum shifts to longer wavelengths. Analysis of [359–361, 434, 435] indicates that luminescent polymer compositions are the sole promising material for light conversion.

The converter, a polymer plate incorporating a luminophor, is an intermediate detector: the light flux coming from the scintillator falls on the larger surface of the plate and, on leaving it in the form of an emission of longer wavelength through the smaller surface, arrives at the light collector (photomultiplier, semiconducting detector). A method for assessing the level of light concentration provided by converters is suggested in [360]. In earlier studies [359], the suggestion was made of applying a multistage conversion system promoting a further light concentration and a further shift to the longer wavelength region. As a result, it becomes possible not only to increase the level of light concentration but also, with the long wavelength light detector, to correlate better the radiation spectrum and the spectral sensitivity of the detector. Refs [359, 434, 435] cover the pioneer applications of fluorescent polymer compositions, along with coloured glass, in converter systems. These references also summarize the basic requirements for such compositions.

The choice of luminophor to be incorporated in the conversion composition depends on its spectral properties, the absorption region which must correspond to the emission region of the scintillation detector, and the fluorescence spectrum which must correspond to the maximum sensitivity of the light detector.

The losses of energy should be minimal during light conversion. The efficiency of conversion is characterized by the transmission factor [360] and depends both on the spectral and luminescent properties of the luminophor and the optical properties of the polymer matrix. The transmission factor is dependent on the natural radiation transmission of the whole composition. High transmission is provided for by selecting a luminophor with a large Stokes shift (minimum reabsorption of natural fluorescence) of the absorption and luminescence spectra. The polymer matrix is also expected to display high transmission due to the chemical structure of the repeat unit, optical uniformity of the polymer block, and high purity of the initial monomer.

The highest transmission over a wide spectral region is displayed by poly(methyl methacrylate) which is why this polymer is used in the production of conversion compositions.

In selecting a luminophor, primary attention should be given to its solubility in the monomer and polymer. In most investigations BBQ has been studied as the luminophor most suitable for the production of conversion compositions; this substance is characterized by an emission in the green spectral region and a large Stokes shift [439].

The luminophor concentration in the converter should be, on the one hand, sufficiently high to ensure the almost total absorption of the light emitted by the scintillator during its transverse passage in the conversion plate, and, on the other hand, it should be sufficiently low for there to almost no absorption of the (long wavelength) converted radiation by the material of the polymer composition during its longitudinal passage. According to [436], at a BBQ concentration in the conversion composition of 90 mg dm^{-3}, the absorption of the scintillation light occurred at 1.34 cm, that is, the absorption was almost total. When the concentration of BBQ was increased from 35 to 300 mg dm^{-3}, the luminescence attennuation wavelength changed from 119 to 69 cm [437].

When producing conversion compositions, special attention should be given to the conditions of polymerization and the nature of the polymerization initiator [438]. Depending on the chemical structure of the luminophor, the peroxide polymerization initiators may form oxidation–reduction systems which leads to a change in the fluorescence region of the luminophor (hypsochromic) and to the inhibition of polymerization [440].

As is the case with plastic scintillators, the main factor determining the optical uniformity of the polymer conversion unit is the temperature of polymerization [404]. The significant light losses occurring during light conversion are caused by low quality machining of conversion plates. In this connection, during the manufacture of conversion devices, particular attention is given to obtaining a high surface finish. Of special importance is the dependence of the conversion efficiency on the surface quality and geometric configuration of the converter [441].

One method to improve the transmission of converters is the development of surface modified converters by the application of thin-film polymer fluorescent varnish coatings on the transparent polymer block or glass.

The application of conversion polymer compositions in scintillation engineering has opened up new opportunities for the development of detectors and instrumentation for ionizing radiation.

7.2 LUMINESCENT SOLAR CONCENTRATORS

Beyond the terrestrial atmosphere at an average distance from the Earth to the Sun, the density of solar radiation energy flux is 1.353 kW m^{-2} in the plane perpendicular to solar rays [442]. Thus, a great amount of energy falls on the Earth at all times. At present, however, the high cost of conversion of solar radiation into other convenient types of energy hampers the advancement of solar energy engineering. One of the possible ways to make it economical is to apply new cheap materials, and technical devices based on them, to improve the efficiency of solar light conversion.

The luminescent solar concentrator (LSC) represents one such device [443, 444].

Commonly used in solar energy engineering are mirror or mirror–lens concentrators. The major advantages of LSCs is that a solar 'tracking' system is not needed, and they intensify both direct and diffuse solar light, whereas conventional concentrators are inefficient in collecting scattered radiation.

The first paper [445] devoted to the application of luminescence for the concentration of solar radiation was published in 1976. As a transparent medium, use was made of glass activated by trivalent neodymium ions absorbing in the spectral region from 500 to 900 nm and emitting with a maximum at 1050 nm with a quantum yield of 50–75%. Then the basic requirement was specified for the spectral overlap of the absorption and emission curves to be minimal. For the concentration of solar energy, the suggestion was made of using mixtures of dyes with different absorption and emission regions.

Further investigations [440] led to specification of the requirements of the materials used in LSCs. At the same time, the possibility was studied of applying LSCs in the intensification of diffuse light [447]. A description was given of energy losses occurring during the repeated reflection of light in the dye–polymer matrix system. Ref. [448] dealt with the conditions providing the optimum efficiency of solar light concentration. The theoretical aspects of light absorption and reflection in LSCs were investigated in parallel [448–455].

In spite of the fact that at present new domains are being opened for the application of LSCs (such as lighting and optical engineering), their main application is to the simplification of conversion of solar radiation into electric or thermal energy [454]. LSCs concentrate by converting solar radiation on the light sensitive surface of the photoelectric converter (PEC) or a special absorber.

In recent years, many electrochemical companies worldwide have been concentrating their efforts on the development and production of water heaters for room heating, hot water supply, and air conditioning, and solar radiation converters, based on photoelectric solar batteries. The world production of solar batteries increases annually by 75%. The best types of PECs show an efficiency of about 27% in converting solar into electrical energy [442], but in terms of cost per kWh they cannot compete with conventional power sources. For this reason, the application of cheap LSCs seems most promising.

7.2.1 Principles of solar light concentration
Generally a LSC is a light-transparent plate incorporating a luminescent dopant which has a broad absorption band in the visible and UV regions. The operation of the LSC is shown in Fig. 7.3.

When passing through the transparent plate, solar light is partly absorbed by the luminophor and converted into luminescent radiation. The portion of the isotropic luminescent radiation, found within the boundaries of two critical cones limited by the angle of incidence $\varphi_0 = \arcsin n$ (n is the refractive index of the matrix), escapes the LSC. The remaining radiation undergoes a total internal reflection from the plate surface and reaches its end faces. For PMMA with a refractive index of 1.49, this fraction makes up 74%.

Normally the luminescent radiation detector is combined with one of the end faces of the plate, while other faces are provided with a mirror coating. Radiant energy is concentrated in LSCs because the smaller end face transmits the radiation originating in the plate with the large area of the illuminated surface.

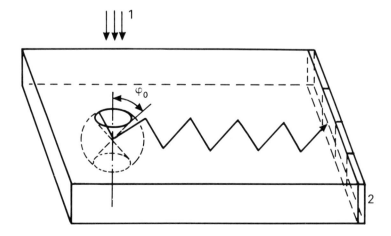

Fig. 7.3. LSC diagram (for clarification, refer to text) 1, solar light; 2, PEC.

The limitation on the efficiency of solar radiation in real LSCs is determined by non-productive losses, which are due to:

(i)   the inadequate spectral absorption of the solar spectrum by the dyes,
(ii)   the magnitude of the quantum yield,
(iii) reabsorption,
(iv)   the absorption of radiation by the matrix material,
(v)   losses by reflection,
(vi)   selection of the shape and area of the plates, and
(vii) the lifetime of the materials employed.

The principal characteristic of the LSC–PEC system is the energy concentration coefficient C_{el} which is determined by relative variations in the outlet electric power of the PEC as a result of introducing the LSC. Generally, it may be represented as follows [455, 456]:

$$C_{el} = W_{LSC}/W_0$$

where W_0 is the electric power of the PEC and W_{LSC} is the electric power of the PEC together with the LSC.

The efficiency of the LSC–PEC system is determined from the following formula [456]:

$$\mu = C_{el}\eta^0_{PEC}/G$$

where G is the ratio of the area of the illuminated plate to that of the light sensitive surface of the PEC (geometric factor) and η^0_{PEC} is the conversion efficiency of the PEC.

Let us consider the general diagram of solar radiation losses in the LSC (Fig. 7.4).

As is well known, a portion of the light flux is reflected at the boundary of two media with different optical densities. The greater the difference between the refractive indices of such media, the larger is the fraction of the reflected light flux. For light falling normally from a vacuum, the values of the reflective and transmitting capacities are described by the Fresnel equation:

$$R = (n-1)^2/(n+1)^2; \quad G = 4n/(n+1)^2$$

where R is the reflective capacity or reflctivity, G is the transmitting capacity, and n is the refractive index of the medium.

For a light beam falling arbitrarily, the Fresnel equation is more complex. The values of R and G depend on the polarization of the incident wave [457]. Calculations show that 4% of light falling normally and 5.8% of light falling at an angle of 50° is reflected from PMMA with $n = 1.5$.

To decrease reflection, antireflective coatings can be applied to the LSC surface. The maximum amount of light collected by the plate depends on the refractive index of the medium. When no antireflective coatings are applied, this maximum is reached at $n = 2$. If antireflective coatings are used, it will be necessary to bear in mind that account is taken in the Fresnel coefficients of the refractive indices and thickness of these coatings. Thus PMMA with a MgF_2 antireflective coating with a thickness of 1.2×10^{-5} m and $n =$

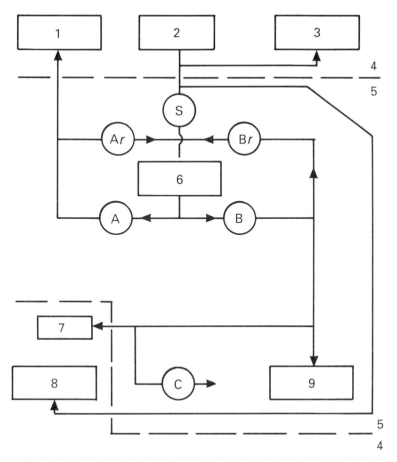

Fig. 7.4. LSC optical losses 1, critical angle losses; 2, incident light; 3, reflected light; 4, air; 5, LSC; 6, dye molecule; 7, PEC; 8, nonabsorbed radiation; 9, transportation losses.

1.38, shows reflection of only 1.5% of the light falling normally and 2.6% of light falling at 50° at a wavelength of 600 nm [450].

On passing through the LSC plate, a fraction of the nonreflected light is partly absorbed by the dye and the matrix material. This fraction is determined by the extinction coefficients of the molecules of the dye and matrix material, and is governed by the Beer–Lambert law:

$$I = I_0 \exp\left\{-\left[\alpha_K\left(\bar{\lambda}\right) + \alpha_M\left(\bar{\lambda}\right)\right]d\right\}$$

where I_0 and I are the intensities of the incident and transmitted light beams, $\alpha_K(\bar{\lambda})$ and $\alpha_M(\bar{\lambda})$ are the extinction coefficients for the average absorption wavelength of the dye and the matrix, and d is the thickness of the sample.

A fraction of solar radiation is lost because of the lack of correspondence between the absorption spectrum of the luminophor and the solar radiation spectrum. As a rule, for an LSC plate containing one luminophor, the amount of absorbed radiation does not exceed

25% [458]. For a noticeable reduction of such losses, it is possible to use several dyes with a significant overlap of their absorption and emission spectra, which optimizes the use of the solar spectrum. This method enables the efficiency of solar concentration to be increased two- and three-fold [452, 453, 459]. It should be noted that the total light flux and spectral distribution of solar radiation are dependent on the weather.

The fraction of solar radiation absorbed by the dye molecules is designated by the letter S in Fig. 7.4.

During the absorption of the photon, the dye molecule passes into its excited singlet state within 10^{-12} s or into the long-lived triplet state via intersystem crossing. There are several ways by which the molecule can relax into the ground state: by fluorescence from the excited singlet state, by phosphorescence from the excited triplet state, by internal conversion through vibrational levels, etc. For molecules of dyes, the processes of internal conversion are not significant, and radiative deactivation becomes predominant. The quantum yields of the dyes used in LSCs are quite high, namely 0.5–1.0 [452].

As discussed above, after shining on the critical cone area, the isotropic luminescent radiation leaves the LSC. In Fig. 7.4, this fraction of radiation is designated by A.

The amount of fluorescent radiation B (Fig. 7.4) collected by total internal reflection increases with the refractive index of the matrix material. At the same time, the losses by the reflection of solar light also increase.

For PMMA with $n = 1.5$, the critical angle is 42°. In view of the fact that the critical angle is dependent only on the refractive index of the matrix, the application of an anti-reflective coating provides a significant reduction in the amount of reflected light on increasing the refractive index of the matrix.

For most luminescent compounds, the absorption and emission spectra overlap, which leads to the repeated absorption of a portion of the luminophor radiation by other molecules of the dye.

The most complex aspect is assessment of the losses occurring in the reabsorption of radiation. In [451, 460–462], the calculation of such losses with regard to specific luminophors doped into the polymer matrix was based on the Monte Carlo simulation method. These calculations show that the level of losses in reabsorption imposes strict limitations on the dimensions of the plate and the luminophor concentration in the LSC.

The light reabsorbed with probability r is emitted at longer wavelengths (red shift), and again this radiation either shines on the critical cone area or undergoes total internal reflection. Owing to the reabsorption, the light quantum emitted by the dye molecule will diminish proportionally to $\exp[-\alpha_K(\overline{\lambda})\ell]$, where ℓ is the path taken by the quantum before it leaves the plate. If the dimensions of the LSC are larger than the average length of the quantum path, then owing to the long optical path, at the plate outlet it will be possible to observe a red shift of the repeatedly absorbed light.

Furthermore, for light leaving the plate at an angle exceeding the critical value, the probability $\overline{r}$ of self-absorption also exists. As a result, the light losses decrease in the LSC by another $A\overline{r}$. For polymer samples of less than 10 cm in length, the light losses in reabsorption make up some 20% [450], whereas longer plates (22.5 cm) suffer losses estimated at approximately 32% [463].

Similar difficulties arise during assessment of the 'transportation losses', that is, the losses occurring in the absorption or scattering of radiation by the matrix material and scattering at the plate surfaces during total internal reflection. Naturally, such losses are dependent on the defects and geometrical parameters of the matrix and, to a large extent, on the LSC material.

A 50 cm long LSC, based on PMMA with a dye emitting at *ca* 700 nm, has an optical density of 0.04 at this wavelength. It can be shown that at an optical density of 0.04, about 20% of transmitted radiation will be lost. On uneven surfaces (irregularities of ~0.05 mm, $d = 2$ mm) the probability of losses by incomplete internal reflection increases by 3% [450].

A further loss of luminescent radiation stems from the reflection of light on the end face C of the plate (Fig. 7.4). When falling on the flat end face of the LSC, some of the quanta are reflected inside the plate.

If the LSC has no mirror coating, only 12% of the dye radiation can fall on the PEC, arranged by one of the faces of the quadrangular plate. This value can be doubled, at the expense of half of the light trapped in the plate, by the introduction of immersion oil between the plate and the PEC [463].

As seen in Fig. 7.4, energy losses occurring in the LSC depend on the properties of the matrix material and the luminophor itself. All the abovementioned optical losses determine the value of the quantum or optical efficiency η_{opt}, which is equal to the ratio of the number of luminescence quanta illuminating the PEC to the number of quanta incident on the LSC.

Generally, the intensification of the light flux incident on the PEC, when a LSC is used, is also dependent on the geometry of the transparent plate. It was found [464] that the optical efficiency is a function of the LSC area which decreases with increasing dimensions of the plate; C_{el} is dependent on the optical path of the luminescent radiation and the LSC's shape. With a similar plate area the C_{el} is higher in pentagonal rather than triangular and quadrangular LSCs [464].

Study of the effect of the thickness of the transparent plate on the parameters of the LSC shows [446, 464, 465] that the production of and operation of LSCs depend on the matrix polymer. Among them the thickness leads to a decrease in η_{opt} at a constant concentration of the dye used. This reduction occurs owing to the following factors. On the one hand, the number of absorbed solar quanta decreases, but, on the other hand, the number of light beam reflections increases in inverse proportion to the plate thickness, which results in increased transportation losses. At the same time, C_{el} can be optimized by appropriate selection of the shape and area of the plate.

Neglecting the above losses, the efficiency of devices based on LSCs will be improved by selection of the optimal LSC radiation detector combination, the efficiency being dependent on the properties of the latter.

7.2.2 Materials for solar concentrators

From the materials at present used in the production of LSCs, the most efficient are polymers activated by organic phosphors and luminescent glasses doped with rare earth ions. For practical use, polymeric LSCs are favoured because their heat conduction is approximately one fifth that of inorganic matrices [453]. The use of polymers does not cause overheating of semiconducting PECs, which is particularly important for their stable operation.

The requirements imposed on the materials used in LSCs can be divided into two groups: requirements of the matrix material and of the dye properties. Along with these requirements, common to all optical polymers (transparency, mechanical and chemical resistance), a number of requirements connected with the production are the possibility to manufacture large plates showing UV radiation resistance and long service life at temperature extremes under field conditions.

Ample evidence based on the photophysical properties of various optical polymers favours an LSC matrix based on PMMA, which shows the best resistance to the above factors. Luminophors are used in greater variety. In addition to the requirements common to all luminophors (high quantum yield, photochemical resistance), special requirements are imposed on LSCs with regard to the physical aspects of their operation:

(i)   effective solar light absorption;
(ii) maximum overlap of the dye emission and PEC sensitivity curves (the maximum spectral sensitivity curve of a silicon and a gallium arsenide based PEC is 0.5–0.95 and 0.6–0.8 μ, respectively);
(iii) minimum overlap of the absorption and luminescence spectra (the ratio between the extinction coefficients at the maximum fluorescence and absorption wavelengths should be approximately 10^{-4} [450]);
(iv) adequate solubility in polymers;
(v)   low concentration and temperature quenching;
(vi) UV resistance (service life of at least 5 years).

In the use of dyes in LSCs, in comparison with other fields of application (lasers, scintillators), the most important of the above requirements is a continuous and long service life. For this reason, the photochemical resistance requirements are the most stringet, this factor being the major one limiting the application of luminophors in LSCs. Thus, the classical laser dye rhodamine 6 G used most frequently in experiments and theoretical calculations of luminescent collectors, shows insufficient light resistance. Unfortunately, a lack of information about the photochemical resistance of commercially available dyes does not permit their immediate application without preliminary investigations.

Study of the photochemical resistance of rhodamine 6 G and coumarine 6 (the formulae of the dyes are given in Table 7.5) exposed to illumination from a xenon lamp over 2500 h allowed assessment of their lifetime on the basis of a 50% drop in their luminescent ability. It reached 2 years for rhodamine 6 G and 6 or 7 years for coumarine [450]. Measurements of the optical density of the test samples (3 by 1 by 0.3 cm) as a function of illumination time show that after an exponential drop lasting for 760 h for coumarine 6 and 1000 h for rhodamine 6 G, the optical density of the samples, activated by coumarine 6, became stabilized, while it continued to decay in the rhodamine 6 G samples. Over a two-week exposure of PMMA samples, the optical density dropped from 3.98 to 2.79, the concentration of rhodamine 590 being 7.6×10^{-5} mol dm^{-3} [450].

Ref. [452] dealt with a large number of LSC samples using various dyes illuminated with solar light for 250 h. For coumarines 460, 500, 540, rhodamines 590, 610, and oxazine 725 incorporated in PMMA, the best results were obtained with coumarin 540 (13% losses), and the worst with oxazine 725 (68% losses). After 243 h, the optical density of LSC samples containing 2.1×10^{-4} mol dm^{-3} of coumarine 540, 8.1×10^{-5} mol dm^{-3} of rhodamine 590, and 2.2×10^{-5} mol dm^{-3} of sulforhodamine 640 decreased by 52, 50, and 51% respectively. It was also shown that protection of the polymeric samples with a glass light filter, absorbing 90% radiation, increases their light resistance by almost 100%.

Study of the kinetics of the optical density and radiative capacity of the samples during

Table 7.5. Dyes used in the manufacture of solar energy concentrators

Conditional name	Structural formula	Description	RMM	λ_{max}^{abs}/nm, ε/dm^{3}mol^{-1}cm^{-1} solvent	λ_{max}^{fl} /nm solvent
4D, 4H pyran, DCM	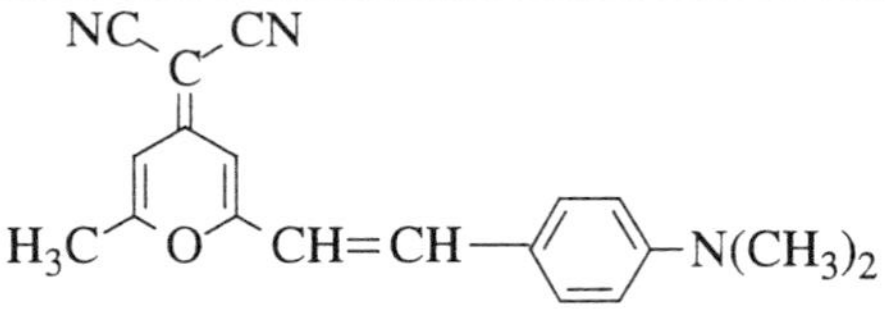	4-Dicyanomethyl-ene-2-methyl-6-(p-dimethylamino-styryl)-4H-pyran	303	472, 43000 ethanol	594 ethanol
C6, coumarin 6	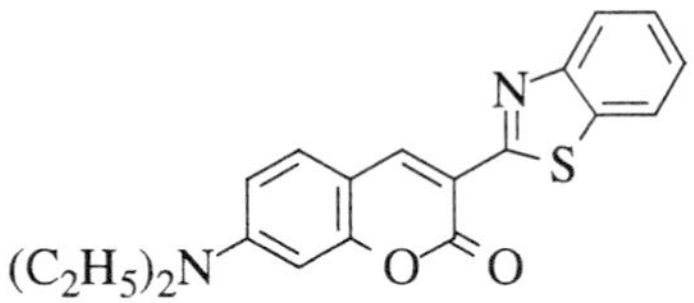	3-(2-Benzothiazolyl)-7-ethylamino-coumarin	350	458, 42000 ethanol	505 ethanol
R6G, rhodamine 6G	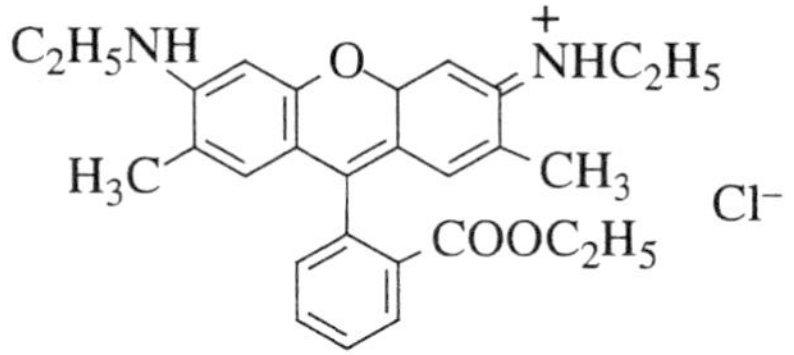	Ethyl ester o(6-ethyl-amino-3-ethylimino-2,7-dimethyl-3H-xan-thene-9-yl)benzoic acid, chloride	479	530, 95000 ethanol	556 ethanol
R64O, R64, rho-damine 640	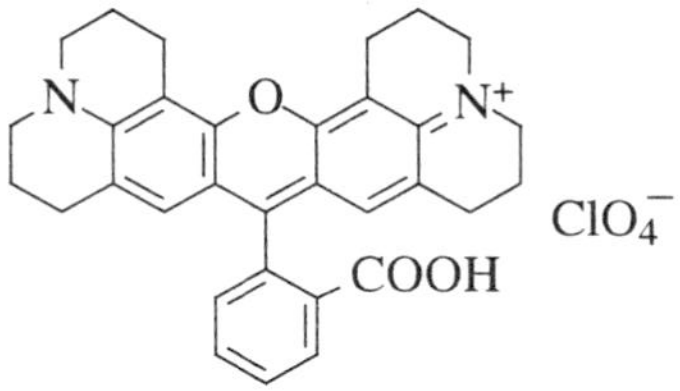	9-(2-Carboxyphenyl)-1,3,6,7,12,13,16,17-octahydro-1H,5H,11H,15H-diquinolizino ⟨1,9-bc; 1,9-hi⟩ xanthi-lium perchlorate	591	568, 95000 ethanol	595 ethanol

Table 7.5. Continued

Conditional name	Structural formula	Description	RMM	λ_{max}^{abs} /nm, ε/dm^3·mol^{-1}·cm^{-1} solvent	λ_{max}^{fl} /nm solvent
C7, coumarin 7	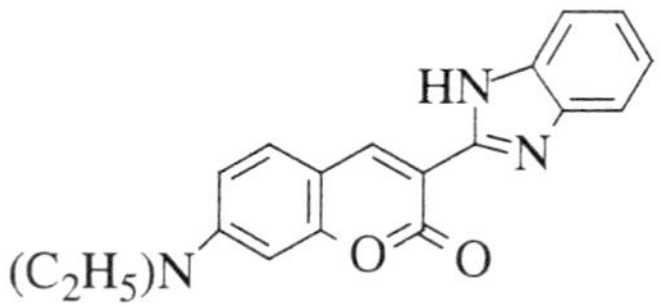	3-(2'-Benzimidazoline)-7-N,N-diethyl-aminocoumarin	333	433, 50000 ethanol	493 ethanol
RB, rhodamine B	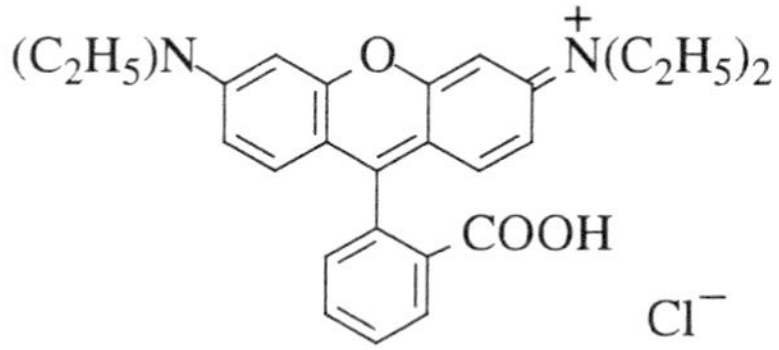	o-(6-Diethylamino-3-diethylimino-3H-xanthene-9-yl)benzoic acid, chloride	479	552, 107000 ethanol	580 ethanol
C307, coumarin 307	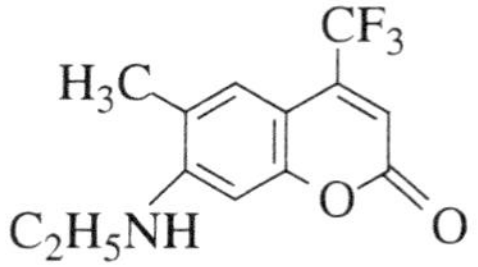	7-Ethylamino-6-methyl-4-trifluoromethyl-coumarin	271	395, 18000 ethanol	490 ethanol
Bis-MSB	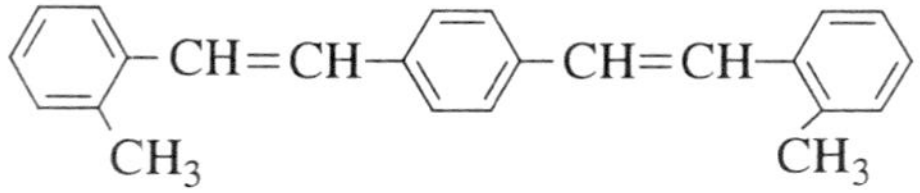	p-Bis-(o-methyl-styryl)benzene	310	350, 48000 cyclohexane	418 ethanol

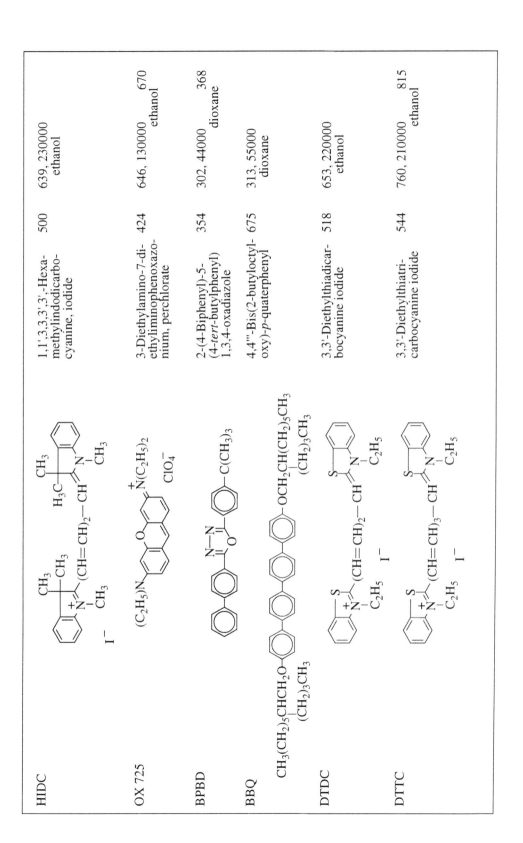

Name	Structure			
HIDC		1,1',3,3,3',3',-Hexamethylindodicarbocyanine, iodide	500	639, 230000 ethanol
OX 725		3-Diethylamino-7-diethyliminophenoxazonium, perchlorate	424	646, 130000 ethanol 670
BPBD		2-(4-Biphenyl)-5-(4-tert-butylphenyl)1,3,4-oxadiazole	354	302, 44000 dioxane 368
BBQ		4,4'''-Bis(2-butyloctyloxy)-p-quaterphenyl	675	313, 55000 dioxane
DTDC		3,3'-Diethylthiadicarbocyanine iodide	518	653, 220000 ethanol
DTTC		3,3'-Diethylthiatricarbocyanine iodide	544	760, 210000 ethanol

Table 7.5. *Continued*

Conditional name	Structural formula	Description	RMM	λ_{max}^{abs} /nm, ε/dm^{3}mol^{-1}cm^{-1} solvent	λ_{max}^{fl} /nm solvent
IR-144	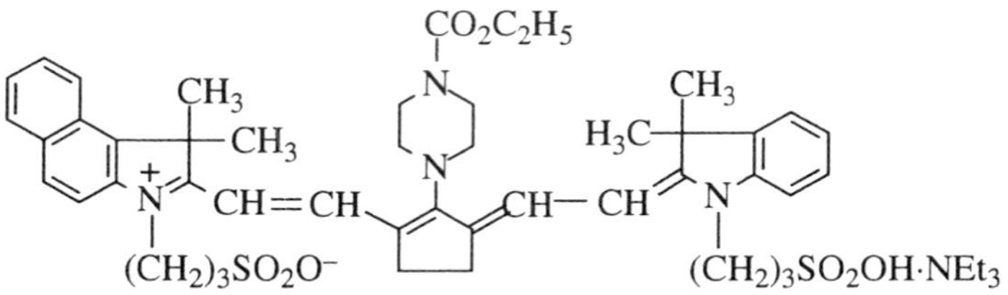	3,3,3',3'-Tetramethyl-1,1'-di-(3-sulfopropyl)-4,5,4',5'-dibenzo-10,12-trimethylene-11-(4-ethoxycarbonyl-piperazino)-indotricarbocyanine betaine, triethylamine compound	1008	750, 140000 ethanol	848 ethanol
DODC	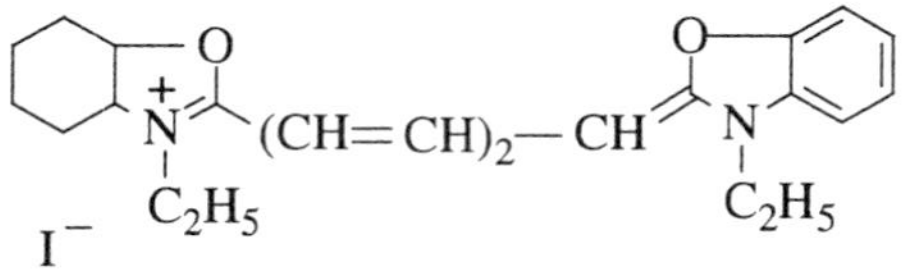	3,3'-Diethyloxadicarbocyanine iodide	486	582, 220000 ethanol	
Nile Blue	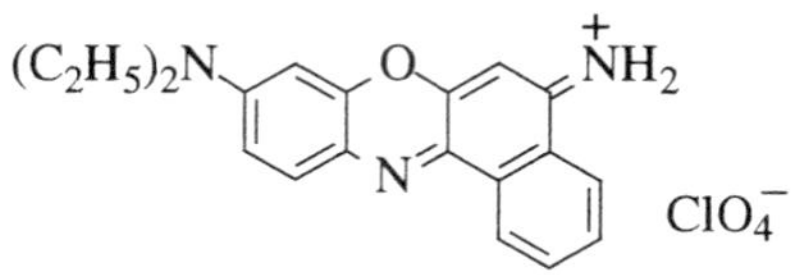	5-Imino-9-diethyl-aminobenzo(a)-phenoxazonium perchlorate	418	633, 78000 ethanol	672
LD-690	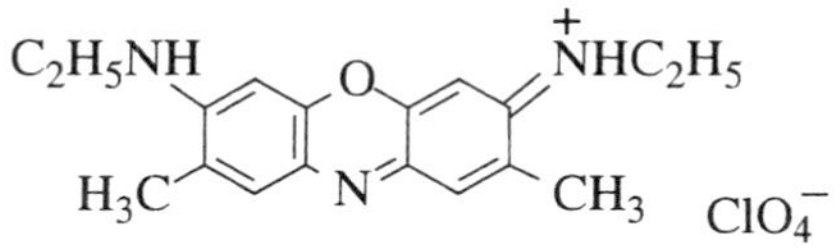	3-Ethylamino-7-ethylimino-2,8-di-methylphenoxazinium perchlorate	396	615, 109000 ethanol	

CAPB 720	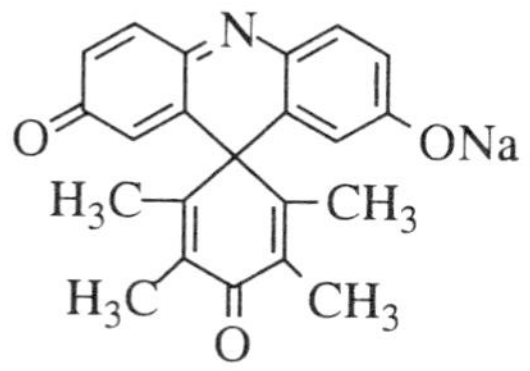	3-Hydroxy-2',3',-5',6'-tetramethyl-spiro⟨carbazin-10,1'-(cyclohexa-2,5-diene)⟩ 7,4'-dyon, Na-salt	367	655, 75000 ethanol
OX 720	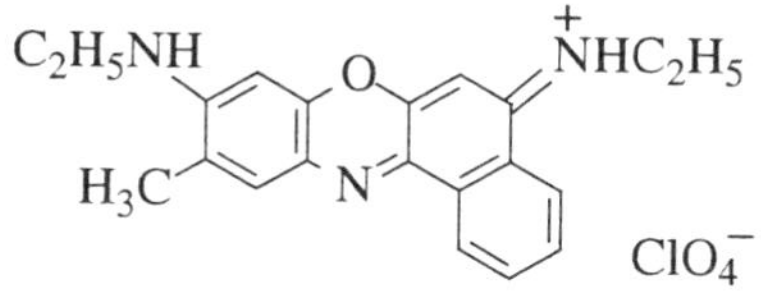	9-Ethylamino-5-ethylimino-10-methyl-5H-benzo(a)phenoxazonium perchlorate	432	620, 83000 ethanol 650
IR-140 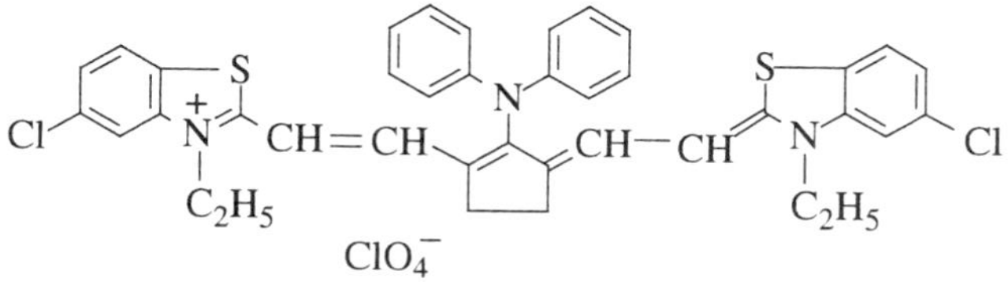		3,3'-Diethyl-5,5'-dichloro-10,12-tri-methylene-11-diphenyl-aminothia-tricarbocy-anine perchlorate	779	810, 150000 ethanol 860
LD-700 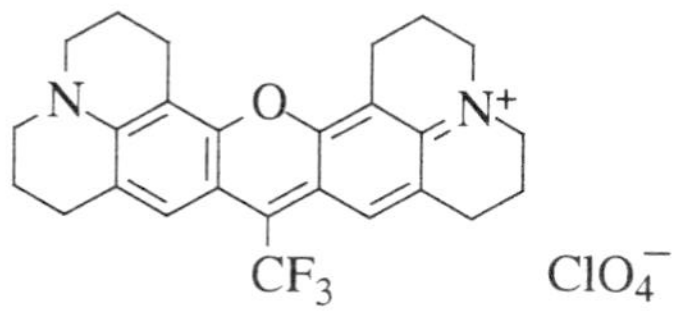		8-(Trifluoromethyl)-2,3,5,6,11,12,14,15-octahydro-1H,4H,10H,13H-diquinolizino [9,9a,1-bc; 9,9a,1'-hi] xanthilium perchlorate	539	643, 93000 ethanol

these tests enabled the authors of [451] to conclude as to the significant effect of the photodegradation products of the dyes on the stability of the LSC operation. After 200 h of exposure of the luminescent plate to illumination, its optical density decreased by almost 92% of its initial value, which explains only half of the total light losses in the LSC. The authors of [451] attribute the remaining losses to the absorption of light by the photoproducts of decomposition of the dyes in their absorption region. It was established that a 500 h illumination in the tough conditions of the accelerated test corresponded to the decay of the efficiency of LSCs exposed to solar illumination for 250 days.

The authors of [466] studied LSCs with the following dyes: rhodamines 6 G and 101, coumarin 6, oxazine, uranine, Na-fluorescein, and KF-241. The best results were obtained with the samples using KF-241, whose absorption and emission spectra are illustrated in Fig. 7.5. The Stokes shift of this dye is 50 nm and the quantum yield is 0.99. It was noted that the efficiency of the LSC based on KF-241 is 1.4 times that of the LSC based on rhodamine 6 G.

The authors of [467] associate the service life of dyes with the lifetime of their fluorescent state (τ_{fl}). Variation in τ_{fl} from 10^{-5}–10^{-9} s increase the 50% efficiency drop period in dyes from 2 days to 40 to 50 years.

The results of operating the LSC–PEC system during daylight (from 8 to 17 h on a summer day in Spain) are given in [466]. The experiment was conducted with use of the BASF KF-241-based LSC. It was shown that within 7 h the value of C_{el} is close to 3.5 and increases up to 4.0–4.5 in the morning and evening hours. Measurements made in poor weather showed that during the day the value of C_{el} varied from 3 to 4.

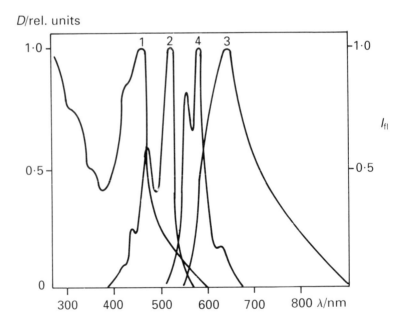

Fig. 7.5. Absorption (1, 2) and emission (3, 4) spectra of tris(2,2'-bipyridyl)ruthenium(II) (1, 3) and the dye KF-241 (2, 4). 1, D, rel. unit; 2, I_{fl}, rel. unit.

Study of samples activated by rhodamine 6 G, B, and fluorescein exposed to temperatures from 25 to 100°C reveals the differences in their thermal resistance [468]. The least resistant is rhodamine B: the optical density of a rhodamine B-based LSC decreased by 37% as a result of exposure of the sample at 80°C for 100 h. Under similar conditions, the losses of the LSC based on rhodamine 6 G and fluorescein were 14 and 10% respectively [469].

As we have already mentioned, overlap of the spectral absorption and luminescence curves of the dye leads to a red shift of its emission maximum which is due to the partial reabsorption of radiation in the concentrator plate, the optical efficiency of the LSC being noticeably reduced [450, 452, 473]. The selection of dyes with a large Stokes shift minimizes the losses. The Stokes shift is associated with different processes occurring in the excited state molecules. The commonest is a change in geometry of the excited molecule, the limiting case being *cis–trans*-isomerization (the *cis*-isomer absorbs, the *trans*-isomer emits).

Ref. [468] deals with testing of DTODC, an efficient new dye, which has been used in LSCs as a PMMA dopant (Table 7.5). It exhibits a large Stokes shift and a fluorescence quantum yield of 0.8. Investigation of its photochemical and spectral luminescent properties showed that as regards stability, it is second to KF-241, but is superior to such dyes as DCM and rhodamine 640. The samples activated by this dye and DCM were characterized by a sharp reduction in optical density in the first hours of exposure to illumination and a dependence of the spectral emission maximum and the quantum yield on the excitation wavelength. The authors explain this deviation by the nonuniformity of the polymerization processes, which manifests itself in local overheating of the polymerizing mass, thus leading to the formation of sites with a different content of residual monomer. When interacting with the dye, the latter forms unstable products.

The Stokes shift increases with increasing polarity of the medium, which is why the addition to the polymerizing mass of a highly polar solvent—dimethyl sulfoxide—yields positive results [470].

Large Stokes shifts, of the order of several thousands of cm^{-1}, are typical of charge transfer complexes (CTC). For the most commonly used CTC components containing aromatic hydrocarbons, the value is of the order of 10^3 cm^{-1} [471].

Large Stokes shifts are also characteristic of excimers and exciplexes, which are complexes containing similar and different molecules in the former and latter cases respectively. An excimer is a complex which comes into existence on the interaction of excited and unexcited molecules and completely dissociated on the loss of electronic excitation. Also known are exciplexes in which a transition metal ion serves as donor or acceptor. The excited molecular associates consisting of a neutral molecule with a positive ion are of interest because of the existence between the components of an electrostatic repulsive, rather than attractive, forces. While both components of the excited complex bear a positive charge, such entities are quite stable, presumably owing to the entropy factor [471].

Ref. [473] deals with the possibility of using excimer emission as one way to reduce the level of reabsorption in LSCs. For instance, the excimer fluorescence yields of 9,10-dimethylanthracene and 9,10-dipropylanthracene are 0.82 and 0.98 respectively. The Stokes shift in these compounds is 6700 cm^{-1}, and the concentration luminescence quenching increases with the viscosity of the medium.

The authors of [473] point to the possibility of using the exciplex emission of chromophore-containing polymers in LSCs. Exciplexes are formed between the chromophoric group and the excited state of the main polymer unit. Since the sensitivity maximum of most PECs is close to the infrared region, the red shift of the excimer and exciplex luminescence increases the efficiency of LSCs in comparison with the emission of the chromophores.

The author of [472] suggests that the polymeric matrix should be doped with soluble inorganic chromophores. This combination allows retention of the advantages of the polymeric matrix used as the base of the LSC and also eliminates the drawbacks of inorganic luminophors such as low extinction coefficients and low quantum yields. Also, in this case, the users are not faced with the problems peculiar to the organic luminophors and stemming from the value of the Stokes shift.

To prepare such compositions, it is necessary that the inorganic salt should be readily soluble in the polymer, and its dissolution should not lead to luminescence quenching. These requirements are met by platinum complexes such as the bis(diphenylphosphino)-imine salt (PPN) of platinum, $(PPN)_4[Pt_2(\mu-P_2O_5H_2)_4]$. A solid poly(vinyl acetate) solution of this complex has its absorption maximum at 376 nm and emission maximum at 513 nm. The Stokes shift ($\Delta\lambda = 137$ nm), and symmetry of the absorption and emission bands are responsible for their minimal overlap. Another advantage of this complex is its excellent stability, which is close to that of inorganic luminophors. The drawbacks of the complex are its poor absorption ($\varepsilon = 110$ dm^3 mol^{-1} cm^{-1} at $\lambda = 452$ nm) and the incompatibility of the spectral characteristics of solar radiation, silicon PECs and of this dye [472, 474].

Ref. [475] covers a series of dyes based on the complexes of ruthenium with 2,2'-bipyridyl and 1,10-phenanthroline and their derivatives. About 30 substituted bipyridyls and 20 substituted phenanthrolines were considered in this study. The spectral luminescent parameters of some complexes are listed in Table 7.6 [475]. The luminescence quantum yield is shown to be considerably affected by substitution of an aryl group for hydrogen in 4,4'-bipyridyl and 4,7-phenanthroline. In this case, the extinction coefficient increases two-fold and the luminescence quantum yield of the liquid and solid solutions increases 9 to 10 times and 2 to 3 times respectively. Fig. 7.5 illustrates the absorption and luminescence spectra of the ternary ruthenium complex based on 2,2'-bipyridyl.

The dyes used in the production of LSCs are listed in Table 7.5. Along with organic luminophors, Table 7.5 also lists crystallophosphors which have advantages of their own.

In most cases, crystallophosphors are UV resistant, have a long service life and a larger Stokes shift than organic luminophors. One of their shortcomings is the poor correspondence of their short wavelength absorption spectrum to the spectrum of solar radiation. The inorganic luminophors with a PMMA matrix listed in Table 7.5 have absorption maxima at 250–335 nm and emission maxima at 440–614 nm. Such compositions are not currently used as LSC dopants, mainly because of their significant light scattering.

With the aim of preparing LSCs absorbing over the broadest possible solar spectral region, the authors of [448, 452] used mixtures of several dyes. The absorption and emission spectra of such compounds were in different frequency ranges. However, not all pairs of dyes are applicable for such purposes, since the association of molecules of different luminophors is likely to cause luminescence quenching. For example, rhodamines 610 and 590 quench the molecular luminescence of oxazines 720 and 750. Good results were shown by mixtures of sulforhodamine 640, rhodamine 590, and coumarine 540. Other compositions of various dyes are listed in Table 7.5.

Table 7.6. Spectral and luminescent properties of ruthenium complexes with bipyridyl and phenanthroline and their derivatives

Compounds	Substituents	λ^{abs}_{max} /nm	ε /10^4 dm^3 mol^{-1} cm^{-1}	λ^{fl}_{max}, solution /nm		φ_n
[Ru(bipyri-dyl)]$_3^{2+}$	–	450	1.43	630	0.089	0.24(PVC)
	4,4'-di(NH$_2$)	504	1.05	705	0.004	–
	4,4'-di(OC$_2$H$_5$)	577	1.30	675	0.020	–
	4,4'-di(CH$_3$)	455	1.70	640	0.086	–
	5,5'-di(CO$_2$C$_2$H$_5$)	495	0.99	720	0.004	–
	4,4'-diphenyl	473	2.80	635	0.306	0.54(PVC)
[Ru(phenan-throline)]$_3^{2+}$	–	445	2.00	595	0.019	0.21(PVC)
	4,7-di(Cl)	456	2.06	668	0.028	
	4,7-diphenyl	463	2.86	618	0.366	0.44(PMMA) 0.70(PVC)
	4,7-di-p-biphenyl	465	3.63	620	0.360	0.40(PVC)

7.2.3 Design and application of LSCs

Along with the search and investigation of new efficient materials to be used in LSCs, the optimal designs of these devices are being developed. The purpose of this design work is to improve the geometrical parameters and optical properties of LSC plates, to use additional elements enhancing the efficiency of LSCs, and to eliminate optical losses, thus increasing the efficiency of the device. The conceptual decisions are also determined by the type of radiation detector: depending on its function, the plates activated by the dye can serve both for the conversion and transmission of light, and also act as light filters.

It can be confidently stated that at present there are no commonly accepted designs ensuring the optimal parameters of LSCs [456]. Certain successful developments are associated with the production of structures composed of several polymeric plates containing a variety of luminophors [449, 453, 459]. This method provides a broader absorption spectrum of solar radiation and, consequently, a higher value of η. A representative diagram of a multilayered LSC is shown in Fig. 7.6. The selection of dyes to be incorporated in the polymeric plates is determined by the coincidence (for pairs of plates) of the emission maximum of the upper plate and absorption maximum of the lower plate. A three-plate LSC described in [449] was activated by luminophors showing an absorption spectrum stretching from the UV region to 550 nm in the first plate, 550 to 900 nm in the second plate, and beyond 900 nm in the third plate. Comparison of these LSCs with single-plate versions absorbing in the region from UV to 530 nm shows that the optical efficiency of the multilayered LSC (η_{opt} = 15.8%) is approximately 1.5 times higher than that of the single plate (η_{opt} = 10%).

Ref. [454] provides data on the application of combined structures composed of two

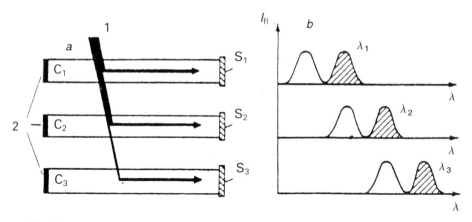

Fig. 7.6. Diagram (a) and spectral characteristics (b) of multilayered LSC: 1, light flux; 2, mirrors.

plates—organic and inorganic. When converting UV into visible radiation, the upper plate based on luminescent glasses (glass activated by uranyl ions, UO_2^{2+}, exhibiting an intense green emission and absorption in the UV region), simultaneously protects from its adverse effects the organic luminophor-based plate (PMMA activated by 16,17-diheptanyloxyviolanthrone).

In multilayered LSCs, provision is made for a mirror coating which is applied to the final plate to increase the amount of energy absorbed at the expense of the double passage of a light beam through the LSC [459]. The upper coating made of Pyrex glass serves as an UV filter and for protection against dust. To decrease the amount of light reflected from the LSC, it is advisable that the illuminated surface should be provided with an antireflective coating [450].

With an increasing number of plates, the geometrical factor remains unchanged, which is an evident drawback of multilayered LSCs. For this reason, the most promising are systems in which the extended absorption band typical of multilayered LSCs should be realized in a single plate [456].

In [464, 473], the suggestion has been made to use diffuse reflective coatings. Thus, replacement of the aluminium coating on the side surfaces of the LSC by several layers of TiO_2 brought about an increase in η_{opt} by 1 or 2% [464]. Additionally, [476] reports a study of the effect of various diffuse screens on the energy concentration coefficient, C_{el}. For this purpose, two side surfaces made from PMMA triangular plates of 43.3 cm^2 with an edge of 10 cm and thickness of 0.3 cm were coated with TiO_2. The experiment compared three different plates activated by rhodamine B. In case (i), the coating was applied directly onto the back surface. In case (ii), a special TiO_2-coated screen was installed at a certain distance behind the plate. In case (iii), use was made of a luminescent screen made from the polymer activated by a luminophor whose emission spectrum corresponded to the absorption region of the LSC. Its back surface was also coated with TiO_2. The corresponding values of C_{el} were 6.3, 7.9, and 9.0 respectively. Without diffuse coatings, the C_{el} of the LSC was 5.6. The authors of [476] explain the difference in the results by the increased optical path taken by light beams in the LSC owing to the changed reflective angle of the normally incident beam.

In [477] the suggestion is made of using, as a reflective luminescent screen, a luminophor coating containing an inorganic luminophor in combination with a polymer.

It was recommended to use as polymers, aromatic polyamides and polyesters containing diphenylanthracene, diethylanthracene, dianilinoanthracene, acridine, stilbene, rhodamine, or diamine phthalimide groups. The luminophor coatings obtained are characterized by increased brightness of emission and resistance to temperatures from 150 to 200°C owing to the thermal resistance of the luminescent polymers used.

Much attention has been given to the investigation of combined LSCs using dye solutions both in liquid and solid matrices and containing polymer films.

The application of thin polymer films in combination with a PMMA light-conducting plate is covered in [467, 478]. The use of a film containing high concentrations of dye mixtures ($n = 1.475$) made it possible to obtain a solar light conversion efficiency of 3.2% in the silicon PEC and 4.5% in the GaAs-based PEC. The advantages of such constructions are as follows: during manufacture, films can be applied both onto the polymer and glass; they make possible a significant increase in the number of optically transparent polymers used in LSCs at the expense of deuterated and fluorinated materials, with which the level of scattering increases after the introduction of dyes in their bulk; they exhibit no radiationless transfer of energy from the dye on the matrix, which permits a high concentration of dye; they make it possible to photostabilize the film by means of a protective coating or UV stabilizer without any losses in matrix transparency; they reduce losses occurring during total internal reflection by applying intermediate films with the required refractive index; and they allow the application of one film layer over another or the use of mixtures of films.

Thin films (25 to 30 μm), absorbing in a broad spectral region like multiplated LSCs, barely increase the thickness of the polymeric base of the concentrator. This is why such designs show no reduction of the light concentration coefficient with an increased number of film coatings [463] and have an optical efficiency η_{opt} of 0.15 [452].

The application of LSCs for the conversion of solar energy into heat was investigated as early as 1977 [449]. This line of investigation was developed later on [479, 480]. The device used in the experiments during direct ($I = 860$ W m^{-2}) and diffuse ($I = 150$ W m^{-2}) solar illumination (Fig. 7.7) allowed attainment of equilibrium temperatures of 555 and 250°C respectively, in the absorber. The absorber was an interferential copper oxide coating applied to a 3 mm copper tube. Contact with the LSC was established by means of a vacuum glass tube which prevented the polymer plate from heating.

The luminescent plate connected with the PEC can function both as a light concentrator and a luminescent light filter re-emitting the absorbed short wavelength solar radiation in the spectral sensitivity region of the converter [456]. The PECs arranged on the reflecting base in the flat layer of the luminescent polymer, absorb, along with the direct solar light, the additional light flux of the luminescence originating in the LSC. Experiments show that the energy efficiency of 20 modules measuring $410 \times 385 \times 8.5$ mm increases by some 25% when the luminescent medium is used.

Noteworthy are cheap self-contained LSC-based devices designated for the lighting of buildings [456, 481]. In such devices, the luminescent plates arranged on the outside of buildings collect and concentrate solar light at the end face of the light duct transmitting radiation inside the building. According to [481], the use of LSCs of area 3.2 m^2 is equivalent to the use of incandescent lamps of 200 to 300 W, permitting an area surface of 60 m^2 to be lit.

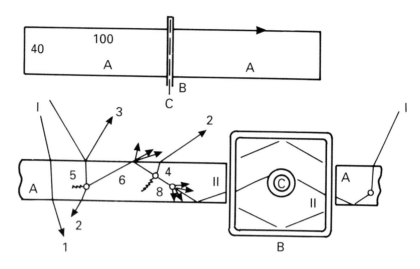

Fig. 7.7. Diagram of LSC-based device for conversion of solar energy into heat: A, plate with dye; B, vacuumized glass tube; C, absorber; D, reflectors; I, solar light; II, luminescent radiation; 1–8, LSC optical losses (see Fig. 7.4).

The great interest shown by many countries in the development and study of LSCs indicates that it is a promising new trend in science and technology. The manifest advantages of these devices give us reason to hope that they will find extensive application in the future.

7.3 MATERIALS FOR LASERS

Transparent polymeric materials have found one of their most promising applications in lasers, in which they can be used as the basis of active elements with generating dyes, antireflective filters for Q-modulation, etc. [13].

The physical principles of generating coherent radiation by dyes in liquid and solid organic solutions are characterized as being similar [13, 482, 483]. The solid-matrix dye lasers make possible combination of the advantages of solid state lasers with the possibility of tuning the radiation over a broad spectral range, which overlaps almost completely the visible region.

7.3.1 Features of polymeric laser elements

The investigation of solid-matrix dyes was undertaken in the USSR in 1968 [484]. The radiation resistance of the first samples was insufficient for practical use. Since then their parameters have been much improved. With the aim of preparing new polymeric media for laser elements, a number of investigations were conducted in the 1970s and 1980s, mainly to eliminate low radiation resistance.

Some advantages of solid polymeric matrices over liquid active media were mentioned in the introduction to this book. Another typical advantage of polymeric materials over other transparent dielectrics—crystals and glasses—is the higher laser resistance of polymeric surfaces as compared with volumes [485]. Such correlation of the damage

Table 7.7. Damage thresholds of volume (I_v) and surface (I_s) of different optical materials

Material	$I_v{}^a$/rel. unit	$I_s{}^a$/rel.unit	$I_v{}^b$/W cm^{-2}
PMMA	1	2	450
Modified PMMA	2.5	5	500
Glass K8	3.5	0.7	440
Sapphire	4.5	0.5	700
Fused quartz	6	1.7	670

^{a} Radiation parameters: $\lambda = 0.69$ μ, $\tau = 20$ ns, $d = 110$ μ.
^{b} $\lambda = 1.06$ μ, $\tau = 10$ ns, $d = 4.6$ μ.

thresholds holds true during different methods of treatment of the polymeric surface: stamping, abrasive treatment with a diamond suspension, diamond paste, rolling, etc. This is very important from the practical point of view since it is the surface resistance of transparent dielectrics which usually represents the limiting factor in their application in laser systems. The surface and volume of the polymer have their own viscoelastic properties: the surface shows a lower limit of forced elasticity, which is usually held responsible for the elevated surface resistance of polymers. Thus, in contrast to fragile dielectrics, in elastic materials such as PMMA, the limiting factor for their application in high powered laser systems is their volume resistance which is, however, several times higher than the surface resistance of glasses and crystalline materials (Table 7.7) [485].

Attempts have been made to use polymers to improve the uniformity of solid state elements based on porous glasses with dyes [486]. The filling of the glass pores with a polymer leads to an increased optical uniformity and reduction of losses through scattering by a factor of 400. At the same time, such systems have a low temperature coefficient of the refractive index, thus $dn/dT \simeq 2 \times 10^{-6}$ K^{-1}. Glass samples filled with a polymer do not change their optical properties over a temperature range from 173 to 423 K. The introduction of a polymer in the porous glass increases the photoresistance of the dye: with the introduction of a polymeric medium, the photodestruction quantum yield of rhodamine 6 G decreases four-fold compared with that of the dye adsorbed on the glass, which is evidently associated with the limited access of aerial oxygen to the dye molecules.

The practical use of laser elements largely depends on the media which determine the service life and the efficiency of conversion of pumping radiation of various intensities. These parameters depend both on the solid state medium and the dye incorporated within it. The critical characteristics of the active solid state medium, such as thermophysical (heat capacity, heat conduction, softening temperature) and mechanical (strength, hardness), are determined solely by the matrix.

For over 20 years of laser history, dozens of polymers and their formulations have been studied as active laser media. These investigations made it possible to identify those materials offering most promise from the practical point of view. The authors of [487] summarized the specifications as regards thermal resistance, mechanical strength, resistance to strong solvents, and optical transparency over a broad wavelength range, and also listed 17 polymers which could be used in high powered lasers.

At present, however, 5 or 6 polymers are used as active laser media among which

PMMA, PS, polyurethane, oxirane polymers and their modifications, and polycarbonates (PC) are distinguished.

PMMA is the most frequently used and most rewardingly studied polymer. Showing the best transparency in the visible spectral region, it has so far ranked first in laser resistance [13].

Being somewhat inferior to PMMA in laser resistance, PS has the best radiation resistance [444]. As regards optical properties, it is, like PMMA, one of the best amorphous polymers.

Laser polymeric elements based on polyurethanes are 'triplexes' in which the light sensitive composition layer is arranged between polished glass, quartz, or other bases. The dense network of hydrogen bonds is responsible for the high mobility of polyurethane macromolecules which, in turn, explains the high radiation resistance of polyurethane matrices [488].

An important advantage of oxirane polymers is their variability in properties over a very wide range. For example, by suitable selection of oxirane polymers through the oligomers and hardener, the refractive index can be gradually varied over the range $n = 1.49–1.61$ [489].

The particular choice of polymeric material is determined by the operating conditions of the laser medium.

7.3.2 Interaction of laser radiation with polymers

The basic requirement imposed on active media, along with optical transparency in the region from the UV to 1.3 μm, is resistance to the laser pumping radiation I_p (or laser resistance). It is usually determined as the intensity at which microdamage of the order of 10^{-2} cm occurs in the bulk of the polymer with a probability $P = 0.5$ [490]. The laser damage threshold depends on the size of the interaction area d having the specified level of intensity I whereas the probability P of damage at fixed dimensions of the laser beam depends on the intensity. In a number of cases, energy density P_d is used instead of intensity [485].

In studies of laser damage under conditions of repeated radiation, basic parameters are introduced to characterize adequately the accumulation process. These are the critical number of pulses averaged over a large number of irradiated points N_{cr} (I) which, at the fixed intensity I or energy density P_d, leads to microdamage of the sample; the threshold intensity with a number N of pulses (I_N) at which damage occurs with a specified probability $P(I)$ during a fixed number of pulses N [491].

The values of the threshold intensity I_p, at which damage occurs, vary for different materials and for samples made from one and the same material, and depend largely on the frequency v, the radiation pulse life τ, and the dimensions of the radiation area d.

Under laser radiation of sufficiently high intensity, polymers develop discretely arranged cracks [492]. The appearance of the cracks is similar in different materials and does not depend on the laser radiation wavelength. The internal surface of the cracks is covered with a thin coating of soot.

Refs [492, 493] provide data on the chemical composition of the gas filling the internal cavity of the cracks formed during laser damage in polymers. Different samples have been used to show that the composition of the gas is independent of the wavelength and intensity of the laser pulse. The results of chemical analysis of the above gas listed in Table 7.8 are indicative of the considerable extent of pyrolysis processes.

Data are available in [493] for, in addition to the gases listed in Table 7.8, the products of laser pyrolysis of PMMA contain both alkanes (C_2H_6, C_3H_8, C_4H_{10}) and alkenes (C_2H_4, C_3H_6).

Table 7.8. Gas content of pyrolysis products of polymers exposed to laser radiation (%)

Polymer	CH_4	H_2	CO_2	CO
PS	91	1.0	1.0	6
PMMA	58	21.0	0.4	20
PMMA (powder)	20–25	45–60	1–3	15

Visible laser damage can occur in the material both during exposure to a single pulse of sufficiently high intensity and a train of pulses of lower intensity.

Of interest are reports [494] that on irradiation of PMMA, PS, and PC with pulses of duration $\tau = 0.001$ to 1 s, starting from a certain value of τ_1, the probability of damage stops growing and, with a further increase in τ to τ_2, it even drops. For PMMA, $\tau_1 = 5$ ms, $\tau_2 > 20$ ms. The authors attribute this to radiation defocusing and note that the value of τ, at which the probability of damage stops growing, is determined by the properties of the material. For PMMA and PS, the constants serving as the quantitative criterion of defocusing are obtained.

Since damage is a mechanical phenomenon, it is necessary to take into account the influence of the polymeric microstructure on the development of laser cracks. In PMMA samples [492] having a microdomain structure (the size of a macrodomain being *ca* 1 µm), laser damage is greater than in samples having a macrodomain structure (*ca* 10 µm), the characteristics of the samples and laser radiation conditions being similar.

7.3.3 Mechanisms of laser damage in polymers

In the application of polymers as a matrix, the mechanism of laser damage is of special importance. However, there is no general concensus on this matter. The discrete nature of laser damage and insignificant absorption coefficient of optically transparent polymeric materials suggest, on the whole, the presence of absorbing defects. Both the chance accumulation of any molecular impurities and any foreign absorbing inclusions may serve as absorbing centres. The importance of inclusions in the mechanism of laser damage in polymeric materials is indicated by the dependence of damage thresholds I_p on the dimensions of the irradiated area of the polymers. With an increased diameter of the irradiated area from 30 to 640 µm, I_p decreases more than 20 times [485]. A study of the dependence of damage thresholds on pulse duration over the range from 10 to 40 ns and radiation wavelength from 1.06 to 0.53 µm, as well as the fluctuations of damage threshold from point to point inside the sample, confirms the conclusion about the effect of absorbing defects on the rate of laser damage in polymers, and, thus, explain its static nature.

So, to ensure a high laser resistance, the initial monomers need to be thoroughly purified. Each process such as distillation, supercooling, or filtration with use of filters with pores of 0.1 µm, increased the damage threshold of PMMA 2.5, 3.2, and 2.5 times respectively. In the initial monomer the damage threshold increased 10.7, 2.8, and 12.0 times respectively [492]. The relationship between the value of the damage threshold and the size of the filter pores (Table 7.9) established in [485] indicates that the former increases monotonically when filters with a pore size decreasing from 16 to 0.22 μm are used.

Table 7.9. Values of energy density P_d for MMA and PMMA in samples of different purity (P_d for K8 glass is taken as unity at same d and λ) [98]

Filter pore size/μm	$d = 110$ μm, $\lambda = 0.69$ μm		$d = 4.6$ μm, $\lambda = 1.06$ μm	
	MMA	PMMA	MMA	PMMA
16	0.19	0.08	0.24	0.24
1.6	—	0.11	0.31	0.31
0.8	—	0.16	0.38	0.38
0.45	—	0.21	0.43	0.43
0.22	>6.0	0.28	0.53	0.53

Summarizing the results of studies of the laser damage mechanism is somewhat complicated by the considerable variety of lasing modes (from continuous lasing to ps pulsed mode), laser radiation wavelengths (from UV to IR), and variations in the purity of samples of one and the same type. Thus, depending on the technology of manufacture, the number of local inclusions in polymers can vary over several orders of magnitude [495].

It would be appropriate to divide the damage occurring in transparent bodies under laser radiation as that occurring in ideally pure media and that caused by impurities. The mechanisms causing damage are different. In pure media, it is optical rupture, which is qualitatively similar to gas rupture; in media containing impurities, it is damage caused by the superheating of impurities [495, 496].

Ref. [492] suggests a multiquantum photodestruction model for polymeric macromolecules, which operates on the general principle of mechanical destruction of a thermofluctuational nature.

Refs [493, 497] deal with the thermochemical mechanism of laser damage according to which gaseous products of pyrolysis are formed in the vicinity of absorbing microinclusions. In this case, a special role is played by soot-like products whose formation results in the growth of absorbing centres. As noted in [491], however, this mechanism, which dominates in laser damage occurring under continuous laser radiation with long pulses (ca 10^{-3} s), does not appear to function if the pulses are short (ns and ps).

The authors of [496] suggest operation of the mechanochemical model of optical rupture in transparent polymeric materials, which is covered in great detail in [485]. This concept explains the following important generality of laser damage in polymers: the dependence of the damage threshold under conditions of single and repeated radiation on the viscoelastic properties of the matrix and the concentration of low molecular dopants.

In essence, this effect implies that macrodamage occurs on the surface or in the bulk of the optical material as a result of its exposure to a train of laser radiation pulses of intensity below the threshold of single-pulse damage.

In polymeric materials, the accumulative effect is particularly strong: it can be observed even when the intensity is below single-pulse damage by a factor of 10^{-2} [485].

Different mechanisms have been suggested to explain the accumulation effect, including (i) the accumulation of gaseous products in the microregions which develop as a result of multiquantum photochemical reactions leading to destruction of the polymer [498], and (ii) the thermomechanical mechanism of accumulation of products such as soot in the polymeric materials [497]. It should be noted that all these mechanisms failed to be adequately confirmed by experiment [499].

7.3.4 Methods for improving laser resistance in polymeric media

Analysis of the accumulation effect [485] offers an explanation for the differences in the characteristics of laser damage occurring in materials with different thermoelastic properties (Table 7.10). Thus, an increase in ambient temperature to the level at which the matrix passes from the glassy into the highly elastic state leads to a significant increase in the damage threshold I_p [13]. This effect is observed under intense laser radiation in the IR, visible, and UV regions and is slightly dependent on the radiation wavelength.

I_p can be increased at the expense of the viscoelastic properties of the matrix by the introduction of plasticizers [485, 490, 496]. Low molecular dopants improve the optical transparency [500] and are responsible for the abrupt reduction in the rate of mechano-chemical reactions and, consequently, for suppression of the accumulation effect. The influence of plasticizers on the rate of destruction of polymeric samples is associated with the transfer of energy by hot radicals to the dopant (oscillation cross-relaxation) [490].

Table 7.10. Laser resistance of polymeric materials at $d = 110$ μm, $\lambda = 0.69$ μm exposed to repeated radiation and fixed radiation intensity at which, for PMM, $N_{dye} = 20$

Polymer	$P_d(1)$, rel. unit	N_{dye}, rad. impulse	Damage type
PMMA:			
initial	0.08	20	Cracks
purified (pore size 0.22 μ)	0.28	90	Cracks
MMA (71%) + MMEG (29%)	0.14	70	Cracks
PMMA + 20% EA	0.11	$>10^4$	None
PTBMA	0.15	40	Cracks
PNBMA	0.10	$>10^3$	Hard spots (~10 μm)
MMEG	0.14	40	Cracks
MMA (50%) + AA (50%)	0.42	30	Cracks
MMA (80%) + DMA (20%)	0.11	$>10^3$	Hard spots (~10 μm)
MMA (35%) + BMA (65%)	0.14	$>10^3$	Hard spots (~10 μm)
PMMA + 30% DBP	0.11	$>10^3$	Same
PS	0.05	5	Cracks
PS + 30% DBP	0.08	$>10^3$	Hard spots (10 μm)

MMEG, monomethacryl ethylene glycol; AA, acrylic acid; DMA, decyl methacrylate; EA, ethanol; DBP, dibutyl phthalate; PTBMA, poly(*tert*-butyl methacrylate); PNBMA, poly(*n*-butyl methacrylate).

An important result of the investigations covered in [485] was the detection of significant improvement in the laser resistance of PMMA with the introduction of plasticizers (Table 7.11) such as ethanol and esters of *o*-phthalic acid:

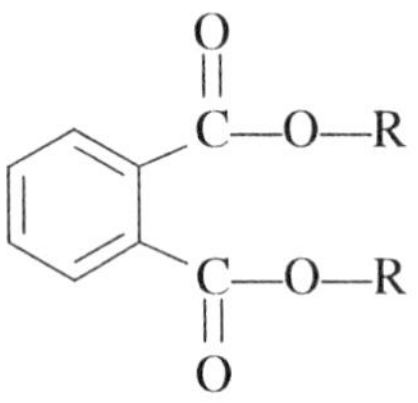

$$
\begin{array}{c}
O \\
\parallel \\
C\!-\!O\!-\!R \\
\\
C\!-\!O\!-\!R \\
\parallel \\
O
\end{array}
$$

where $R = CH_3$, C_4H_9, C_5H_{11}.

Temperature studies of the laser resistance of PMMA samples with a variety of dopants, introduced in equal quantities and ensuring similar viscoelastic properties of polymers, under repeated radiation show that the initiation of microdamage of less than 10 μm in size and its subsequent growth to macrosize do not depend on the structure and macrocharacteristics of the plasticizers. It has been established, for example, that the thresholds of multipulse damage of PMMA containing 20% dimethyl phosphate and 20% ethanol differ more than 5 times.

Ref. [501] deals with PMMA samples containing propyl alcohol and dibutyl phthalate as dopants. It was found that the glass transition temperature, laser resistance, and mass of the samples did not change on exposure to successive doses of γ-radiation.

Table 7.11. Laser damage thresholds of polymers exposed to repeated radiation $P_d(200)$ at $\lambda = 0.69$ μm, $d = 110$ μm [$P_d(200)$ of KB glass is taken as unity]

Polymer	$P_d(200)/10^{-3}$		$I_d(N)^a/°C$	$T_c/°C$
	55°C	20°C		
PMMA	4.0	4.0	120	119
PTBMA	4.0	11.2	50	53
PNBMA	5.6	16.0	20	20
PMMA + 20% DMP	4.0	4.0	65	61
PMMA + 20% DBP	4.0	6.4	63	53
PMMA + 20% EA	8.0	20.8	60	51

[a] Temperature at which laser resistance of polymers has been recorded to increase ten-fold.

The authors of [490] studied the action of various low molecular alcohols on the laser resistance of PMMA. It was shown that the introduction of butanol and hexanol leads to a considerable enhancement of the laser resistance of the polymer (Fig. 7.8). At the same time, the introduction of other dopants such as butyronitrile and cyclohexanol caused only a slight increase in the resistance of polymers. This difference was explained by the absence of resonance in the vibrational spectra of the carbonyl group of PMMA and molecules of the above compounds. Their relatively weak influence on laser resistance is associated by the authors of [490] with variations in the viscoelastic properties of the matrix stemming from the plasticizing effect produced by these dopants.

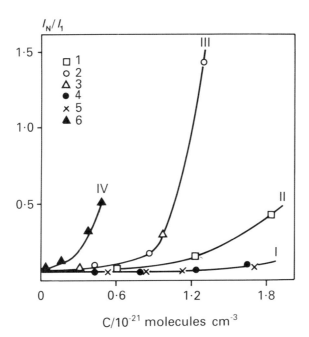

Fig. 7.8. I_N/I_1 ($N = 200$) for PMMA (I, II, III) and PMMA + 5% propanol (IV) versus concentration of dopant C: 1, propanol; 2, butanol; 3, hexanol; 4, butyronitrile; 5, cyclohexanol; 6, water.

The same reference gives data on the effect of adding a water–propanol (about 1%) mixture to PMMA. The addition of low concentrations produced a significant increase in the multipulse damage threshold. Similar experiments conducted on PMMA containing DBP showed that the introduction of water has no bearing on the laser resistance of the polymer. These results confirm the hypothesis about the crucial role of mechanochemical reactions in the accumulation effect and so allow a systematic search for polymers featuring the best laser resistance.

It would seem that the most promising way, though yet unrealized, to improve the laser resistance of PMMA matrices is to introduce plasticizer moieties directly into the polymeric chain, for instance, by the copolymerization of methyl methacrylate with the methacryl derivatives of polyhydric alcohols:

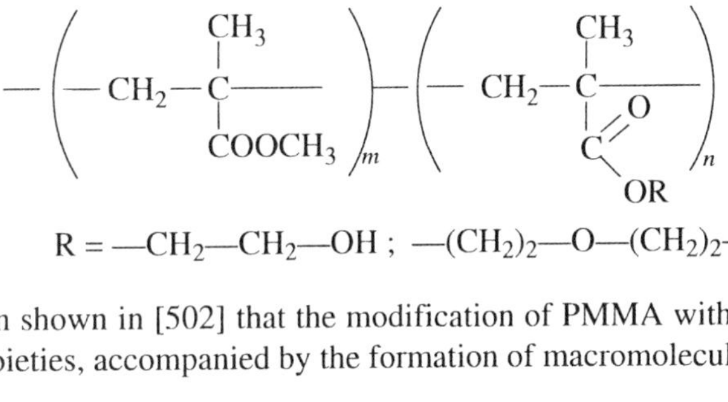

$$-\left(-CH_2-\underset{\underset{COOCH_3}{|}}{\overset{\overset{CH_3}{|}}{C}}-\right)_m\left(-CH_2-\underset{\underset{\underset{OR}{C}}{|}}{\overset{\overset{CH_3}{|}}{C}}\overset{O}{\underset{}{\diagdown}}\right)_n$$

$$R = -CH_2-CH_2-OH ; \ -(CH_2)_2-O-(CH_2)_2-OH$$

It has been shown in [502] that the modification of PMMA with diaminophenylbenzimidazole moieties, accompanied by the formation of macromolecules such as:

$(l/m = 1/500$ to $1/2500)$,

imparts to it better UV resistance without affecting its uniformity and high light transmission. Earlier similar results were obtained by introducing hydroxybenzophenone
moieties into the side chain of PMMA [21].

Refs [503, 504] compare the laser resistance of a solid solution of the laser dye 4-fluoro-
methyl-7-aminocoumarin in PMMA and the copolymer of MMA with 4-fluoromethyl-7-
(N-allylamino)coumarin on exposure to the radiation of a neodymium laser with $\lambda =$
1060 nm and a radiation energy density of 14.7 J cm^{-2}. It was shown that the laser resistance of the copolymer is increased over that of the solid solution by 20–30%, the concentration of the luminophoric centres being the same. During longitudinal pumping by
light with $\lambda = 355$ nm, the copolymer showed a lasing action threshold of 5 MW cm^{-2}. At a
pumping power of 65 MW cm^{-2} it had an electrical power efficiency of 6.0% (a lasing
action quantum yield of 7.7%). The service life was estimated by a 50% reduction in the
lasing action efficiency after 80 pulses at a pumping power of 30 MW cm^{-2}. It is worth
noting that at the same pumping power, the corresponding solid solution of the
luminophor in PMMA exhibits no lasing properties, which is explained by the authors of
[504] by the presence in the test samples of associates of aminocoumarine with a low fluorescence quantum yield.

Another method of enhancing the laser resistance of polymeric matrices used in active
laser media, along with technological uniformity (absence of dust particles, bubbles, etc.)
and the introduction of plasticizers, is a significant improvement of the structural uniformity of the polymeric systems and a more uniform distribution of the dye in the matrix.

7.3.5 Laser dyes for solid state elements
The number of dyes available as active dopants for solid state lasers has notably increased
the number of light transmitting polymers and currently totals about 700 [505]. The structure and optical properties most dyes used in active polymeric media are listed in Table 7.5.

The spectral luminescent and lasing properties of the active elements are determined
both by the properties of the luminophors and the polymeric matrix into which they are
incorporated, and also by their interaction in the process of operation. The polymeric base
affects the location of the absorption and luminescence maxima of the luminophor [506,
507]. The action produced by the polymeric medium on the laser characteristics of rhodamine 6 G is covered in [507]. Thus the lasing efficiency of rhodamine 6 G in a copolymer
of PMMA + 10% MAA (36%) exceeds the efficiency of conversion of rhodamine 6 G in
PMMA (25%). However, in one and the same matrix, for example in PMMA modified with
dye 11B and with 6 G, the lasing efficiency reaches 52% and only 24% respectively [508].

It should be noted that the photochemical resistance of most dyes depends on the
nature of the polymeric medium. For instance, rhodamine 6 G incorporated in PMMA

shows better photoresistance in comparison with rhodamine 6 G present in an oxirane matrix, and, conversely, rhodamine C doped into an oxirane matrix has a higher radiation resistance compared with a PMMA matrix activated by the same dye [509]. It also explains the fact that laser radiation has been obtained with PMMA on the basis of no more than 10 dyes. Accordingly the photoresistance of laser dyes should be investigated with reference to their host media.

The authors of [488] studied the photobleaching of dyes doped into polymeric matrices, which enabled them to identify the main causes of photodecomposition:

(i)   interaction of organic dye molecules with free radicals;
(ii) interaction with atmospheric oxygen dissolved in the matrix.

After photopolymerization, the solid matrix always contains unreacted molecules of the initiator which form free radicals during subsequent illumination within the absorption region (for example during operation of the laser element). These free radicals attack and destroy the closely packed molecules of the dye. An effective method for eliminating this channel of photodecomposition is neutralization of the unreacted molecules of initiator.

On exposure of organic dye molecules to light with a maximum corresponding to the long wavelength absorption band (500 to 1000 nm), photoaddition of molecular oxygen takes place. Photo-oxidation and competing processes can be represented as follows:

$$A^{S_0} + h\nu_1 \rightarrow A^{S_1}$$
$$A^{S_1} \rightarrow A^{T_1}$$
$$A^{S_1} \rightarrow A^{S_0} + h\nu_2$$
$$A^{S_1} + {}^3O_2 \rightarrow A^{T_1} + {}^1O_2$$
$$A^{T_1} + {}^3O_2 \rightarrow A^{S_0} + {}^1O_2$$
$${}^1O_2 + A^{S_0} \rightarrow AO_2$$

where A is the dye molecule and AO_2 are products of photo-oxidation.

The quantum efficiency of photo-oxidation depends on the molecular structure of the organic dyes. Thus for polymethine dyes with similar terminal heterocycles, differing in the length of the polymethine chain, elongation of the conjugated chain leads to an increased fraction of decomposed molecules [488]. Study of this process revealed the formation of a variety of intermediate products, for example dyes with shortened polymethine chains, and the subsequent formation of colourless dyes. Those results show that the photoresistance of organic dyes can be improved appreciably by the removal of atmospheric oxygen from the matrix and neutralization of the molecules of initiator.

The authors of [510] suggest two mechanisms of photodestruction of xanthene dyes in polymeric matrices—radical and anionic. In the former case, following extensive population of the first excited singlet level of the dye molecule, higher singlet levels become highly populated as a result of a two-quantum process. Interaction of the dye molecules with the surrounding polymeric macromolecules (or products of their destruction) gives rise to radiationless deactivation of excited electronic states through the highly excited state of the matrix molecules, with the formation of free radicals. Since the ionization energies of radicals are much less than those of molecules, the former are effective electron donors. A neutral radical is formed as a result of combination of the excited dye molecule with the electron derived from the free radical.

In the anionic mechanism, the electron is transferred from the anion of the dye itself to the active cation. When $I < I_{dye}$, the main mechanism of photodestruction is the radical route, but when $I \sim I_{dye}$, the anionic mechanism prevails.

The kinetic features of photodestruction of rhodamine 6 G in PMMA are covered in [511]. Analysis of the dependence of the kinetic photodestruction curves of rhodamine 6 G in PMMA on the radiation dose received on exposure to light of different spectral composition shows that the molecular photodestruction of rhodamine 6 G proceeds independently; that is, the molecules of dye present in the polymeric matrix are nonequivalent in their light resistance. This nonequivalence, and also the fact that the destruction products of the dye do not greatly alter the spectral characteristics of the system, enabled the authors to suggest the 'kinetic stabilization' method. This is based on selection of the molecular concentration for the initial system which should exceed the optimal concentration by a certain value. The dye concentration is further brought towards the optimal level in the process of preliminary radiation of the system under spectral conditions similar to the operating ones. The light resistance of the system prepared in such a way will be improved by the increased fraction of dye molecules showing the best light resistance. Such results have been obtained with low intensity near UV radiation.

The processes of photodecomposition in intense light fluxes are noticeably different, mainly owing to saturated absorption. On the basis of studies into the kinetics of photochemical decolorization of rhodamine 6 G and the polymethine dye PK-560 incorporated into a polyurethane matrix and excited by the second harmonic pulses of the neodymium laser, it has been established [512] that decomposition of the dye is described by first order kinetics, the probability of decomposition being dependent on the intensity of the exciting light. Those results are explained by the authors of [512] on the basis of the three-level model of vibronic compounds. Some practical conclusions have been formulated for the partial suppression of photochemical bleaching of dyes in polymeric media.

To improve the photoresistance of the laser system, the polymeric matrix, that is poly(methyl methacrylate) and rhodamine 6 G, was doped with the fluorinated monomer or a photostabilizer, that is, phenyl salicylate [506]. The fluorinated group present in the polymer decreases its lasing efficiency in comparison with that of the active element based on the copolymer of MMA and 10% MAA. The introduction of the photostabilizer in PMMA does not decrease the lasing efficiency. The lasing stability of the sample is, however, appreciably decreased.

The lasing properties of PAE largely depend on their methods of manufacture. The manufacture of PAE is based on radical polymerization of a monomer solution of luminophor. Almost no literature is available on the modes of polymerization. We emphasize the need for careful purification of the luminophor and repeated distillation of the monomer. Purification of the monomer by filtration is covered in [507, 499]. An important factor is the initiation of polymerization, thus the initiator should not affect the optical properties of the luminophor. Azobisisobutyronitrile proved to be the most suitable oxygen-free initiator of polymerization [513]. The need to degas the initial solutions stems from the inhibition of MMA polymerization by atmospheric oxygen.

The temperature conditions of polymerization should ensure optical uniformity of the polymeric matrix. Ref. [514] deals with the action of exciting light on the polarization characteristics of lasers based on active polymeric elements. It is shown that the lasing parameters are most affected by the intrinsic anisotropy of the polymeric matrix developed during polymerization.

Active polyurethane based elements have been prepared by photochemical polymerization. The authors of [488] note the photodecomposition of luminophors (polymethine dye PK-890) during polymerization. During the polymerization of rhodamine 6 G in the presence of benzophenone (photoinitiator), the optical density of the activator was virtually unchanged.

The polymeric laser elements based on polyurethanes are essentially 'triplexes' in which the light-sensitive layer is arranged between bases (machined glass, quartz, etc.).

Active elements based on PMMA [13] and oxirane polymers [516–518] are prepared from polymeric blocks in the form of cylinders and disks.

Stringent requirements placed upon active laser elements significantly constrain their methods of manufacture. The well developed methods of treating polymeric materials (injection moulding, extrusion moulding, stamping) are not well suited for the manufacture of polymeric laser elements because they cannot ensure high optical uniformity and purity of the material as well as the absence of internal stress. For such purposes, preferable are optical mould polymerization, the distribution of the polymeric solution over the optical surface and subsequent removal of the solvent, radical bulk polymerization of the initial monomeric composition, and subsequent precision machining of the material obtained [499].

7.4 MATERIALS FOR THE PAINT AND VARNISH INDUSTRIES

Fluorescent polymeric materials are widely used in the production of fluorescent pigments and paints. Such pigments appeared for the first time in the early 1940s. Fluorescent pigments are used for advertisement posters and placards, signals and road symbols, firefighting equipment, indicators, motor vehicle bodies, sides of motor boats, inflatable dinghies, for marking locomotives, civil aircraft, etc. ([5], p. 173).

Fluorescent pigments are also extensively used for decorative purposes, such as combining two patterns on one surface. One pattern is usually made in nonfluorescent paints, and the other in luminescent paints (preferably in colourless paints emitting as different colours). Under UV irradiation, the viewer can see a picture different from that seen in daylight. This method is successfully used in theatres, enabling one to avoid changing the scenery.

Fluorescent paints, emitting during excitation by sunlight and showing no emission when the excitation ends, are classified as 'ordinary' and 'daytime' [519]. The emission of ordinary fluorescent paints occurs as a result of excitation only by UV. The low intensity of UV radiation near the Earth's surface is responsible for the insignificant increase in the brightness of such paints in comparison with nonfluorescent ones. Daytime fluorescent paints are excited over a wide range of wavelengths, from 320 to 500 nm, that is, including short wavelength visible rays. As a result of absorption of these rays, the brightness of daytime fluorescent pigments greatly exceeds that of ordinary paints.

The brightness of paints depends not only on the radiation of the absorbed fraction of incident light but also on reflection of the unabsorbed portion. Consequently, the fuller the reflection of the unabsorbed fraction, and the radiation of the absorbed fraction of incident light by the coating, the brighter is the coating. This is why fluorescent paints are usually applied on white coatings with a high reflective capacity.

The optical properties of surfaces painted with fluorescent pigments are described within the terms of the colorimetric characteristics applied to ordinary paints. They are: colour tone, colour purity, and brightness coefficient [520]. The colour tone depends on the wavelength of monochromatic radiation which should be added to white for reproduction of the test radiation.

The most extensively used bases for daytime fluorescent pigments and paints are thermoplastic melamine- and urea-formaldehyde resins modified with p-toluene sulfamide (MTCP-resins) [5], p. 174. The advantage of these resins is their brittleness, which allows very fine pigments to be obtained. They are colourless and sufficiently transparent to light and the luminescence emitted by the luminophors incorporated in daytime fluorescent pigments. In addition, such resins have a higher light resistance than any other resins.

Ref. [5] p. 173 covers daytime fluorescent pigments which are solid solutions of low molecular luminophors in polycondensation polymers. Much attention is given to polymeric resins chemically bonded to the luminophor molecules. Pigments of this type possess certain advantages over traditional pigments because the luminophoric components do not migrate from the polymeric matrix and are resistant to the washing-out action of solvents. Such systems are discussed briefly below.

Bright fluorescent pigments have been prepared by interaction of the condensation products of epichlorohydrin and aryl imides of 4-aminonaphthalic acid (I) with polycaproamide or MTCP-resins (Inventor's Certificate 306160, USSR).

$$ClCH_2-CHOH-CH_2-NH-NH-\underset{\text{I}}{\text{[naphthalimide ring]}}N-R \qquad (R=C_6H_5)$$

It should be noted that the imide derivatives of 4-aminonaphthalic acid proved to be effective luminescent components of daytime fluorescent pigments [521].

The pigments based on MTCP-resin have a golden yellow colour and were used in the manufacture of paints for polyethylene printing [522]. These pigments found wide application in the production of enamel paints and water soluble colours, but because of their comparatively low dispersity they proved to be of little use in the manufacture of fluorescent temperas. In ref. [523] such paints were prepared with the use of polyester resins such as xylitol phthalic, trimethylolpropane phthalic, or glyptal. For example, the pigment containing 0.9% luminophor I (R = $C_{10}H_7$) in glyptal resin is superior in brightness to that based on the same resin and 3-methoxybenzathrone and used in industry.

Yellow-lemon pigments with a high brightness coefficient (116%) and good colour purity (90%), fluorescing with a maximum at 480–500 nm have been prepared by interaction of the anhydride of phenyloxazolyl naphthalic acid (II) with MTCP-resins [5]. On addition to the above polymer of rhodamine 6 G, whose absorption spectrum overlaps the emission spectrum of luminophor II, red fluorescent pigments (fluorescence maximum at 610 nm) are obtained. The authors of [524] suggest using 4-(4,5-benzimidazolyl-2)-naphthalic anhydride (III) for the reaction with MTCP-resins.

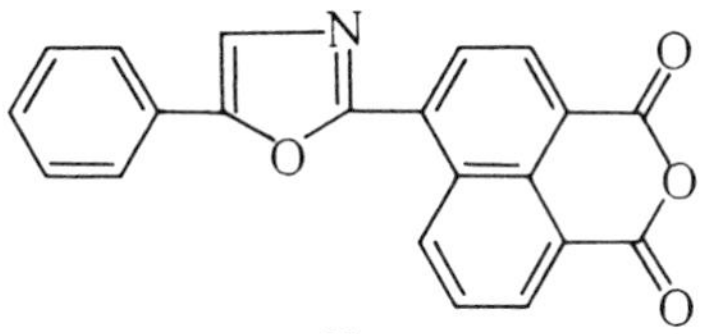

II

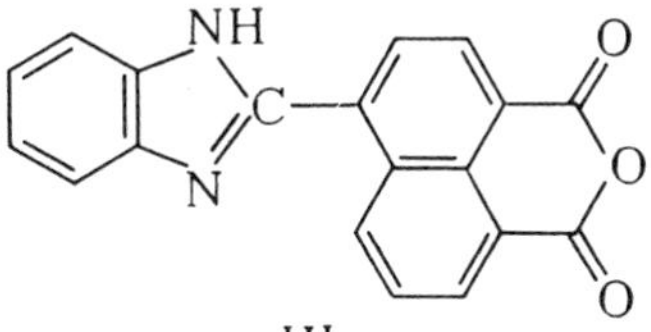

III

By adding rhodamine C or 6 G to the pigment formed, and dispersing it in varnish CBM-31, the authors obtained red enamel paints with an adequate brightness coefficient and high colour purity.

Migration resistant luminescent dyes were prepared for polyethylene and poly(vinyl chloride) by adding the derivatives of 4-hydroxy-3-carboxyphenyl imides of naphthalic acid (IV) to the reaction mass during the synthesis of trimethylolpropanephthalic and MTCP-resins [525].

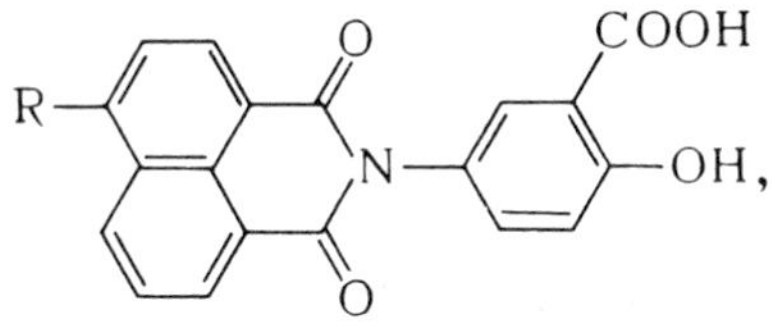

IV

In the preparation of fluorescent pigments, resins with luminophoric dopants were precipitated on Al_2O_3. More intense and brighter paints were obtained on the basis of MTCP-resins. A certain reduction in the brightness and emission intensity of trimethylol-propanephthalate-based pigments seems to stem from the partial formation of nonfluo-rescent dye complexes containing hydroxy- and carboxy-groups in the *ortho*-position mutual with Al^{3+}.

The preparation of daytime fluorescent polymer based pigments emitting in the violet and blue regions presents a number of difficulties associated also with the selection of appropriate luminescent components and the search for appropriate nonluminescent dyes since only some of them, in combination with the selected luminophor, make possible the preparation of fluorescent paints of high colour purity ([5] p. 178). Noteworthy in this respect is [526] which describes daytime fluorescent pigments based on MTCP-resins chemically bonded to the amine derivatives of anthraquinone diazoles. In solution, these dyes fluoresce in organic solvents with a maximum from 596 to 630 nm. They can impart a bright colour to pigments in daylight without addition of nonluminescent dyes, which permits the preparation of bright violet pigments in a variety of tints.

Of great importance for the practical use of painted polymeric materials is their light fastness [527, 528]. A low resistance to the action of sunlight has long delayed the wide application of polymeric coatings containing daytime fluorescent pigments.

Even the best coatings began to lose brightness after a month-long exposure to light. Studies to improve light fastness have been carried out along the following lines: (i) to prepare luminophors showing better light fastness, (ii) to select polymeric bases enhancing the light fastness of pigments, (iii) to select light stabilizers protecting the fluorescent coating.

Light stabilizers have been used in the preparation of protective varnishes which exhibit high transparency to visible radiation but absorb most UV radiation. The application of polyacrylic resin containing 4 or 5% 2,2-hydroxy-5-methylphenyl benzotriazole proved to be the best. The use of this varnish in the protection of daytime fluorescent pigments showed that, after a 3- to 4-month exposure to atmospheric conditions, the paints did not noticeably fade or become dull [519]. Without a protective coating, the appearance of the same fluorescent pigments degraded very much after 2 or 3 months of tests.

Besides benzotriazole derivatives, protective varnishes are doped with other stabilizers such as 2-hydroxy-4-methoxybenzophenone and free aromatic ketones.

A number of reports [529–531] deal with the synthesis and properties of daytime fluorescent pigments. Thus, Industrial Colours Ltd are engaged in the production of polyacrylic resin based pigments combining high colour purity and a satisfactory brightness coefficient [529]. Film-forming compositions are used in the preparation of paints and inks for stencilling, lithography, and offset printing.

An important application of daytime fluorescent pigments is the production of luminescent dyed plastics and fibres. Poly(methyl methacrylate) plastics dyed with such luminophors as rhodamine C are used to prepare luminescent scales in instrument making. The production, properties, and fields of application of such materials are covered in ([5] p. 182). The method of structural dyeing of polycaproamide by the introduction, before polycondensation, in ε-caprolactam of the anhydride mixture of percarbonic acids and o-diamines is an example of preparation of fluorescent fibres for the textile industry (Inventor's Certificate No. 203156, USSR). The derivatives of naphthoylene benzimidazole formed in the process of synthesis are chemically bonded with the polymeric chain.

The bright yellow fluorescent polyurethanes, which can be used for decorative purposes, are prepared by the interaction of oligomeric polyesters with the products of reaction between diisocyanates and 4-aminonaphthalimides of type V (US Patent No. 3096333).

$$O{=}C{=}N{-}\langle\ \rangle{-}NH{-}CO{-}NH{-}$$

V

In a number of cases, the dyeing of polymers with daytime fluorescent pigments can lead to the enhancement of their light fastness. The mechanism of light-protective action of luminophors seems to consist of the absorption from solar light of UV radiation, responsible for polymer destruction, and emission of less active light by molecules of the luminophor [57].

7.5 NEW FIELDS OF APPLICATION

The new fields of application of fluorescent polymers which are yet to be widely adopted are the preparation of photoluminescent materials (luminescent photolayers)

[532–536] and luminescent probes, often in association with fibre-optic sensors [537], for the medicobiological study of cell membranes ([8] p. 1188).

The suggestion has been made to prepare luminescent photolayers on the basis of polymers with light-sensitive molecular dopants which, when exposed to radiation, form luminescent compounds [532, 533]. Under the action of light, a latent image is formed on the irradiated areas of such materials. To visualize it, the exposed material should be additionally irradiated by light of wavelength close to the maximum of the absorption band of the luminescent compound formed.

Such luminescent photolayers are exemplified by polymeric compositions doped with nonluminescent dimers of aromatic compounds which, under radiation, dissociate into monomeric molecules luminescing with a high quantum yield. In [534] anthracene, sodium 2-anthracene sulfonate, the propyl ester of 2-anthracene sulfonic acid and acridicinium perchlorate were used as dimers. The photochemical decomposition of the dimers was studied in poly(methyl methacrylate) films exposed to UV with a wavelength of 275 nm. Using an anthracene dimer, the fluorescence of the monomer formed (λ_{max} 405 and 455 nm) was recorded on excitation with UV at 340 and 345 nm. The kinetics of monomer accumulation were studied and sensitograms permitting estimation of the photographic speed of the photolayers were obtained based on the dependence of the fluorescence intensity on the irradiation time of the monomeric anthracene. In a number of experiments, the photographic speed reached 300 cm^2 J^{-1}. The opinion of the authors of [534] is that this value can be increased 20 times and more for the complete absorption of incident light by a poly(methyl methacrylate) film doped with an anthracene dimer. In practice, the absorbing excitation light and luminescent impurities contained in both the dimer and polymeric matrix considerably reduce the photographic speed of the photolayer.

The luminescent layers based on the dimers of aromatic compounds retain high values of photographic speed after storage for several years. It has been noted in [534] that the photographic speed displayed by such layers is at least an order of magnitude higher than that of conventional photochromic materials.

Luminescent photolayers can be prepared by another method. The authors of [535] suggest using thin films of polycrystalline anthracene sprayed on various bases including polymeric ones. The initial layer is luminescent. However, when exposed to light at a wavelength exceeding 400 nm, anthracene undergoes either photodimerization or photo-oxidation with the nonluminescent image being formed in the irradiated areas. At exposures of over 10^{-1} J cm^{-2}, a yellow visible image can be obtained. In this example the image develops during exposure, and the excitation of luminescence is not required for its visualization.

During the application of anthracene photolayers, absorption spectra can be used to obtain the photographic characteristics. It is expedient to apply this method in reading coded information and for long radiation exposures [535]. For visual control, it is more convenient to excite the luminescence to develop the recorded image.

Anthracene photolayers exhibit a reasonably high level of resolution, that is, no fewer than 400 lines mm^{-1} and the image they carry is stable over time. When kept in storage without any precautions taken against daytime diffuse light, atmospheric moisture, and variation in room temperature, they do not lose their properties for several years. A detailed study of the photographic characteristics of such layers showed that on irradiation with light of 9×10^{-2} W cm^{-2}, luminescence quenching becomes noticeable when the radiation time is equal to or exceeds 1 s. The image obtained is, however, spontaneously written off within 24 hours at room temperature. With increase of the exposure time to several minutes, the image is no longer written off [535]. The drawbacks of anthracene photolayers include the insufficient photographic speed which equals *ca* 0.05 cm^2 J^{-1}, and reduction of the image contrast range in the process of reading. The latter is associated with the actinic nature of the light exciting the image.

Depending on the intensity of the luminescent image obtained on luminescent photo-layers of any type, it is possible to use the following methods of reading:

(i) Direct viewing during which the exposed photolayer is lit by a light source through a light filter.
(ii) Photographing of the image on a conventional light-sensitive film.
(iii) Monitoring of the image through an electro-optical transducer which permits intensification of the signal 10^5 or 10^6 times.
(iv) Raster reading by means of a photomultiplier tube accumulating the signal at a high resolution level.
(v) Contact printing through a light filter on conventional photomaterials.

Since one of the basic requirements imposed on luminescent photolayers is a high quantum yield of the primary photochemical event initiated by the exposure light, the suggestion was made to use aromatic and heteroaromatic azides as additives to polymeric matrices [532, 536]. These compounds are able to dissociate under the action of UV with a quantum yield close to unity, the compounds formed luminescing with a high quantum yield. Table 7.12 features the azides used and the region of luminescence of their solid solutions in the polystyrene film after exposure.

Table 7.12. Heteroaromatic azides used in manufacture of luminescent photo-layers and colour of image obtained

Compound	Exposure/nm	Colour
4-Azidophthalimide	375	Green
3-Azidophthalimide	3⁻5	Blue
N-Phenyl-3-azidophthalimide	370	Bluish-green
2,3-Diazidoquinoxaline	320	Blue

The most extensively studied are the luminescent photolayers based on 4-azidophthal-imide and polymers such as polystyrene, poly(methyl methacrylate) poly(vinyl ethylal), and cellulose esters [532]. It has been found that the behaviour of 4-azidophthalimide is virtually independent of the nature of the polymeric matrix; in all cases, the photodisso-ciation quantum yield reaches 0.9–1.0, and the luminescence spectrum of the photolysis products is characterized by a structureless band with λ_{max} *ca* 500 nm. However, this type of polymeric matrix determines the amount of luminescent product formed, which indicates the involvement of the polymer in the further conversion of intermediate products formed during photolysis of the azide:

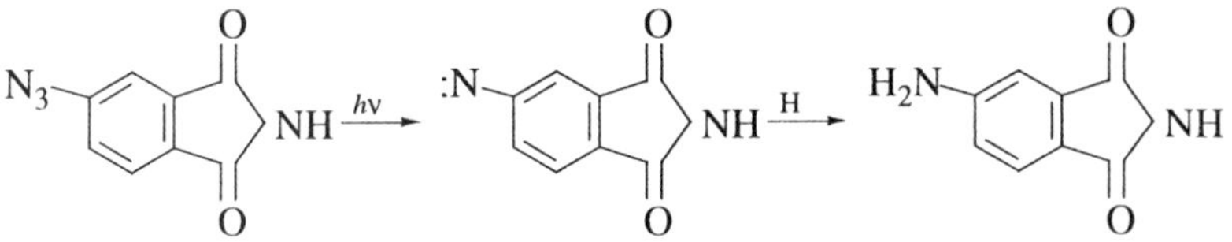

The nitrene, into which the azide is converted on irradiation, abstracts two hydrogen atoms from the polymeric matrix forming 4-aminophthalimide whose luminescence quantum yield is close to unity. This mechanism is confirmed by the similarity of the absorption spectra of the photolysis product of 4-azidophthalimide and pure 4-aminophthalimide.

Polystyrene or poly(methyl methacrylate) films containing 0.1–1% (by mass) of heteroaromatic azides form the basis for the production of luminescent photolayers applicable to recording images by the contact method from a slide or a stencil [532]. This image is reproduced by light of wavelength 400–420 nm in the form of a luminescent image without irradiation of the nonexposed sites. This suggested positive luminescent material has a photosensitivity of up to 10^{12} quantum cm^{-2} and a resolution of up to 400 lines mm^{-1}.

As is well known [538], nitrenes formed in the initial stage of azide photolysis are incorporated deeply in double bonds which can be used to photocrosslink polymers. On this basis, the author of [532] studied the possibility of fixing the image in the luminescent layers formed by 4-azidophthalimide and poly(vinyl ethylal) or polystyrene applied to a cellulose triacetate base. After image-wise exposure, the irradiated sites of the photolayers developed a luminescent image. For fixation, it was necessary to expose the image to radiation for a couple of minutes (the light-sensitive layer being 5–10 μm thick). To remove 4-azidophthalimide from the nonexposed sites, the photolayers were rinsed with methanol or ethanol. In such cases, the character of the absorption spectrum of the irradiated sites of the photolayers and the luminescent image were virtually unchanged. The clean sample becomes insensitive and can be stored at room temperature and under moderate lighting conditions for 3 years without changing the luminescent image, which can be read over a broad range of wavelengths (from 250 to 420 nm). The long-term storage within this timescale of the nonexposed photolayers in darkness, and regular examination of their spectral characteristics, showed that the samples do not lose their sensitivity.

One may conclude that luminescent photolayers based in polymeric matrices on solid solutions of heteroaromatic azides have a satisfactory light sensitivity and high resolution. They offer the advantage over other non-silver photomaterials with the possibility of intensifying the image by the further prolonged accumulation of the signal or by means of electronics. It is stated in [532] that the new material can be used for microfilming and soundtrack recording in cinematography.

As has already been noted, the second of the new trends of application of fluorescent polymers is their employment as luminescent probes both in analytes and in the investigation of biological membranes. The polymeric structure of these probes can contribute to their increased selectivity with regard to separate cell components in comparison with low molecular luminophors. By analogy with low molecular luminescent probes [279, 282], polymeric probes can be arbitrarily divided into three large classes:

(i)   Water soluble polymers, such as derivatives of poly(vinyl alcohol) containing luminophoric groups in their side chain. The distinctive feature of these probes is their increased sensitivity to the membrane potentials and acidity gradient existing in the cell. The distribution of water soluble probes between the membrane material and ambient aqueous medium is shifted towards the latter.

(ii) Hydrophobic polymeric probes such as the copolymers of polystyrene with vinyl-substituted aromatic hydrocarbons. The introduction of additional *tert*-butyl substituents into polymers will, apparently, solubilize them in the lipid membrane component. The fluorescence parameters of such probes, their intensity of polarization, and lifetime depend on the physical characteristics of the probe microenvironment in the membrane. Fluorescent probes of this type are a sensitive tool designed for investigating the physical properties of the internal contents of the membrane.

(iii) Polymeric probes distributed between the aqueous and membrane phases in a measured proportion. The salt complexes of such polymers as poly(vinyl amine) and 1-dimethylaminonaphthalic acid can probably be used as such probes. It should be assumed that such macromolecules will fluoresce with a high quantum yield in the membrane bound state and will serve as a sensitive indicator of the transmembrane potential, ionic membrane transport, membrane protein conformation, etc. This is why almost any structural membrane changes occurring under the action of chemical or physical factors, including heating or cooling, the action of ferments, and medical preparations and hormones have a bearing on the fluorescence of such probes.

The possibilities of using luminescent probes in the investigation of biological membranes are covered in greater detail in [282, 537].

The more general topic of applying fluorescent polymeric fibres as fibre optic chemical sensors (FOCS) has been reviewed [537]. One approach depends on fluorescence quenching by the analyte at the remote end of the fibre of an optical signal generated in the fluorescent polymeric fibre at the opposite (observer) end; molecular oxygen and protons can be determined in this way. A useful set of papers on FOCS covering chemical and biochemical systems appears in [282A], which also includes developments in near-infrared fluorescence probes. The latter are based on cyanine and naphthalocyanine dyes, of which chloroaluminium 2,3-naphthalocyanine tetrasulfonic acid is typical [539].

References

[1]   Terenin A. N. *Photonics of dye molecules*, Nauka, Leningrad (1967), 616 p.

[2]   Meister T. G., *Electronic spectra of polyatomic molecules*, LGU Publ., Leningrad (1969), 206 p.

[2A] Sinha H. K., Dogra S. K., *J. Chem. Phys.* (1986), **102**, 337–347.

[2B] Acuna A. U., Amat F., Catalan J., Costela A., Figuera J. M., Munoz J. M., *Chem. Phys. Lett.* (1986), **132**, 567–569.

[2C] Costela A., Amat F., Catalan J., Douhal A., Figuera J. M., Munoz J. M., Acuna A. U., *Opt. Commun.* (1987), **64**, 457–460.

[3]   Nurmukhametov R. G., *Absorption and luminescence of aromatic compounds*, Moscow, Khimia (1971), 216 p.

[4]   McGlynn S. P., Azumi T., Kinoshita M., *Molecular spectroscopy of the triplet state*, New Jersey, Prentice Hall, Inc., Englewood Cliffs (1969), 448 p.

[5]   Krasovitsky B. M., Bolotin B. M., *Organic luminophors*, Moscow, Khimia (1984), 336 p.

[6]   Parker C. A., *Photoluminescence of solutions*. Amsterdam, Elsevier (1968), 510 p.

[7]   Murrell J. N., *The theory of the electronic spectra of organic molecules*, London, Chapman and Hall (1971), 328 p.

[8]   Barashkov N. N. *Russ. Chem. Rev.* (1985), **54**, 690–709.

[9]   Sommersall A. C., Guillet J. E. *J. Macromol. Sci.-Rev.* (1975), Pt. C. **13**, p. 135–188.

[10] Krejcoves J., Drobnik J., Kalal J. *Chem. listy* (1979), **73**, 363–388.

[11] Oster G., Nishijima Y., *Fortschr. Hochpolymer Forsch*, (1964), **3**, 313–331.

[12] Gunder O. A., *Polymeric systems and their scinillation properties*, Moscow (1975), 60 p.

[13] Mnuskin V. E., Trinchuk B. F., Tokareva A. N., Solid–dye Lasers. *Electr. Technika*, Ser. II (1987), 3–52.

[14] Polymer Photophysics: *Luminescence, energy migration, and molecular motion in synthetic polymers*, Ed., Phillips D, Chapman and Hall, London (1985), 437 pp.

[15] Guillet J., *Polymer photophysics and photochemistry*. Cambridge, Cambridge University Press (1985).

[15A] Winnik M. A., Ed., *NATO Asc Ser.*, *Ser. C* (1986), **182** (Photophysical and photochemical tools in polymer science: conformation, dynamics, morphology), D. Reidel, Dordrecht, 642 pp.

[15B] Rabek J. F., Ed., *Photochemistry and photophysics*, *Vol. 1* CRC, Boca Raton, Florida (1990), 192 p.

[15C] Tsuchida E., Ed., *Macromolecular Complexes*: *Dynamic interactions and electronic processes*, VCH, Weinheim (1991), 400 p.

[16] Allen N. S., *Photochemistry*, Ed., D. Bryce–Smith, Royal Society of Chemistry, London (1988), **19**, 459–506.

[17A] Allen N. S., *Photochemistry*, Ed., D. Bryce–Smith, Royal Society of Chemistry, London (1989), **20**, 455–503.

[17B] Allen N. S., *Photochemistry,* Ed. D. Bryce–Smith, A. Gilbert, Royal Society of Chemistry, London (1990), **21**, 483–543.

[17C] Allen N. S., Edge M., *Photochemistry,* Ed. D. Bryce–Smith, A. Gilbert, Royal Society of Chemistry, London (1991), **22**, 411–501.

[17D] Allen N. S., Edge M., *Photochemistry,* Ed. D. Bryce–Smith, A. Gilbert, Royal Society of Chemistry, London (1992), **23**, 403–468.

[17E] Allen N. S., *Photochemistry,* Ed., D. Bryce–Smith, A. Gilbert, Royal Society of Chemistry, London (1993), **24**, 403–509.

[18] Cuniberti C., Perico A. *Progr. Polymer Sci.* (1984), **10**, 271–316.

[19] Anufrieva E. V., *Modern Physical methods for polymer research* (Ed. G. L. Slonimsky), Khimia, Moscow (1982), 77–91.

[19A] Bokobza L., *Prog. Polym. Sci.* (1990), **15**, 337–360.

[19B] Itagaki H., Horie K., Mita I., *Prog. Polym. Sci.* (1990), **15**, 361–424

[20] McKellar J. F., Allen N. S., *Photochemistry of man-made polymers*, Applied Science Publ. London (1979).

[21] Shlyapintokh V. Ya., *Photochemical conversions and stabilization of polymers*, Moscow, Khimia (1979), 344 p.

[22] Birks J. B., Conte J. C. *Proc Roy Soc. (London)* (1968), Ser. A, **303**, 85–95.

[23] Galanin M. D. *Moscow, AN SSSR Publ.* (1960), **12**, 3–53.

[24] Fox R. B. *Pure Appl. Chem.* (1973), **34**, 235–241.

[25] Powell R. C. *J. Chem. Phys.* (1971), **55**, 1871–1877.

[26] Faidysh A. N., Slobodnyak V. V., Yaschuk V. N. *AN SSSR, Ser. Phys.* (1978), **42**, 318–322.

[27] Vala M. T., Haebig J. *J. Chem. Phys.* (1965), **43**, 886–897.

[28] Klopffer W. *Organic molecular photophysica*, Ed. J. B. Birks. N. Y.: Wiley–Interscience (1973), 357–402.

[29] Yoon D. Y., Sundararajan P. R., Flory P. J. *Macromolecules* (1975), **8**, 776–783.

[30] Birks J. B. *Rep. Progr. Phys.* (1975), **38**, 903–974.

[31] Kuzmin M. G. *JVHO imeni D. I. Mendeleeva* (1974), **19**, 362–375.

[32] Parker C. A. *Advances in Photochemistry*, Ed. W. A. Moyes, G. S. Hammond, J. N. Pitts. H. Y.: Intern. (1964), **2**, 453 p.

[33] Biteman V. B., Gunder O. A., Petrova I. B., Senchishin V. G. *J. Prikladn. Spectroscopy* (1980), **33**, 723–726.

[34] Biteman V. B., Gunder O. A., Senchishin V. G. *Vysokomol. Soed.* (1984), Ser. B, **26**, 86–88.

[35] MacCullum J. R., *Polym. Prepr. Am. Chem. Soc. Div. Polym. Chem.* (1986), **27**, 69.

[36] Fraser I. M., MacCullum J. R., *Eur. Polym. J.* (1987), **23**, 171.

[37] Qian R., Ti C., *Polym. Commun.* (1986), **27**, 169.

[38] Jiang Y. C., Lucki J., Rabek J. F., Ranby B., *Macromol. Chem., Rap. Commun.* (1986), **7**, 563.

[39] Allen N. S., McKellar J. F., *Photochemistry of dyed and pigmented polymers*, Applied Science Publ. London (1980), 284 p.

[40] Watanabe A., Matsuda M., *Macromolecules* (1986), **19**, 2253.

[41] Sienicki K., Bojarski C., *Polym. Photochem.* (1986), **7**, 243.

[42] Kauffman H. F., Weixelbaumer W. D., Burbaumer J., *Polym. Prepr., Am. Chem. Soc., Div. Polym. Chem.* (1986), **27**, 364.

[43] Klopffer W., *Polym. Prepr., Am. Chem. Soc., Div. Polym. Chem.* (1986), **27**, 346.

[44] Phillips D., *Polym. Prepr., Am. Chem. Soc., Div. Polym. Chem.* (1986), **27**, 368.

[45] Schiveizer K. S., *Polym. Prepr., Am. Chem. Soc., Div. Polym. Chem.* (1986), **27**, 134.

[46] Birks J. B., *Photophysics of aromatic molecules*. London, Wiley–Interscience (1970), 704 p.

[47] Ermolaev V. L., Bodunov E. N., Sveshnikova E. B., Shakhverdov T. A., Nauka Publ., *Radiationless electronic excitation energy transfer.* (1977), 311 p.

[48] Agranovich V. M., Galanin M. D., *Electronic excitation energy transfer in condensed media.* Moscow, Nauka, (1978), 383 p.

[49] Rozman I. M. *AN SSSR. Ser. Phys.* (1972), **36**, 922–928.

[50] Rozman I. M. *AN SSSR. Ser. Phys.* (1973), **37**, 502–507.

[51] Doktorov A. B., Burstein A. I. *JETP.* (1975), **68**, 1349–1362.

[52] Gosele U., Hauser M., Klein U.K.A., Frey R. *Chem. Phys. Lett.* (1975), **34**, 519–522.

[53] Heisel F., Laustriat C. *J. Chim. Phys. Phys. Chim. Biol.* (1969), **66**, 1895–1902.

[54] David C., Piens M., Geuskens G. *Europ. Polymer. J.* (1973), **9**, 533–541.

[55] Biteman V. B., Gunder O. A., Senchishin V. G. *Zh. Prikl. Spectr.* (1978), **28**, 823–826.

[56] Biteman V. B., Gunder O. A., Senchishin V. G. *Zh. Prikl. Spectr.* (1979), **30**, 1059–1063.

[56A] Ghiggino K. P., Smith T. A. *Prog. React. Kinet.* (1993), **18**, 375–436.

[56B] Soutar I., *Polymer International* (1991), **26**, 35–49.

[57] Rabek J. F., Ranby B. G., *Photodegradation, photo–oxidation and photostabilization of polymers: Principles and applications.* London, Wiley (1975), 573 p.

[58] Stepanov B. I., Gribkovsky V. P., *Introduction to luminescence theory*, Minsk, Izd. AN SSSR (1963), 443 p.

[59] Strokach N. S. *et al.*, *Vibronic spectra of polyatomic molecules* Moscow, Nauka (1982), 143 p.

[60] Nurmukhametov R. N., Plotnikov V. G., Shigorin D. N. *Russ. J. Phys. Chem.* (1966), **40**, 622–625.

[61] Shigorin D. N. *Russ. J. Phys. Chem.* (1980), **54**, 1095–1105.

[62] Shigorin D. N. *Russ. J. Phys. Chem.* (1983), **57**, 501–509.

[63] Nakaya T., Totomoto T., Imoto M. *Bull. Chem. Soc. Jap.* 1967, **40**, 691–692.

[64] Sommersall A. C., Guillet J. E. *Macromolecules*. (1973), **6**, 218–233.

[65] Hawkins E. G. *J. Chem. Soc.* (1957), 3858–3862.

[66] Krakovyak M. G., Anufrieva E. V., Ananyeva T. D. *et al. Vysokomol. Soed.* (1975), Ser. A, **17**, 1983–1988.

[67] Krakovyak M. G., Anufrieva E. V., Skorokhodova S. S. *Vysokomol. Soed.* (1972), Ser. A, **14**, 1127–1131.

[68] Meyer G. *Bull. Soc. Chim. France*. (1969), 3629–3632.

[69] Tanikawa K., Ishizuka T., Suzuki K., *et al.*, *Bull. Chem. Soc. Jap.* (1968), **41**, 2719–2722.

[70] Tanikawa K., Ishizuka T., Suzuki K., *et al.*, *Bull. Chem. Soc. Jap.* (1983), **56**, 1529–1534.

[71] Imai K., Iton Y., Furue M., Nozakura S. *J. Polymer Sci., Polymer Chem. Ed.* (1983), **21**, 2439–2449.

[72] Fieser M., Fieser L., *Reagents for Organic Synthesis*, N. Y, Wiley (1974), **4**, 724 p.

[73] Homer R. B., Shinitzky M. *Macromolecules* (1968), **1**, 469–472.

[74] Milner R. S., Fedosyuk M. I. *Monocrystals, scintillators and organic luminophors*. Kharkov, BNII monocryst. (1969), **5**, Part 1, 147, 148.

[75] Ibemesi J. A., Kinsinger J. V., El–Bayomi M. A. *J. Polymer Sci., Polymer Chem. Ed.* (1980), **18**, 879–890.

[76] Tazuke S., Banba F., *Macromolecules* (1976), **9**, 451–455.

[77] Etienne A., Arcos J. C. *Bull. Soc. Chim. France* (1951), 731–733.

[78] Tazuke S., Banba F. *J. Polymer Sci., Polymer Chem. Ed.* (1976), **14**, p. 2463–2478.

[79] Tazuke S., Hayashi N. *Polymer J.* (1978), **10**, 443–450.

[80] Tazuke S., Tanabe T. *Macromolecules* (1979), **12**, 848–853.

[81] Stewart F.H.C. *Austr. J. Chem.* (1960), **13**, 483–487.

[82] Horhold H. H., Wolf H. *Faserforschung u. Textiltechnik* (1978), **29**, 393–395.

[83] Reznikova N. I., Veselova T. V. *Zh. Org. Khim.* (1970), **6**, 840–845.

[84] Gupta V. S., Kraft S. C., Samuelson J. C., Mays R. L. *J. Chromatog.* (1967), **26**, 158–163.

[85] Ioffe I. S., Otten V. F., *JOKh* (1961), **31**, 1511–1516.

[86] Hayashi Y., Tazuke S., *Macromol. Chem.* (1973), **171** (1), 57–68.

[87] Matsuyama Y., Tazuke S. *Macromol. Chem.* (1975), **176** (6), 1657–1667.

[88] Levshin L. V., Nizamov N. N. *Izv. AN SSSR. Ser. Phys.* (1970), **34**, 599–603.

[89] Nurmukhametov R. N., Kunavin N. I., Khachaturova G. T., *Izv. AN SSSR Ser. Phys.* (1978), **42**, 517–523.

[90] Kunavin N. I., Nurmukhametov R. N., *Zh. Prikl. Spectr.* (1977), **26**, 1120–1124.

[91] Feofilov P. P. *Polarized luminescence of atoms, molecules and crystals*, Moscow, Fizmagiz (1959), 288 p.

[92] Levshin L. V., Mitsel Y. A., Nizamov N. N. *Zh. Prikl. Spectr.* (1969), **11**, 509–514.

[93] Levshin L. V., Akvarova D. M. *Vestnik MGU. Ser. Phys.* (1964), **2**, 16–20.

[94] Berlin A. A., Geiderikh M. A. *et al.*, *Khimiya polysopr. sistem*, Moscow, Khimiya Publ. (1972), 272 p.

[95] Gachkovsky V. F. *J. Str. Chem.* (1967), **8**, 69–75.

[96] Pivovarov A. P., Gak Y. V., Lukovnikov A. F., *Vysokomol. Soed.* (1971), Ser. A, **13**, 2110–2120.

[97] Yanari S. S., Bovey F. A., Lumry R. *Nature*. (1963), **200**, 242–244.

[98] Harrash L. A. *J. Chem. Phys*. (1972), **56**, 385–387.

[99] Roberts A. J. *J. Photochem*. (1981), **17**, 11–14.

[100] Halary J. L., Monnerie L., *NATO Advanced Science Institutes*, Ser. C. Netherlands (1986), 589 p.

[101] Frank C. W., Gelles R. (in [100]), 561 p.

[102] Nagamov N. A., Sivokhin V. S., Batrak G. V. *et al.*, *Vysokomol. Soed*, Ser. B, **28**, (1986), 13.

[103] Polacki Z. *J. Photochem*. (1985), **28**, 135–137.

[104] Brechtbuhler T., Magat M. *J. Chim. Phys*. (1950), **47**, 676–684.

[105] Cherkasov A. S., Voldaikina K. G., *Vysokomol. Soed*., (1963), **5**, 79–86.

[106] Cherkasov A. S., Voldaikina K. G. *Vysokomol. Soed. (Carbon–chain compounds)* (1963), **4**, 179–185.

[107] Cherkasov A. S., Voldatkina K. G. *Vysokomol. Soed*. (1965), **7**, 175–179.

[108] Anufrieva E. V., Volkenstein M. V., Koton M. V., *Zh. Fiz. Khim*. (1957), **31**, 1532–1549.

[109] Andreeschev E. A., Baroni E. E., Viktorov E. S. *et al.*, *Vysokomol. Soed*. (1963), **5**, 1482–1487.

[110] Cherkasov A. S., Voldaikina K. G. *Spectroscopy of polymers*. Ed. by M. Volkenstein, Kiev, Nauk. Dumka (1968), 126–132.

[111] Wahl Ph., Meyer G. Parrod J., Anchet J. C. *Europ. Polymer J*. (1970), **6**, 585–608.

[112] Gunder O. A. *Excimer Formations in Polymer Scintillation Systems*, *Cherkassy* (1974), **14**, 14C85.

[113] Gunder O. A., Petrova I. B. *Scintillators and Organic Luminophors*, *Kharkov* (1973), **2**, 111–116.

[114] Andreeschev E. A., Rozman I. M., *Optics and Spectr*. (1960), **8**, 828–831.

[115] Grigoryeva V. I., Gunder O. A., Krasovitsky B. M. *et al.*, *J. Prikl. Spectr*. (1968), **5**, 884–887.

[116] Gunder O. A., Petrova I. B., *Izv. AN SSSR, Ser. Phys*. (1972), **36**, 1134–1136.

[117] Alekseeva T. A., Dmitrievskaya L. I., Grigoryeva B. I. *et al.*, *Vysokomol. Soed*. (1968), Ser. B, **10**, 226–228.

[118] Iinuta F., Mikawa H., Shirota Y. *Macromolecules* (1981), **14**, 1747–1751.

[119] Tazuke S., Takasaki R. J. *Polymer Sci., Polymer Chem. Ed*. (1983), **21**, 1529–1534.

[120] Ushiki H., Horie K., Okamoto A., *Muta I. Polymer J*. (1981), **13**, 191–200.

[121] Burkart R. D., Burrows J. A. J., *Polym. Prepr, Am. Chem. Soc., Div. Polymer Chem*. (1987), **28**, 66.

[122] Torkelson J. M., Gilbert S. R., *Polym. Prepr. Am. Chem. Soc. Div. Polym. Chem*. (1986), **27**, 437.

[123] Coulter D. R., Gubta A., Yavronian A. *et al.*, *Macromolecules* (1986), **19**, 1227.

[124] Banks E., Okamoto Y., Ueba Y., *J. Appl. Polymer Sci*. (1980), **25**, 359–368.

[125] Anufrieva E. V., Gotlib Y. A., Krakovyak M. G. *et al.*, *Vysokomol. Soed*. (1972), Ser. A, **14**, 1430–1449.

[126] Krakovyak M. G., Milovskaya E. B., Rudkovskaya G. D. *et al.*, *Vysokomol. Soed*. (1980), Ser. A, **22**, 143–150.

214 **References**

[127] Barashkov N. N., Sakhno T. V., Alexeev N. N. *et al.*, *J. Prikl. Spectr.* (1990), **53**, 386.

[128] Kosfeld R., Marsch K., *Adv. Chem. Ser. Multicom. Polymer System.* Washington D. C. (1971), 581–588.

[129] Phillips D., Roberts A. J., Soutar I., *J. Polymer Sci., Polymer Lett. Ed.* (1980), **18**, 123–129.

[130] Anderson R. A., Reid R. F., Soutar I. *Europ. Polymer J.* (1980), **16**, 945–950.

[131] Aspler J. S., Guillet J. E. *Macromolecules* (1979), **12**, 1082–1088.

[132] Bratchkov C., Sinigersky V., Markova L. *et al.*, *Europ. Polymer J.* (1985), **21**, 569–572.

[133] Bouderska H., Bratchkov C., Kolova V. *et al.*, *Europ. Polymer J.* (1978), **14**, 619–621.

[134] Bairamov Yu. Yu. *Effect of quantity and distribution of chromophor groups on photostability of some macromolecules*, Moscow (1974), 21 p.

[135] Amerik Y. B., Bairamov Yu. Yu., Krencel B. A., *J. Prikl. Spectr.* (1973), **18**, 1023–1027.

[136] Wang F. W., Lowry R. E., Grant W. H. *Polymer* (1984), **25**, 690–692.

[137] Barashkov N. N., Malysheva L. I., Sakhno T. V. *et al.*, Kharkov (1990), 19 (*Reports of Union Conference on organic luminophors*).

[138] Khakhel O. A., Nurmukhamedov R. N., Barashkov N. N. *et al., J. Prikl. Spectr.* (1991), **55**, 503.

[139] Slobodyanik V. V., Faidysh A. N., Yaschuk V. N. *et al.*, *J. Prikl. Spectr.* (1982), **36**, 309–316.

[140] Sasski T., Yamamoto M., Nishiyima Y., *Macromol. Chem. Rap. Comm.* (1986), **7**, 345.

[141] Chin G., Winnik M. A., Croucher M. D., *Colloid Polymer Sci.* (1986), **264**, 25.

[142] Kim N., Webber S. E. *Macromolecules* (1985), **18**, 741–746.

[143] Roberts A. J. *J. Chem. Soc. Faraday Trans. 1* (1981), **77**, 2725–2734.

[144] Webber S. A. Avots–Avotins P. E., Deumie M. *Macromolecules* (1981), **14**, 105–110.

[145] Kainitsky A. Y., Slobodyanik V. V., Yaschuk V. N. *All-Union conference on molecular luminescence and its applications.* Kharkhov (1982), 102.

[146] Galli G., Solaro R., Chiellini E. *Polymer* (1981), **22**, 1088–1095.

[147] Keyanpour–Rad M., Ledwith A., Hallam H. *et al.*, *Macromolecules* (1978), **11**, 1114–1118.

[148] Tazuke S., Inoue T., Tanabe T. *et al.*, *J. Polymer Sci., Polymer Lett. Ed.* (1981), **19**, 11–13.

[149] Ito S., Yamamoto M., Nishiyima Y., *Koenshu–Kyoto Daizaku, Nippon Kagaku Seri Kenkyusho*, (1986), **43**, 25.

[150] Kalechits I. I., Kuzmin G. M., Zubov V. P. *et al.*, *Mekh. Kompoz. Mater.*, (1986), **2**, 331.

[151] Yokoyama M., Tamamura T., Nakano T., Mikawa H. *Chem. Lett.* (1972), 499–502.

[152] Kaoru I., Yoshihiro I., Masaoki F., Shun–Ichi N. *J. Polymer Sci., Polymer Chem. Ed.* (1983), **21**, 2439–2449.

[153] Hargreaves J. S., Webber S. E. *Macromolecules* (1984), **17**, 235–240.

[154] Altomare A., Carlini C., Gardelli F. *Macromolecules* (1985), **18**, 729–734.

[155] Phillips D., Roberts A. J., Soutar J. *Polymer* (1981), **22**, 427–429.

[156] Gupta A., Liang R., Moacanin J. *et al.*, *Europ. Polymer J.* (1981), **17**, 485–490.

[157] Burkhart R. D., Aviles R. G., Magrini K. *Macromolecules* (1981), **14**, 91–95.

[158] Fitzgibbon P. D., Frank C. W. *Macromolecules* (1981), **14**, 1650–1658.

[159] Bai F., Chang C. H., Webber S. E., *Macromolecules* (1986), **19**, 2484–2494.

[160] Burkhart R. D., Haggquist G. W., Webber S. E., *Macromolecules* (1987), **20**, 3012.

[161] Abuin E. B., Lissi E., Gargallo L. *et al.*, *Europ. Polymer J.* (1980), **16**, 1023–1025.

[162] *Organic semiconductors*. Ed. V. A. Kargin, Moscow, Nauka (1968), 546.

[163] Gachkovsky V. F. *Vysokomol. Soed.* (1970), Ser. B **12**, 411–413.

[164] Wegner G., *Macromol. Chem.* (1972), **154**, 35.

[165] MacCallum J. R., Hoyle C. E., Guillet J. E. *Macromolecules* (1980), **13**, 1647–1650.

[166] Guillet J. E., Hoyle C. E., MacCallum J. R. *Chem. Phys. Lett.* (1978), **54**, 337–339.

[167] Samedova T. G., Gavrilenko I. F., Karpacheva G. P. e*t al.*, *Izv. AN SSSR, Ser. Chem.* (1970), **3**, 682–683.

[168] Davydov B. E., Karpacheva G. P., Samedova T. G. *et al.*, *Europ. Polymer J.*, (1971), **7**, 1569–1574.

[169] Samedova T. G., Karpacheva G. P., Davydov B. E. *Europ. Polymer J.*, (1972), **8**, 599–611.

[170] Silin E. A., Ekmane A. Y., Khutareva G. V. *et al.*, *Vysokomol. Soed.* (1968), Ser. A., **10**, 1786–1793.

[171] Bendersky V. L., *J. Str. Chem.* (1963), **4**, 415–420.

[172] Cherkashin M. I., *Investigations into π-conjugated systems*. Moscow (1970), 78 p.

[173] Cherkasov Y. A., Cherkashin M. I., *Advances in Chemistry* (1979), **48**, 1119–1149.

[174] Bhattacharjee H. R., Preziosi A. F., Patel G. N. *J. Chem. Phys.* (1980), **73**, 1478–1480.

[175] Misin V. M., Cherkashin M. I. *Advances in Chemistry* (1985), **54**, 956–1008.

[176] Bykov A. N., Kirillova T. M., Lits N. P. *Vysokomol. Soed.* (1963), **5**, 428–431.

[177] Kharitonova V. P., Bykov A. N., Alexandriysky S. S. *Ser. Khimiya i khim. tekhnologiya* (1965), **8**, 297–300.

[178] Vinogradova S. V., Antonova–Antipova I. P. *Progress of Polymeric Chemistry*. Ed. Korshak V. V., Moscow, Nauka (1969), 375–396.

[179] Biswas M., Das S. K. *Polymer* (1982), **23**, 1713–1726.

[180] Tazuke S., Matsuyama Y. *Macromolecules* (1975), **8**, 280–283.

[181] Matsuyama Y., Tazuke S. *Polymer J.* (1976), **8**, 481–483.

[182] Tazuke S., Nagahara H., Matsuyama Y. *Macromol. Chem.* (1980), **18b**, 2199–2206.

[183] Hayashi Y., Tazuke S. *Macromol. Chem.* (1973), **171**, 57–68.

[184] Sato K., Nayashi N., Tazuke S. *J. Polymer Sci., Polymer Lett. Ed.* (1977), **15**, 671–674.

[185] Tazuke S., Tanabe T. *Macromolecules* (1979), **12**, 853–862.

[186] Tazuke S., Hayashi N. *Polymer J.* (1978), **10**, 443–450.

[187] Suzuki Y., Tazuke S., *Macromolecules* (1981), **14**, 1742–1747.

[188] Suzuki Y., Tazuke S. *Macromol. Chem.* (1984), **185**, 521–529.

[189] Tazuke S., Yuan H. L., Iwaya Y. *et al.*, *Macromolecules* (1981), **14**, 267–270.

[190] Yuan H. L., Tazuke S. *J. Polymer Sci.*, *Polymer Lett. Ed.* (1982), **20**, 81–84.

[191] Ibemesi J. A., Kinsinger J. B., El–Bayomi M. A. *J. Macromol. Sci. Chem.* (1980), Pt. A, **14**, 813–822.

[192] Ledwith A. *Macromol. Chem. Suppl.* (1981), **5**, 42–57.

[193] Nichimoto S., Izukawa T., Hatura Y., Kagiya T. J. *Polymer Sci. Polymer Lett. Ed.* (1984), **22**, 199–202.

[194] Hogle C. E., Kim K. J., *Macromolecules* (1980), **20**, 597.

[195] Egerton P. L., Trigg J., Hyde E. M. *et al.*, *Macromolecules* (1981), **14**, 100–104.

[196] Tatsi G. V., Pereyaslova D. G., Levchenko N. F., *et al.*, *Scintillators and organic luminophors.* Kharkov (1975), **4**, 65–69.

[197] Konalev S. E., Krasovitsky B. M., Popova N. A. *Scintillators and organic luminophors.* Kharkov (1973), 3–8.

[198] Barashkov N. N., Gorbunov L. A., Alexandrov V. N., Hetmanchuk Y. P. *J. Prikl. Spectr.*, (1989), **51**, 709.

[199] Barashkov N. N., Vysotsky V. N., Grigorieva I. N., *Khimicheskiye Volokna* (1987), **6**, 38.

[200] Barashkov N. N., Vysotsky V. N., Grigorieva I. N. *et al.*, *Khimicheskiye Volokna* (1991), **1**, 44.

[201] Horhold H. H. Z. *Chem.* (1972), **12**, 41–45.

[202] Pebalk A. V., Kardash I. E., Pravednikov A. N. *Vysokomol. Soed.* (1981), Ser. A, **23**, 1926–1930.

[203] Pebalk A. V., Barashkov N. N., Kozlov Y. A. *et al.*, *Vysokomol. Soed.* (1981), Ser. A, **23**, 2705—2712.

[204] Kozyreva E. F. *Vysokomol. Soed.* (1979), Ser. A, **21**, 855–860.

[205] Kozyreva E. F. *All–Union Conference on molecular luminescence and its application,* Kharkov (1982), 114 p.

[206] Kozyreva E. F. *Vysokomol. Soed.* (1975), Ser. B, **17**, 651–654.

[207] Misurkin I. A., Ovchinnikov A. A. *Advances in Chemistry* (1977), **46**, 1834–1870.

[208] Allen N. S. McKellar J. F. *Macromol. Chem.* (1979), **180**, 2875–2882.

[209] D'Alelio G. *Macromol. Sci.–Chem.* (1967), Pt. A, **1**, 1161–1165.

[210] Vasilenko N. A., Nurmukhametov R. N., Belaits I. L., Pravednikov A. N. *Russ. J. Phys. Chem.* (1977), **51**, 941–944.

[211] Uno A., Kondo T., *Polymer J.* (1974), **6**, 267–334.

[212] Barashkov N. N., Ryzhakova N. V., Nurmukhametov R. N. *Vysokomol. Soed.* (1984), Ser. **B**, 26, 356–359.

[213] Barashkov N. N., Sakhno T. V., Kuzmin N. I., *J. Prikl. Spectr.* (1990), **53**, 514.

[214] Barashkov N. N. *Preparation of polymers with prescibed spectral luminescent properties by structural chemical modification of the molecular chain.* Moscow (1990), 68 p.

[215] Barashkov N. N., Semenova L. I., Nurmukhametov R. N. *Vysokomol. Soed.* (1983), Ser. A, **25**, 1090–1094.

[216] Yakovlev Y. Y., Barashkov N. N., Nurmukhametov R. N. *et al.*, *Vysokomol. Soed.*, Ser. A (1991), **33**, 1033.

[217] Nurmukhametov R. N., Barashkov N. N., Semenova L. I., Shablygin M. V. *Russ. J. Phys. Chem.* (1988), **62**, 649–651.

[218] Nurmukhametov R. N., Semenova L. I., Nekrasova L. P., Barashkov N. N., *Russ. J. Phys. Chem.*, (1988), **62**, 40–43.

[219] Kojima T. *J. Polymer Sci.*, *Polymer Phys. Ed.* (1980), **18**, 1685–1695.

[220] Voischev V. S., Sazhin B. I., Mikhantiev B. I., Yakubovich V. S., *et al.*, *Vysokomol. Soed.* (1973), Ser. B, **15**, 775–778.

[221] Barashkov N. N., *Structurally–stained polymers and materials*, Moscow, Khimiya (1987), 80 p.

[222] Barashkov N. N., Sakhno T. V., Semenova L. I., *J. Prikl. Spectrosc.*, **51** (1989), 152.

[223] Wachsman E. D., Frank C. W., *Polymer* (1988), **29**, 1191.

[224] Hasegawa H., Mita I., Kochi M., Yokota R., *J. Polym. Sci., Polym. Lett. Ed.* (1989), **27**, 263.

[225] Arjavalingam G., Hougham G., Lafemina J. P., *Polymer* (1990), **31**, 840.

[226] Feofanov B. N., Siling S. A., Barashkov N. N. *et al.*, *Vysokomol. Soed.*, Ser. B (1988), **30**, 286.

[227] Siling S. A., Barashkov N. N., Feofanov B. N. *et al.*, *Vysokomol. Soed.* Ser. A (1989), **31**, 839.

[228] Barashkov N. N., Kozlov Y. A. *VII Republican Conference of Young Scientists.* Riga, Zinatne (1981), 31, 32.

[229] Snyder R. W., Thomson B., Bartges B. *et al.*, *Macromolecules* (1989), **22**, 4167.

[230] Bellus J. D., *Adv. Photochem.* (1971), **8**, 109.

[231] Naboikin E. N., Krainov I. P., Distanov B. G. *III All–Union Conference on Organic Luminophors*, Kharkov (1980), 48 p.

[232] Wahl Ph. Thèse presentées à la faculté des science de l'Université de Strasbourg. Chartres (1963), 36 p.

[233] Krakovyak M. G., Luschik V. B., Sycheva E. A. *et al.*, *Vysokomol. Soed.*, Ser. B (1986), **28**, 289–294.

[234] Krakovyak M. G., Luschik V. B., Ananieva T. D. *et al.*, *Vysokomol. Soed.*, Ser. A (1987), **29**, 598.

[235] Anufrieva E. V., Luschik V. B., Nekrasova T. N. *et al.*, *Zh. Prikl. Spectr.* (1987), **46**, 931.

[236] Krakovyak M. G., Anufrieva E. V., Skorokhodov S. S. *Vysokomol. Soed.* (1969), Ser. A, **11**, 2499–2504.

[237] Lushchik V. B., Krakovyak M. G., Skorokhodov S. S. *Vysokomol. Soed.* (1980), Ser. A, **22**, 1904–1908.

[238] Haraoubia R. *Macromol. Chem.* (1982), **183** (10), 2383–2397.

[239] Fenio J. C. *IUPAC Macro. Florence. Intern. Symp. Macromol.* (1980), **4**, 123–126.

[240] Mataga N., Kubota T. *Molecular interactions and electronic spectra.* N. Y.: Marcel Dekker (1970), 504 p.

[241] Anufrieva E. V., Volkenstein M. V., Krakovyak M. G. *et al.*, *DAN SSSR* (1968), **182**, 361–363.

[242] Elmgren H. *J. Polymer Sci.*, *Polymer Lett. Ed.* (1980), **18**, 815–822.

[243] Kainkova T. V., Barashkov N. N., Nekrasov V. V. *IV All–Union Conference on Organic Luminophors*, Kharkov (1984), 69.

[244] Cuniberti C., Perico A. *Europ. Polymer J.* (1980), **16**, 887–893.

[245] Barashkov N. N., Sakhno T. V., Semenova L. I. *J. Prikl. Spectr.*, **45** (1986), 331.

[246] Anufrieva E. V., Gromova R. A., Krakovyak M. G. *et al.*, *Vysokomol. Soed.* (1984), Ser. A, **26**, 1276–1281.

[247] Krakovyak M. G., Anufrieva E. V., Ananieva T. D. *et al.*, *Vysokomol. Soed.* (1976), Ser. A, **18**, 1494–1497.

[248] Herkstroeter W. G. *J. Polymer Sci.*, *Polymer Chem. Ed.* (1984), **22**, 2395–2412.

[249] Qui T. C., Chang T., Hanc C., Nishijima Y., *Polymer* (1986), **27**, 1705.

[250] Bootsma J. P. C., Challa G., Muller F. *J. Polymer Sci.*, *Polymer Chem. Ed.* (1984), **22**, 705–709.

[251] Bootsma J. P. C., Challa G., Visser A. J. *et al. Polymer* (1985), **26**, 951–956.

[252] Iizawa T., Nishikubo T., Takahashi E., Hasegawa M., *Makromol. Chem.* (1983), **184**, 2297–2312.

[253] Pazkowski J., Neckers D. C., *Macromolecules* (1985), **18**, 1245.

[254] Gupta S. N., Linden S. M., Wrzyszynski A., Neckers D. C., *Macromolecules* (1988), **21**, 51.

[255] Barashkov N. N., Sakhno T. V., Semenova L. I., *J. Prikl. Spectr.* (1988), **49**, 161.

[256] Selb J., Jerome R., *Eur. Polym. J.* (1987), **23**, 603.

[257] Ushiki H., Kano Y., Akiyama S., Kitazaki Y., *Eur. Polym. J.* (1986), **22**, 381.

[258] Ueba Y., Banks E., Okamoto Y. *J. Appl. Polymer Sci.* (1980), **25**, 2007–2017.

[258A] Higashiyama N., Nakamura H., Mishima T., Shiokawa J., Adachi G., *J. Electro-chem. Soc.* (1991) **138**, 594–598.

[259] Sumi K., Furue M., Nazakura S. *J. Polymer Sci.*, *Polymer Chem. Ed.* (1984), **22**, 3779–3788.

[260] Sakhno T. V., Barashkov N. N., Danilevsky I. P. *et al.*, *All–Union Conference on Polymeric Optical Materials* (1991), 65.

[260A] Baxter S. M., Jones W. E., Danielson E., Worl L., Strouse G., Younathan J., Meyer T. J., *Coord. Chem. Rev.* (1991), **111**, 47–71.

[261] Shore V. G., Pardee A. B. *Arch. Biochem. Biophys.* (1956), **60**, 100–103.

[262] Burshtein E. A. *Natural luminescence of proteins*. Itoghi Nauki Tekh.: Biofiz. (1977) 7, 190 pp.

[263] Burshtein E. A., *Luminescence of protein chromophores*. Itoghi Nauki Tekh.: Biofiz. (1976) 6, 213 pp.

[264] Haggis G. H. *Introduction to molecular biology*. London, Longman, Green and Co., (1964), 401 p.

[265] Teale F. W. *Biochem. J.* (1960), **76**, 381–385.

[266] Pechere J.–F. Demaille J., Capony J.–P. *Biochim. Biophys. Acta* (1971), **236**, 391–393.

[267] Chernitsky E. A. *Luminescence and structural lability of proteins in solution and cells*. Minsk: Nauka i Technika (1972), 277 p.

[268] Steiner R. F., Kolinsky R. *Biochemistry* (1968), **7**, 1014–1016.

[269] Morawetz H. (in ref. [100]), 1.

[270] *Luminescent antibodies*. Ed. Meisel M. N., Moscow: Medicina (1972), 144 p.

[271] *Luminescent antibodies in microbiology*. Ed. Levina E. N., Moscow: Medgiz (1962), 240 p.

[272] Marshall J. D., Eveland W. C., Smith S. W. *Proc. Soc. Exp. Biol. Med.* (1958), **98**, 898–900.

[273] Yudenfriend S. *Fluorescence analysis in biology and medicine*: trans. from English. Moscow, Mir (1965), 484.

[274] Barsky V. E., Ivanov V. B., Sklyar Y. E. *et al.*, *Izv. AN SSSR, Ser. Biol.* (1968), 744–747.

[275] Freishtat D. M. *Reagents and preparations for microscopy*. Moscow, Khimiya (1980), 480.

[276] Barsky V. E., Ivanov V. B. *Electronic and fluorescence microscopy of the cell*. Moscow, Nauka (1974), 157–163.

[277] Virnik A. D., Chekalin M. A. *Zh. Prikl. Khim.* (1962), **35**, 588–593.

[278] Vladimirov Y. A. *Photochemistry and luminescence of proteins*. Moscow, Nauka (1965), 232 p.

[279] Vladimirov Y. A., Dobretsov G. E. *Fluorescent probes in study of biological membranes*, Moscow, Nauka (1980), 320 p.

[280] Barashkov N. N., *Luminescence in public health*, Mir, Moscow, (1988), 151 p.

[281] Priezzhev A. V., Tuchin V. V., Shubochkin L. P. *Laser diagnostics in biology and medicine*, Moscow, Nauka (1989), 240 p.

[282] Dobretsov G. E., *Fluorescent probes in a study of cells, membranes and lipoproteins*. Nauka, Moscow (1989), 274 p.

[282A] *Proc. Advances in Fluorescence Sensing Technology*, eds. Lokowicz J. R., Thompson R. B., Katzir A., S. P. I. E., Washington (1993).

[282B] Hamasaki K., Ueno A., Toda F., *J. Chem. Soc., Chem. Commun.* (1993), 331–333.

[283] Gachkovskii V. F., *Zh. Strukt. Khim.* (1968), **9**, 1018–1023.

[284] Bondareva L. V., Litovchenko G. D., Mikhailov N. V., *Vysokomol. Soedin., Ser B* (1973), **15**, 96–99.

[285] Loutfy R. O., *Macromolecules* (1981), **14**, 270–275.

[286] Kalechits I. I., Kuz'min M. G., Zubov V. P., Lachinov M. B., Kabanov V. A., *DAN SSSR*, (1979), **244**, 1157–1160.

[287] Kalechits I. I., Zubov V. P., Kuz'min M. G., Kabanov V. A., Lachinov M. B., *Vysokomol. Soedin., Ser. B*, (1979), **21**, 447–452.

[288] Kalechits I. I., Kuz'min M. G., Zubov V. P., Kabanov V. A., *DAN SSSR*, (1981), **256**, 407–410.

[288A] Kotch T. G., Lees A. J., Fuerniss S. J., Papathomas K. I., *Chem. Mater.* (1991) **3**, 25–27.

[289] Barashkov N. N., Sakhno T. V., Alekseev N. N. *et al.*, *Zh. Prikl. Spektrosk.* (1990) **53**, 386–391.

[290] Rameesdonk H. J. V., Vos M., Verhoeven J. W. *et al.*, *Polymer* (1987), **28**, 951.

[291] Leicht R., Fuhrmann J., *Polymer Bull.* (1981), **4**, 141–148.

[292] Barashkov N. N., Sakhno T. V., Semenova L. I. *Zh. Prikl. Spectr.* (1991), **54**, 335.

[293] Papisov I. M., Nekrasova N. A., Pautov V. D. *et al.*, *DAN SSSR* (1974), **214**, 861–864.

[294] Hemker D. J., Garza V., Franc C. W., *Macromolecules* (1990), **23**, 4411.

[295] Krakovyak M. G., Ananyeva T. D., Anufrieva E. V. *et al.*, *Vysokomol. Soed., Ser. A*, (1984), **26**, 2071–2076.

[296] Scarlata S. F., Ors J. A., *Polym. Commun.* (1986), **27**, 41.

[297] Wang F. W., Lowry R. E., Fanconi B. M., *Polymer* (1986), **27**, 1529.

[298] Wang F. W., Wu E. S., *Polym. Comm.* (1987), **28**, 73.

[299] Sung C. S. P., Pyun E., Sun H. L., *Macromolecules* (1986), **19**, 2922.

[300] Sung C. S. P., Chin I. J., Yu W. C., *Macromolecules* (1985), **18**, 1510.

[301] Sung C. S. P., Mathisen R., Filardi C., *Amer. Chem. Soc., Polym. Prepr.* (1986), **27**, 308.

[302] Zamotayev P. V., Litsov N. I., Kagan A. A., *Vysokomol. Soed.* Ser. B, (1982), **24**, 577–580.

[303] Slobodyanik V. V., Faidysh A. N., Yashuk V. N. *et al.*, *Vysokomol. Soed.* Ser. A (1978), **20**, 2264–2269.

[304] Slobodyanik V. V., Faidysh A. N., Yashuk V. N. *et al.*, *Izv. AN SSSR*, Ser. Phys. (1978), **42**, 318–322.

[305] Allen N. S., McKellar J. F., Phillips G. O. *et al.*, *J. Polymer Sci., Polymer Chem. Ed.* (1974), **12**, 2647–2650.

[306] Allen N. S., McKellar J. F., Phillips G. O., *J. Polymer Sci., Polymer Lett. Ed.* (1974), **12**, 477–479.

[307] Allen N. S., McKellar J. F., Phillips G. O., *J. Polymer Sci., Polymer Chem. Ed.* (1974), **12**, 1233–1241.

[308] Bauer D. R., *Polymer Deg. and Stabil.* (1987), **19**, 97.

[309] Smirnova E. I., Vechkanov G. N., Alekseyeva L. N. *et al.*, *Khim. Volokna* (1982), 39–40.

[310] Allen N. S., McKellar J. F., Phillips G. O., *J. Polymer Sci., Polymer Lett. Ed.* (1974), **12**, 253–255.

[311] Allen N. S., McKellar J. F., Chapman C. B., *J. Appl. Polymer Sci.* (1976), **20**, 1717–1719.

[312] Allen N. S., McKellar J. F., Wilson D., *J. Photochem.* (1976), **6**, 73–76.

[313] Allen N. S., Harrison M. J., *Europ. Polymer J.* (1985), **21**, 517–526.

[314] Sharf H. D., Dieris C. D., Leismann H., *Angew. Makromol. Chem.* (1979), **79**, 193–206.

[315] Gachkovsky V. F., *Vysokomol. Soed.*, (1965), **7**, 2009–2014.

[316] Baumbach D. O., *J. Polymer Sci., Polymer Lett. Ed.* (1982), **20**, 117–121.

[317] Heintz E., *J. Phys. Radium.* (1954), **15**, 219–225.

[318] Abuin F., Lissi E., Gargalo L. *et al.*, *Europ. Polymer J.* (1984), **20**, 105–107.

[319] Char K., Frank C. W., Gast A. P., Tang W. T., *Macromolecules* (1987), **20**, 1833.

[320] Nogue Y., Hisamune T., Goto T., Tsuruta H., *J. Phys. Soc. Japan* (1986), **55**, 4053.

[321] Utena Y., Yajuma H., Ishii T., Handa T., *Eur. Polym. J.* (1987), **23**, 643.

[322] Skorokhodov S. S., Anufrieva E. V., Koltzov A. I. *et al. J. Polymer Sci.*, Pt. C (1973), 1583–1586.

[323] Anufrieva E. V., Volkenstein M. V., Gotlib Yu. Ya. *et al.*, *DAN SSSR* (1970), **194**, 1108–1110.

[324] Ghiggino K. P., Amer. Chem. Soc., *Polym. Prepr.* (1986), **27**, 331.

[325] Soutar I., Toynbee J., Amer. Chem. Soc., *Polym. Prepr.* (1986), **27**, 338.

[326] Ushiki H., Tanaka F., Mita I., *Eur. polym. J.* (1986), **22**, 827.

[327] Nekrasova T. N., Ptitsyn O. V., Shikanova M. S., *Vysokomol. Soed.* Ser. A (1968), **10**, 1530–1538.

[328] Anufrieva E. V., Gromova R. A., Kondratjeva E. V. *et al.*, *Vysokomol. Soed.* Ser. B (1976), **18**, 916–918.

[329] Anufrieva E. V., Pautov V. D., Freidzon Ya. S. *et al.*, *Vysokomol. Soed.* Ser. A (1977), **19**, 755–758.

[330] Anufrieva E. V., Pautov V. D., Freidzon Ya. S. *et al. DAN SSSR* (1984), **278**, 383–386.

[331] Anufrieva E. V., *Spectroscopic methods for polymer research*, Moscow, Znanie (1975), 35–62.

[332] Sommersall A. C., Dan E., Guillet J. E., *Macromolecules* (1974), **7**, 223–244.

[333] Gachkovsky V. F., *Zh. Strukt. Khim.* (1967), **8**, 362–364.

[334] Konev S. V., Katibnikov M. A., *DAN SSSR* (1961) **163**, 472–475.

[335] Binana–Limbele W., Zana R., *Macromolecules* (1987), **20**, 1331.

[336] Moisya E. G., Egorov Yu. P., *Zh. Prikl. Spektrosk.* (1964), **1**, 363–366.

[337] Moisya E. G., Mozdor E. V., Arjev I. A. *et al.*, *Spectroscopy of polymers*. Ed. M. V. Volkenstein, Nauk. Dumka Publ., Kiev (1968), 60–63.

[338] Nishijima Y., Onogi Y., Asai T., *J. Polymer Sci.*, Pt. C (1966), 237–240.

[339] Menzheres G. Ya., Moisya E. G., *Composite polymer materials* (1983), 14–17.

[340] McGraw G. E., *J. Polymer Sci.*, Pt. A–2 (1970), **8**, 1323–1337.

[341] Kaplanova M., Kudlacek L., Herner M., *Sb. Ved. Pr. VSCHT, Pardubice* (1981), **43**, 43–51.

[342] Azarov V. Yu., *Composite Polymer Materials*, (1986), 36–39.

[343] Brestrin Yu. V., *Acta Polymerica*, (1980), **31**, 646–653.

[344] Monnerie L., *Amer. Chem. Soc. Polymer Prepr.* (1981), **22**, 96–97.

[345] Perico A., Gruenza M., *J. Chem. Phys.* (1986), **84**, 510.

[346] Jarry J. P., Monnerie L., *J. Polymer Sci.*, *Polymer Phys. Ed.*, (1980), **18**, 1879–1890.

[347] Bokobza L., Pham Van Cang C., Giordano C. *et al.*, *Polymer* (1988), **29**, 251–255.

[347A] Freeman B. D., Bokobza L., Sergot P., Monnerie L., De Schryver F. C., *J. Lumin.* (1991) **48–49** (Pt. 1), 259–264.

[347B] Bur A. J., Lowry R. E., Roth S. C., Thomas C. L., Wang F. W., *Macromolecules* (1991) **24**, 3715–3717.

[348] Jarry J. P., Monnerie L., *J. Macromol. Sci.*, Pt. B, (1980), **18**, 637–639.

[348A] Fofana M., Veissier V., Viovy J. L., Monnerie L., *Polymer* (1989) **30**, 51–57.

[348B] Halary J. L., Leviet M. H., Kwei T. K., Pearce E. M., *Macromolecules* (1991) **24**, 5939–5942.

[348C] Halary J. L., Larbi F. B. C., Oudin P., Monnerie L., *Makromol. Chem.* (1988) **189**, 2117–2124.

[348D] Halary J. L., Monnerie L., *NATO ASC Ser.*, *Ser. C* (1986), **182** (Photophys. Photochem. Tools Polym. Sci.: Conform, Dyn., Morphol.) 589–610; Monnerie L., *ibid*, 371–396.

[348E] Bahar I., Mattice W. L., *J. Chem. Phys.* (1989) **90**, 6783–6790.

[348F] Bahar I., Mattice W. L., *J. Chem. Phys.* (1989) **90**, 6775–6782.

[349] Henry R., Soutar I., *J. Polymer Sci.*, *Polymer Phys. Ed.*, (1980), **18**, 1021–1034.

[350] Kirsh Yu. E., Pavlova N. R., Kabanov V. A., *DAN SSSR*, (1974), **218**, 863–866.

[351] Peterlin A., *J. Polymer Sci.*, Pt. B (1972), **10**, No. 2, 101–105.

[352] Winnik M. A., Hua M. H., Hougham B. *et al.*, *Macromolecules*, (1984), **17**, 262–266.

[353] Pavlov E. A., Starodubtsev S. G., Pavlov A. A. *et al.*, *Vysokomol. Soed. Ser. A*, (1984), **26**, 1432–1436.

[354] Kalechits I. I., Zubov V. P., Kuzmin M. G. *et al.*, *Vysokomol. Soed. Ser. A*, (1984), **26**, 2128–2136.

[355] Izumrudov V. A., Savitsky A. P., Zezin A. B. *et al.*, *DAN SSSR* (1983), **272**, 1408–1412.

[356] Izumrudov V. A., Bronich T. K., Zezin A. B. *et al.*, *DAN SSSR* (1984), **278**, 404–408.

[357] Webber S. E., *Macromolecules* (1986), **19**, 1658.

[357A] Okamoto Y., Kido J., Ref. [15C] p. 143–173.

[358] Vyasemsky V. O., Lomonosov I. I., Pisarevsky L. N. *et al.*, *Scintillation methods in radiometry*. Moscow, Atomizdat (1961), 429 p.

[359] Keil G. J. *Appl. Phys.* (1969), No. 9, 3544–3547.

[360] Tsyrlin Y. A., Sokolovskaya T. I., Pomerantsev V. V. *et al.*, *Scintillators and organic luminophors*. Kharkov (1974), **3**, 113–118.

[361] De Marzo C., De Palma M., Distante A. *et al.*, *Nucl. Instr. Meth. Phys. Res.* (1983), **217**, No.3, 405–417.

[362] Levin M. B., Cherkasov A. S., Bakranov V. K., *Optico–mechan. Ind.* (1988), **3**, 47–55.

[363] Burakov V. S., Vasiliev N. N., Gorelenko A. Y. *DAN BSSR* (1982), **26**, 12, 1085–1087.

[364] Dyumaev K. M., Manenkov A. A., Malyukov A. N. *et al. Quant. Elect.*, (1983), **10**, 4, 810–818.

[365] Shorr M. G., Torney F. Y., *Phys. Rev.* (1950), **80**, 474–479.

[366] Gunder O. A. *Luminescent Dopants in Plastic Scintillators*. Moscow, 1978, 24 p.

[366A] Bross A. D., Pla-Dalmau A., Spangler C. W., *Radiat. Phys. Chem.* (1993) **41**, 379–387.

[367] Sandler S. R., Loshaek S. *J. Chem. Phys.* (1961), **34**, No. 2, 439–444.

[368] Gunder O. A., Korunova A. F. *Scintillators and organic luminophors*. Kharkov, *VNII Monocryst.* (1974), **3**, 73–82.

[369] Bezugly V. D., Grachev N. M., Petrova I. B. *Scintillators and scintillation materials*. Kharkov, *KGU Publ.* (1963), **3**, 80–81.

[370] Bezugly V. D., Grachev N. M., Dykhanova A. S. *Prib. i Tehn. Exper.* (1963), **1**, 163–165.

[371] Koton M. M. *JTF.* (1956), **26**, 8, 1741–1743.

[372] Gunder O. A., Grachev N. M., Belyaev V. A. *et al.*, *Vysokomol. Soed.*, Series B (1968), **10**, 852–855.

[373] Bezugly V. D., Chernobuy A. V., Dmitrievskaya L. I. *et al.*, *Scintillators and scintillation materials*. Kharkov, KGU Publ. (1963), **3**, 72–79.

[374] Chernobuy A. V., Gunder O. A., Milner R. S. *et al.*, *Prib. i Tehn. Exper.* (1967), **6**, 58–62.

[375] Grachev N. M., Bezugly V. D., Dykhanova A. S. *Prib. i Tehn. Exper.* (1964), **6**, 61–62.

[376] Malinovskaya S. A. *Plastic scintillators based on acrylic polymers*, Moscow, (1970), 19 p.

[377] Gunder O. A., Malinovskaya S. A., Duyche A. R. *et al.*, *Prib. i Tehn. Exper.* (1969), **3**, 66–69.

[378] Anderson R. A., Davidson K., Soutar I., *Eur. Polym. J.* (1989), **25**, 745.

[379] Gunder O. A., Malinovskaya S. A., Teslya L. E. *Prib. i Tehn. Exper.* (1971), **5**, 73–79.

[380] Aleshin V. I., Bakalayarov A. M., Balysh A. Y. *et al.*, *Prib. i Tehn. Exper.* (1977), **4**, 68–70.

[381] Markley F. W. *Mol. Cryst.* (1968), **4**, 1–4, 303–317.

[382] Kumasaki H. *Jap. J. Appl. Phys.* (1974), **1**, 195–196.

[383] Bezugly V. D., Mukhina S. A. *Prib. i Tehn. Exper.* (1967), **2**, 82–84.

[384] Inagaki I., Fakashima R. *Nucl. Instr. Meth. Phys. Res.*, (1982), **201**, 2, 511–517.

[385] Kilimov A. P., Grachev N. M. *Prib. Tehn. Exper.* (1963), **3**, 175–176.

[386] Vershinina S. P., Volosyuk G. P., Tsyrlin Y. A. *et al.*, *Atom. Energ.* (1969), **26**, 4, 341–344.

[387] Kempe R. *Kernergie*, (1962), **5**, No. 6, 487–490.

[388] Chernobuy A. V., Kolesnikov L. N. *Prib. i Tehn. Exper.* (1964), **2**, 120–121.

[389] Baroni E. E., Silin S. F., Lebsadze T. N. *et al.*, *Atom. Energ.* (1964), **17**, 6, 497–500.

[390] Petrova I. B., Volosyuk G. P., Gunder O. A. *et al.*, *Scintillators and organic luminophors*, Kharkov, (1975), **4**, 105–110.

[391] Volosyuk G. P., Petrova I. B., Prokofieva L. M. *et al.*, *Monocrystals, scintillators and organic luminophors.* Cherkassy, ONIITEHIM (1972), **6**, part 1, 293–298.

[392] Frolova A. V., Kolotilova V. G., Gunder O. A. *et al.*, *Medical Radiology.* (1968), **13**, 2, 43–47.

[393] Volosyuk G. P., Petrova I. B., Prokofieva L. S. *et al.*, *Optika i Spektr.* (1971), **30**, 3, 466–471.

[394] Biteman V. B., Gunder O. A., Petrova I. B. *et al.*, *Methods for preparation and study of monocrystals and scintillators*, Kharkov, *VNII Monocryst* (1980), **5**, 161–163.

[395] Cho Z. H., Tsai C. M., Eriksson L. A. *IEEE Trans. Nucl. Sci.* (1975), **NS–22**, 1, 72–80.

[396] Petrova I. B., Volosyuk G. P., Prokofieva L. S. *Plastic scintillators with naphthyl derivatives of metals*, Moscow, NIITEHIM, (1982), 7 p.

[397] Charlesby A., *Atomic radiation and polymers.* Oxford, Pergamon Press, (1960), 522 p.

[398] Brooks F. D. *Nucl. Instrum. Meth.* (1979), **162**, Pt. 2, 1–3, 477–505.

[399] Barony E. E., Viktorov D. V., Rozman I. M. *et al.*, *Nukl. Electron.* (1962), **1**, 131–138.

[400] Tsyrlin Y. A., Chernikov V. V., Timofeeva T. V. *Prib. i Tehn. Exper.* (1965), **4**, 72–74.

[401] Tsyrlin Y. A., Zalyubovsky I. I., Sokolovskaya T. I. *et al.*, *Prib. i Tehn. Exper.* (1968), **1**, 55–57.

[402] Berlman I. B., Grismore R., Oltman B. G. *Trans. Faraday Soc.*, (1963), **59**, 2010–2015.

[403] *Interaction of ultra–short pulses with matter.* Ed. M. D. Galanin, Moscow, Nauka (1984), 96 p.

224 References

[404] Gunder O. A., *Physico–chemic. principles of manufacture of plastic scintillators*, Moscow, NIITEHIM (1979), 57 p.

[405] Walker J. K. *Nucl. Instrum. Meth.* (1964), **1**, 131–134.

[406] Albikov Z. A., Veretennikov A. I., Kozlov O. V. *Detectors of ionizing pulse radiation.* Moscow, Atomizdat (1978), 63–68.

[407] Tsyrlin Y. A. *Light collection in scintillation counters*, Moscow, Atomizdat (1975), 265 p.

[408] Bengtson B., Moszynski M. *Nucl. Instrum. Meth.* (1978), **1**, 2, 221–231.

[409] Moszynski M., Bengtson B. *Nucl. Instrum. Meth.* (1979), **1**, 1–31.

[410] Kunze R., Langkau R. *Nucl. Instrum. Meth.* (1971), **4**, 667–668.

[411] Kelly T. M., Merrigan J. A., Lambrecht R. M. *Nucl. Instrum. Meth.* (1973), **2**, 233–235.

[412] Volkov N. G., Gunder O. A., Lyapidevsky V. K. *et al., Prib. i Tehn. Exper.* (1973), **6**, 188–189.

[413] Andreeshchev E. A., Silin S. F., Kovyrzina K. A. *et al., Prib. i Tehn. Exper.* (1983), **3**, 52–54.

[414] Lions P. B. *IEEE Trans. Nucl. Sci.* (1977), **1**, 177–181.

[415] Beriman I. B., Lutz S. S., Flournoy Y. M. *Nucl. Instrum. Meth. Phys. Res.* (1984), **1**, 78–82.

[416] Gunder O. A., Koval L. P., Grigorieva V. I. *et al., Monocrystals, scintillators and organic luminophors.* Cherkassy, ONIITEHIM (1976), **6**, part 1, 79–82.

[417] Hayen J., *Z. Phys.* (1968), **210**, No. 2, 182–192.

[418] Hardwick R., Schwartzenbach V. *Adv. Radiat. Res., Phys. Chem.* (1973), **2**, 547–553.

[419] Bezugly V. D., Nagornaya L. D. *Atom. Energ.* (1964), **1**, 67–70.

[420] Beregovenko E. D., Gorbachev V. M., Uvarov N. A. *Atom. Energ.* (1973), **2**, 124–126.

[421] Klein J., Gresset J., Heisel F. *et al., Intern. J. Appl. Radiat. Isotop.* (1967), **6**, 399–406.

[422] Oldham G., Ware A. R. *Radiat. Eff.*, (1975), **1–2**, 95–97.

[423] Gunder O. A., Koba V. S., *Khimiya Vysok. Energ.* (1974), **1**, 83–84.

[424] Malinovskaya S. A., Gunder O. A., Generalova V. V. *et al., Monocrystals, scintillators and organic luminophors*, Kharkov, VNII *Monocryst.*, (1969), **5**, part 1, 158–164.

[425] Gunder O. A., Koba V. S., Eckerman V. M. *Radiochemistry* (1976), **6**, 913–914.

[426] Gorbachev V. M., Kuzyanov V. V., Peshkova Z. I. *et al., Atom. Energ.*, (1975), **5**, 427–429.

[427] Gunder O. A., Koba V. S. *Radiochemistry* (1969), **1**, 119–122.

[428] Bezugly V. D., Semenenko M. G., Vlasov V. G. *et al., Scintillators and Scintillation Materials*, Kharkov, KGU (1963), **3**, 43–53.

[429] Bezugly V. D., Zaplesnichenko G. P. *Prib. I Tehn. Exper.* (1966), **1**, 186–188.

[430] Heisel F., Laustriat C., Coche A. *Intern. J. Appl. Radiat. Isotop.*, (1964), **2**, 89–93.

[431] Koton M. M., Sivogranova K. A., Tolstikova Z. D. *et al., Plastics* (1960), **2**, 48–52.

[432] Novakova O., Nejedla A. *Jad. Energ.* (1971), **17**, Pt. 10, 333–336.

[433] Solomonov V. M., Gunder O. A. *Monocrystals, scintillators and organic luminophors*, Cherkassy, ONIITEHIM (1972), **6**, part 1, 92–102.

[434] Keil G. *Nucl. Instrum. Meth.* (1970), **89**, 111–123.

[435] Keil G. *Nucl. Instrum. Meth.* (1970), **83**, 145–147.

[436] Barish B. *IEEE Trans. Nucl. Sci.* (1978), **1**, 532–536.

[437] Eckardt V., Kalbach R., Manz A. *et al.*, *Nucl. Instrum. Meth.* (1978), **155**, 389–398.

[438] Malinovskaya S. A., Kornilovskaya D. D., Skripkina V. T. *Preparation and Study of Optical and Scintillation Materials*, Kharkov, VNII *Monocryst.* (1984), **12**, 132–134.

[439] Allemand L., Auronet C., Beauval I. I. *et al.*, *Nucl. Instrum. Meth.* (1979), **164**, 93–95.

[440] Titskaya V. D. *Study into polymerization of scintillation systems*, Moscow (1981), 22 p.

[441] Biteman V. B., Pomerantsev V. V., Kutsina L. M. *et al.*, *Radiation converters*, Cherkassy, ONIITEHIM (1974), 8 p.

[442] Mak–Veig D., *Application of solar energy*, Moscow, Energoizdat (1981), 216 p.

[443] Barashkov N. N., Ishchenko A. A., Krainov I. P. *et al.*, *Zh. Prikl. Spectr.* (1991), **55**, 897.

[444] Barashkov N. N., Sakhno T. V., *Optically transparent polymers and materials*, Moscow, Khimiya (1992), 80 p.

[445] Weber W. H., Lambe J. *Appl. Opt.* (1976), **5**, 2299–2300.

[446] Levitt J. A., Weber W. H. *Appl. Opt.* (1977), **16**, 2684–2689.

[447] Goetzberger A. *Appl. Phys.* (1978), **16**, 399–404.

[448] Swartz B. A., Cole T. C., Zewail A. H. *Opt. Lett.* (1977), **2**, 73–75.

[449] Goetzberger A., Greubel W. *Appl. Phys.* (1977), **14**, 123–139.

[450] Batchelder J. S., Zewail A. H., Cole T. *Appl. Opt.* (1979), **18**, 3090–3110.

[451] Goetzberger A., Wittwer V. *Solar Cells* (1981), **4**, 3–23.

[452] Batchelder J. S., Zewail A. H., Cole T. C. *Appl. Opt.*, (1981), **21**, 3733–3754.

[453] Wittwer V., Heidler K., Zastrow A. *et al.*, *J. Luminesc.*, (1981), **24–25**, 873–876.

[454] Reisfeld R., Jorgensen C. K. *Struct. Bonding*, Berlin, (1982), **15**, 1–36.

[455] Lempicki A. *Appl. Opt.* (1983), **8**, 1160–1164.

[456] Levin M. B., Cherkasov A. S., Baranov V. K. *Optiko–Mech. Ind.*, (1988), **3**, 47–55.

[457] Born M., Wolf E. *Principles of optics*, 2nd ed., Elmsford N. Y. Pergamon (1964).

[458] Boer S. *Solar Energy* (1977), **19**, 525–537.

[459] Drake J. M., Lesiecki M. L., Sansregret J. *et al.*, *Appl. Opt.* (1982), **16**, 2945–2952.

[460] Thomas W. R. L., Drake J. M., Lesiecki M. L. *Appl. Opt.* (1983), **21**, 3440–3450.

[461] Carrascosa M., Unamura S. *Appl. Opt.* (1983), **20**, 3236–3241.

[462] Dobro L. F., Popov V. V., Romanov A. D. *et al.*, *Zh. Prikl. Spectr.*, (1983), **1**, 142–145.

[463] Levin M. B., Starostina G. P., Sherkasov A. S. *Zh. Prikl. Spectr.* (1987), **46**, 432–437.

[464] Roncali J., Carnier F. *Appl. Opt.* (1984), **16**, 2809–2817.

[465] Sakuta K. *Bull. Electrotechn. Lab.* (1986), **8**, 787–796.

[466] Sidrach de Cardona M., Carroscosa M., Mesequer F *et al.*, *Appl. Opt.* (1985), **13**, 2028–2032.

[467] Herman A. M. *Solar Energy* (1982), **4**, 323–329.

[468] Meseguer Rico F., J. Jaque F., Cusso F. *J. Power Sources*, (1981), **6**, 383–388.

[469] Mugnier J., Dordet Y., Pouget J. *et al.*, *Solar Energy Mater.* (1987), **2**, 65–75.

[470] San R. E., Buar G., Kelker H. *Appl. Phys.* (1980), **23**, 369–376.

[471] Kapinus E. I. *Photonics of molecular complexes*, Kiev, Naukova Dumka (1988), 256 p.

[472] Roundhill D. M., *Solar Energy* (1986), **36**, 297–299.

[473] Heidler K. *Appl. Opt.* (1981), **5**, 773–777.

[474] Olson R. W., Loring R. F., Fayer M. D. *Appl. Opt.* (1981), **17**, 2940–2943.

[475] Cook M. J., Thompson A. J. *Chem. Brit.* (1984), **20**, 914–917.

[476] Roncali J., Garnier E. *Solar Cells* (1984), **13**, 133–143.

[477] Barashkov N. N., Sakhno T. V., Nurmukhametov R. N. *et al.* Author's certificate No. 1368324, USSR (1988).

[478] Boling N. C., Rapp. C. F. *Conf. Rec. 13th IEEE. Photovolt. Spec. Conf.* Washington (1978), p. 690.

[479] Wittwer V., Stahl W., Goetzberger A. *Solar Energy Mater.* (1984), **11**, 187–197.

[480] Stahl W., Wittwer V., Goetzberger A. *Solar Energy* (1986), **1**, 27–35.

[481] Bornstein J. G. *Proc. Soc. Photo–Opt. Instrum. Eng.* SPIE, (1984), **502**, 138–145.

[482] Bass M., Deich T., Veber M. *Usp. Phys. Nauk*, (1971), **3**, 521–573.

[483] Denisov L. K., Kozlov N. A., Uzhinov B. M. *Organic compounds—active laser media*, Moscow, CNII Electronika (1980), 61 p.

[484] Kotsubanov V. D., Malkes L. Y., Naboikin Y. V. *et al.*, *zv. Akad. Nauk. SSSR, Phys. Series* (1968), **32**, 1466–1470.

[485] Dyumaev K. M., Manenkov A. A., Maslyukov A. P. *et al.*, *Sov. J. Quant. Electronics* (1983), **4**, 503–507.

[486] Zemsky V. I., Kolesnikov Y. L., Meshkovsky I. K. *Pisma v JTF* (1986), **12**, 331–335.

[487] O'Connel R. M., Saito T. T. *Opt. Eng.* (1983), **4**, 393–399.

[488] Bezrodny V. I., Przhonskaya O. V., Tikhonov E. A. *et al.*, *Quant. Electronics*, (1982), **12**, 2455–2464.

[489] Gueidur S. A., Morozov A. G., Sidyakova V. G. *5th All–Union Conf. 'Laser Optics'*, Leningrad, (1987), 227.

[490] Dyumaev K. M., Manenkov A. A., Maslyukov A. P. *et al.*, *AN SSSR, Phys. Series*, (1987), **8**, 1387–1398.

[491] Manenkov A. A., Prokhorov A. M. *Usp. Phys. Nauk*, (1986), **1**, 170–211.

[492] Novikov N. P. *Structure and properties of polymeric materials*, Riga, Zinatne (1979), 160–189.

[493] Butenin A. V., Kogan B. D. *Quant. Electronics* (1986), **10**, 2149–2151.

[494] Glauberman G. Y., Pilipetsky N. F., Savenin S. Y. *Quant. Electronics* (1989), **6**, 1221–1225.

[495] Delone N. B. *Interaction of laser radiation with matter*, Moscow, Nauka (1989), 280 p.

[496] Kusakawa H., Takahashi K., Ito K. *Appl. Phys.* (1969), **10**, 3954–3958.

[497] Butenin A. V., Kogan B. Y. *Quant. Electronics* (1976), **3**, 1136–1140.

[498] Novikov N. P. *Usp. Phys. Nauk* (1981), **26**, 1676.

[499] Manenkov A. A., Matyushin G. A. Nechitailo V. S. *et al.*, *Quant. Electronics* (1983), **12**, 2426–2432.

[500] Speranskaya T. A., Tarutina L. I. *Optical properties of polymers*. Leningrad, Khimiya (1976), 136 p.

[501] Belichenko A. S., Dyumaev K. M., Manenkov A. A. *et al.*, *AN SSSR* (1986), **1**, 89–92.

[502] Barashkov N. N., Muravieva T. M., Gorbunov L. A. *et al.*, *Light–resistant poly(methyl methacrylate) with luminophoric fragments in the chain*. Patent application No. 4665307/05 of 31.03.89 MKI CO8F120/14.

[503] Yaroslavtsev V. T., Barashkov N. N., Klimenko V. G. *et al.*, *VII All–Union Coord. Conference 'Photochemistry of dye-based laser media'*, Lutsk (1990), 88.

[504] Yaroslavtsev V. T., Muravieva T. M., Barashkov N. N. *et al.*, *V All–Union Conference on optical polymeric materials*, Leningrad (1991), 87.

[505] Denisov L. K., Dyachkov A. I., Kristaleva V. N. *et al.*, *Plastics* (1987), **12**, 22–23.

[506] Naboikin Y. V., Ogurtsova L. A., Podgorny V. P. *et al.*, *Optics Spectr.* (1970), **28**, 528–532.

[507] Rodchenkova V. V., Tsogoeva S. A. *IV All–Union Conference on organic luminophors*, Kharkov, VNII Monocrystals (1984), 138.

[508] Manenkov A. A., Nechitailo V. S. *Quant. Electr.*, (1980), **3**, 616–619.

[509] Bermas T. B., Paramonov Y. M., Barashkov N. N. *et al.*, *VII All–Union Conference 'Photochemistry of dye-based laser media'*, Lutsk (1990), 13–14.

[510] Gromov D. A. Dyumaev K. M., Manenkov A. A. *et al.*, *AN USSR, Phys. Series* (1984), **7**, 1364–1369.

[511] Mardaleishvili I. R., Anisimov V. M. *Zh. Prikl. Spectr.* (1986), **4**, 581–584.

[512] Bondar M. V., Przhonskaya O. V., Tikhonov E. A. *Quant. Electr.* (1985), **6**, 1242–1247.

[513] Vasiliev N. N., Gorelenko L. Y., Kalosha I. I. *et al.*, *Zh. Prikl. Spectr.* (1985), **1**, 51–55.

[514] Burakov V. S., Vasiliev N. N., Gorelenko A. Y. *et al.*, *Zh. Prikl. Spectr.* (1985), **1**, 35–40.

[515] Bezrodnyi V. I., Bondar M. V., Kozak G. Y. *et al.*, *Zh. Prikl. Spectr.* (1989), **50**, 711.

[516] Geidur S. A., Morozov A. G., Sidyakova V. P. *et al.*, *Optics Spectr.*, (1988), **64**, 1148.

[517] Bortkevich A. V., Geidur S. A., Karapetyan O. O. *et al.*, *Zh. Prikl. Spectr.* (1989), **50**, 210.

[518] Bermas T. B., Zaitsev Y. S., Kostenich Y. V. *et al.*, *Zh. Prikl. Spectr.* (1987), **47**, 569.

[519] Denker I. I., Kalinina E. P. *Varnish materials and their applications* (1963), **2**, 18–21.

[520] Sharonov V. V. *Light and colour*. Moscow, Phyzmatgyz, (1961), 311 p.

[521] Smith T. *Pigm. Resin Technol.* (1981), **11**, 13–15.

[522] Krasovitsky B. M. *Scintillators and organic luminophors*, Kharkov, *VNII Monocryst.* (1972), **3**, 3–16.

[523] Ionchev V. D. *Daytime fluorescent pigments and paint compositions for polyethylene printing*, Moscow, (1971), 15 p.

[524] Pereyaslova D. G., Kuznetsov A. M., Tatsky G. V. *et al. Scintillators and organic luminophors*, Kharkov, *VNII Monocryst.* (1973), **2**, 71–74.

[525] Karmilova L. I., Yermolenko I. G., Serdechnaya T. *A. All–Union Conf. on molecular luminescence and its applications*, Kharkov, *VNII Monocryst.* (1982), 118.

[526] Slezko G. F. *Chemistry and technology of 1,3,5-triarylpyrazolines*, Kharkov (1971), 22 p.

[527] Krichevskii G. E., Gombkoto Y. *Light-fastness of dyed textiles*, Moscow, Leg. Ind. (1975), 168 p.

[528] Barashkov N. N., *Polymeric composites: preparation, properties, application*, Moscow, Nauka (1984), 129 p.

[529] Barker M. Polymers, *Paint Colour J.* (1979), **4008**, 913–917.

[530] Dane C. D. *Chem. Brit.* (1977), **9**, 335–340.

[531] Dane C. D. *Brit. Ink. Maker* (1977), **1**, 11–13.

[532] Smirnov V. A. *Photodissociation of aromatic and heteroaromatic compounds*, Moscow (1984), 36 p.

[533] Zweig A., *Pure Appl. Chem.* (1973), **33**, 389.

[534] Kuzmin M. G., Sadovsky N. A., Kozmenko M. V. *Intensification processes in photographic data recording systems*, Minsk, Nauka i Tehnika (1981), 178–180.

[535] Kondratenko P. A., Kurik M. V., Sandul G. A. *et al.*, *Methods of data recording on silverfree media*, Kiev, Vyshcha Shkola, (1974), **5**, 81–94.

[536] Alfimov M. V., Nazarov V. B., Smirnov V. A. *et al.*, *All–Union Conf. on Luminescence*, Leningrad (1981), 205.

[537] Arnold M. A., *Anal. Chem.* (1992) **64**, 1015A–1025A.

[538] *Photochemical processes in layers*. Ed. Yeltsov A. V., Leningrad (1978), 232 p.

[539] Casay G. A., Czuppon T., Lipowski J., Patonay G., Ref. [282A], 324–336.

Index